Portable Electronics
World Class Designs

DATE DUE

JUN 0 1 2015	

Newnes World Class Designs Series

Analog Circuits: World Class Designs
Robert A. Pease
ISBN: 978-0-7506-8627-3

Embedded Systems: World Class Designs
Jack Ganssle
ISBN: 978-0-7506-8625-9

Power Sources and Supplies: World Class Designs
Marty Brown
ISBN: 978-0-7506-8626-6

FPGAs: World Class Designs
Clive "Max" Maxfield
ISBN: 978-1-85617-621-7

Digital Signal Processing: World Class Designs
Kenton Williston
ISBN: 978-1-85617-623-1

Portable Electronics: World Class Designs
John Donovan
ISBN: 978-1-85617-624-8

RF Front-End: World Class Designs
Janine Sullivan Love
ISBN: 978-1-85617-622-4

For more information on these and other Newnes titles visit: www.newnespress.com

Portable Electronics
World Class Designs

John Donovan

with

Cheryl Ajluni
Alan Bensky
Thomas Coughlin
Rick Gentile
K.F. Ibrahim
Keith Jack
Bruce Jacob
David Katz
Spencer W. Ng
Steve Rackley
Douglas Self
Findlay Shearer
Marc T. Thompson
David T. Wang

ELSEVIER

AMSTERDAM • BOSTON • HEIDELBERG • LONDON
NEW YORK • OXFORD • PARIS • SAN DIEGO
SAN FRANCISCO • SINGAPORE • SYDNEY • TOKYO

Newnes is an imprint of Elsevier

Newnes

Newnes is an imprint of Elsevier
30 Corporate Drive, Suite 400, Burlington, MA 01803, USA
Linacre House, Jordan Hill, Oxford OX2 8DP, UK

Library of Congress Cataloging-in-Publication Data
Application submitted

British Library Cataloguing-in-Publication Data
A catalogue record for this book is available from the British Library.

ISBN: 978-1-85617-624-8

For information on all Newnes publications
visit our Web site at www.elsevierdirect.com

09 10 11   10 9 8 7 6 5 4 3 2 1

Printed in the United States of America

Working together to grow
libraries in developing countries

www.elsevier.com | www.bookaid.org | www.sabre.org

ELSEVIER   BOOK AID
International   Sabre Foundation

Contents

Preface

Not long ago traveling humans only carried their necessities with them: food, clothing, passports, kids. If you wanted to talk with someone, you found a phone booth; to listen to music or catch up on the news, turn on the car radio; to see a movie, check the listings in the paper and drive to the neighborhood theater.

That was then, this is now. Somewhere along the line the definition of "necessities" changed to include music. Sony's introduction of the transistor radio in the 50s—the product on which they launched the company—was radical. Suddenly every teenager on the face of the earth had to have one. The sound was tinny and you needed to carry spare batteries, but transistor radios were "cool." Even in the 50s, being cool could sell products whose technology could stand some improvement.

In the early 80s the transistor radio gave way to the cassette-based Walkman, which several years later became the Diskman, moving to the new CD music format. You had to carry along a lot of CDs in a "CD wallet," but that was a small price to pay to be able to listen to the latest rock band while attempting to focus on your homework.

Now of course everything is digital and therefore portable—including the latest flash-based Walkman. Now you can carry around dozens of videos, hundreds of high-resolution photos, and thousands of songs, emails and documents on a thumbnail-sized memory card in your cell phone. If you want a larger screen or better sound, you can also carry a personal media player (PMP). If you're an inveterate web surfer and really want a larger format, one of the new mobile Internet devices (MIDs) may be just your cup of tea. But notice that the option here is to scale *up*, not down. Portable consumer electronics devices have gotten so conveniently small that ergonomic considerations, rather than technical ones, dictate the form factor. The overarching engineering challenge now is how to make all this advanced functionality run for an acceptable length of time from a tiny battery.

Consumer electronics is driving the entire electronics and semiconductor industries worldwide. The Consumer Electronics Association (CEA) estimates the size of the consumer electronics industry in the US at $170B in 2008. Worldwide the CEA predicts the consumer electronics industry will hit $700B in 2009, which if it were a country would make it the 19th largest country in the world, just behind Australia and ahead of the Netherlands.

The CEA estimates that the average American home has 20 consumer electronics devices in it, most of them portable (at least if you have kids). Teenagers have found they can't live without their iPods and iPhones. Their parents can't live without their Blackberries. Portable consumer electronics devices are no longer luxuries, they're necessities.

Cell phones have led the charge, with sales of 1.3B units in 2008 according to ABI Research—and a growth rate of 13% that shows no sign of slowing down. In 2008 for the first time there were more cell phone subscribers worldwide than landline users. The explosion in cell phone use has been helped in large part by the rapid growth of a middle class in India, China and other developing countries which, instead of burying miles of landline cables, are focusing instead on a next-generation cellular infrastructure. ABI predicts that by 2009 more people will access the Internet from their cell phones than from computers. Cell phones have become the ultimate converged devices, handheld microcomputers that compute, communicate and entertain all at the same time.

Cell phones are where engineers are wrestling with all the hardest design problems and coming up with innovative solutions. Adaptive voltage and frequency scaling, fine-grained clock gating and low-power design flows are all energy-saving techniques that were developed for the portable space. Multi-band, multi-protocol RF front ends are appearing for cell phones that feature various combinations of OFDM, 802.11g/n, Bluetooth, NFC, Mobile WiMAX/MIMO and Wireless USB. A few years ago this wasn't possible—or even considered possible—but demand from the portable consumer electronics market made it happen. Rather, some very smart portable designers did.

The same holds true for displays. Over the past few years TFT LCDs have increased rapidly in resolution, at the same time becoming more power efficient. Organic LEDs (OLEDs), less power hungry still, have become common as the outer screens in clamshell-design handsets; if their proponents can overcome outstanding yield and durability issues, they may become ubiquitous. Meanwhile, other firms are actively developing screens based on active-matrix OLEDs (AM-OLEDs) as well as digital light processing (DLP), surface-conduction electron-emitter display (SED) and MEMS technologies. It remains to be seen which, if any, of these new display types will wind up in your next cell phone. But the competition is driving a wave of innovation, and all of us will benefit from the results.

Portable Electronics: World Class Designs is intended for engineers who design leading-edge portable consumer electronics products—cell phones, portable media players, digital cameras, camcorders: basically anything with a battery and a CPU. The book is a snapshot of a moving target, the fast-paced world of portable design. To ensure that it remains useful to you over time, we've focused on basic challenges, technologies and techniques—primarily system-level design, portable power and RF issues. These have proven to be the most vexing issues for readers of *Portable Design* magazine, and ones that aren't about to go away any time soon.

Wherever possible we've tried to provide useful ideas as well as more detailed material that will prove useful in your everyday work, including schematics, code snippets, tables, tips and tricks. We've emphasized system-level design, design methodologies and tools, all of which are critical for complex, cutting-edge, low-power designs, which is what you're doing. We hope that *Portable Electronics: World Class Designs* becomes a fixture on your bookshelf and proves to be as useful as we've tried to make it.

Now go off and turn that bright idea you've been mulling over into a world-class design. And have fun doing it.

About the Editor

John Donovan is Editor-in-Chief of *Portable Design* magazine. John has spent 25 years writing about technology, from both sides of the inbox: 12 doing mostly semiconductor PR and an equal amount as an editor at *EDN Asia*, *Portable Design*, *Circuits Assembly Asia* and *PC Fabrication Asia* as well as a columnist for *Electronic Business Asia*. John has written one previous book, dozens of manuals and hundreds of articles, the latter having appeared (in addition to his magazines) in *EE Times*, *EDN*, *Asian Electronics Engineer* and the *Asian Wall Street Journal*.

John has a B.A. in English Literature from U.C. Berkeley—which he earned while working his way through school as a microwave technician—and an MBA in Marketing from San Francisco State University. He is a member of the IEEE, Association for Computing Machinery (ACM), Audio Engineering Society (AES) and the Society for Information Display (SID). His favorite pastimes include skiing, diving, ham radio and playing with his kids.

About the Contributors

Cheryl Ajluni (Chapter 6) is a contributing editor and freelance technical writer specializing in providing technology-based content for the Internet, publications, tradeshows and a wide range of high-tech companies. She writes a monthly wireless e-newsletter, regularly moderates technology webinars and produces podcasts on a range of topics related to mobile technology.

Alan Bensky (Chapter 5), MScEE, is an electronics engineering consultant with over 30 years of experience in analog and digital design, management, and marketing. Specializing in wireless circuits and systems, Bensky has carried out projects for varied military and consumer applications and led the development of three patents on wireless distance measurement. Bensky has taught electrical engineering courses and gives lectures on radio engineering topics. He is the author of *Short-range Wireless Communication, Second Edition*, published by Elsevier in 2004, and has written several articles in international and local publications. He also wrote a recently published book on wireless location technologies and applications. Bensky is a senior member of IEEE.

Thomas Coughlin (Chapter 8), President, Coughlin Associates is a widely respected storage analyst and consultant. He has over 30 years in the data storage industry with multiple engineering and management positions at many companies. Tom has over 60 publications and six patents to his credit. Tom is also the author of *Digital Storage in Consumer Electronics: The Essential Guide*, which was published by Newnes Press in March 2008. Coughlin Associates provides market and technology analysis (including reports on several digital storage technologies and applications and a newsletter) as well as Data Storage Technical Consulting services. Tom is active with IDEMA, SNIA and the IEEE. Tom is the founder and organizer of the Annual Storage Visions Conference, a partner to the annual Consumer Electronics Show as well as the Creative Storage Conference that was recently held during the 2008 NAB. Tom is also an organizer for the Flash Memory Summit and the Data Protection Summit.

Rick Gentile (Chapter 1) joined Analog Devices Inc. (ADI) in 2000 as a Senior DSP Applications Engineer. He currently leads the Applications Engineering Group, where he is responsible for applications engineering work on the Blackfin, SHARC and TigerSHARC processors. Prior to joining ADI, Rick was a Member of the Technical Staff at MIT Lincoln

Laboratory, where he designed several signal processors used in a wide range of radar sensors. He received a B.S. in 1987 from the University of Massachusetts at Amherst and an M.S. in 1994 from Northeastern University, both in Electrical and Computer Engineering.

K. F. Ibrahim (Chapter 12) is the author of *Newnes Guide to Television and Video Technology* and *DVD Players and Drives*.

Keith Jack (Chapter 11) is Director of Product Marketing at Sigma Designs, a leading supplier of high-performance System-on-Chip (SoC) solutions for the IPTV, Blu-ray, and HDTV markets. Previously, he was Director of Product Marketing at Innovision, focused on solutions for digital televisions. Mr. Jack has also served as Strategic Marketing Manager at Harris Semiconductor and Brooktree Corporation. He has architected and introduced to market over 35 multimedia SoCs for the consumer markets, and is the author of *Video Demystified*.

Bruce Jacob (Chapter 7) is an Associate Professor of Electrical and Computer Engineering at the University of Maryland, College Park. He received his Ars Baccalaureate, cum laude, in Mathematics from Harvard University in 1988, and his M.S. and Ph.D. in Computer Science and Engineering from the University of Michigan in 1995 and 1997, respectively. In addition to his academic credentials, he has extensive experience in industry: he designed real-time embedded applications and real-time embedded architectures in the area of telecommunications for two successful Boston-area startup companies, Boston Technology (now part of Comverse Technology) and Priority Call Management (now part of uReach Technologies). At Priority Call Management he was employee number 2, the system architect, and the chief engineer; he built the first working prototype of the company's product, and he built and installed the first actual product as well.

David Katz (Chapter 1) has over 15 years of experience in circuit and system design. Currently, he is Blackfin Applications Manager at Analog Devices, Inc., where he focuses on specifying new processors. He has published over 100 embedded processing articles domestically and internationally, and he has presented several conference papers in the field. Additionally, he is co-author of *Embedded Media Processing* (Newnes 2005). Previously, he worked at Motorola, Inc., as a senior design engineer in cable modem and automation groups. David holds both a B.S. and M. Eng. in Electrical Engineering from Cornell University.

Spencer W. Ng (Chapter 7) is a senior technical staff member with Hitachi Global Storage Technologies. Prior to that, he was a research staff member with IBM Almaden Research Center. Before joining IBM, he was a member of technical staff with Bell Labs in Naperville, Illinois. He has been a researcher in the field of storage for over twenty years, and is the holder of about twenty issued U.S. patents. For the past fifteen years, he is heavily involved in studying and developing various techniques for improving disk drive performance.

Steve Rackley (Chapter 4) is the author of *Wireless Networking Technology.* He is an Independent Consultant.

Douglas Self (Chapter 10) has dedicated himself to demystifying amplifier design and establishing empirical design techniques based on electronic design principles and experimental data. His rigorous and thoroughly practical approach has established him as a leading authority on amplifier design, especially through the pages of *Electronics World* where he is a regular contributor.

Findlay Shearer (Chapters 2 and 3) holds a B.S.E.E. from Glasgow Caledonian University and a M.S.S.E. and M.B.A. from University of Texas at Austin. He currently leads the Technology Product Management team for Cellular Products at Freescale Semiconductor, Inc.

Marc T. Thompson (Chapter 9) specializes in custom R/D, analysis, and failure investigations into multi-disciplinary electrical, magnetic, electromechanical and electronic systems at Thompson Consulting, Inc. (Harvard MA). Dr. Thompson received the BSEE in 1985, the MSEE in 1992, and the Ph.D. degree in Electrical Engineering in 1997, all from the Massachusetts Institute of Technology. Dr. Thompson is also Adjunct Professor of Electrical and Computer Engineering at Worcester Polytechnic Institute. At W.P.I., Dr. Thompson teaches 2 different graduate design courses. He has also taught for University of Wisconsin-Madison, covering classes in electric motors, electromechanical systems, power electronics and magnetic design. Dr. Thompson is author of a textbook entitled *Intuitive Analog Circuit Design*, published in 2006 by Elsevier Science/Newnes. Another text entitled *Power Quality in Electronic Systems*, co-authored with Dr. Alexander Kusko, was published by McGraw-Hill in 2007. Dr. Thompson has contributed 3 chapters to *Analog Circuits (World Class Designs)*, published by Elsevier in 2008.

David T. Wang (Chapter 7) received his PhD from the University of Maryland in 2005. David's primary research interest is into power efficient, high performance memory systems that use commodity DRAM devices. As part of his research, David has collaborated with memory system architects and design engineers, and presented his work to JEDEC in support of proposals for future generation DRAM device specifications. David is presently working as an architect for MetaRAM, a memory systems startup in the Silicon Valley.

CHAPTER 1

System Resource Partitioning and Code Optimization

David Katz

Rick Gentile

In portable designs the two biggest concerns are performance and power; cell phone users want to watch streaming high-definition video while still having their tiny batteries last all day. High performance and low power are diametrically opposed, of course, and the battle is ultimately won or lost at the architectural level.

The architectural model determines the hardware/software partitioning of the design. Partitioning involves deciding which functions to implement in software, what to implement in hardware, even which loops of an algorithm could most benefit from hardware acceleration. Decisions made at this stage have huge impacts on both performance and power that can only be fine tuned at later stages of the design.

System-level design involves bringing together two very different disciplines. Hardware engineers are typically a lot more conversant with register structures than data structures, and the reverse is true for their software colleagues. The two groups have very different training and interests, and traditionally they've worked in isolation from each other, which leads to a lot of late-stage integration surprises.

The old serial style of system design—where software development waits until it has a stable hardware platform on which to run—is no longer viable in light of the very short market windows for portable consumer devices. Software engineers need a detailed knowledge of what they can and can't rely on hardware to do for them; and hardware engineers need to let their software colleagues know how to optimize their code to run most efficiently on their hardware— and then to customize their hardware for the resulting software.

The process of system-level design can't be reduced to a chapter, but this chapter is a good first approximation. In their book Embedded Media Processing, *Katz and Gentile deconstruct the problems involved in designing high-performance, low-power, small-footprint, low-cost multimedia devices, which covers the great majority of portable consumer designs. In this chapter they focus on processor architectural features that architects and software engineers need to understand to make intelligent choices early on. They cover event generation and handling, instruction pipelines, native addressing modes, and a good deal of detail on external memory flows and data memory management.*

This chapter is a good place to start when considering your next portable design.

—John Donovan

1.1 Introduction

In an ideal situation, we can select an embedded processor for our application that provides maximum performance for minimum extra development effort. In this utopian environment, we could code everything in a high-level language like C, we wouldn't need an intimate knowledge of our chosen device, it wouldn't matter where we placed our data and code, we wouldn't need to devise any data movement subsystem, the performance of external devices wouldn't matter… In short, everything would just work.

Alas, this is only the stuff of dreams and marketing presentations. The reality is, as embedded processors evolve in performance and flexibility, their complexity also increases. Depending on the time-to-market for your application, you will have to walk a fine line to reach your performance targets. The key is to find the right balance between getting the application to work and achieving optimum performance. Knowing when the performance is "good enough" rather than optimal can mean getting your product out on time versus missing a market window.

In this chapter, we want to explain some important aspects of processor architectures that can make a real difference in designing a successful multimedia system. Once you understand the basic mechanics of how the various architectural sections behave, you will be able to gauge where to focus your efforts, rather than embark on the noble yet unwieldy goal of becoming an expert on all aspects of your chosen processor.

Here, we'll explore in detail some Blackfin processor architectural constructs. Keep in mind that much of our discussion generalizes to other processor families from different vendors as well.

We will begin with what should be key focal points in any complex application: interrupt and exception handling and response times.

1.2 Event Generation and Handling

Nothing in an application should make you think "performance" more than event management. If you have used a microprocessor, you know that "events" encompass two categories: interrupts and exceptions. An interrupt is an event that happens asynchronous to processor execution. For example, when a peripheral completes a transfer, it can generate an interrupt to alert the processor that data is ready for processing.

Exceptions, on the other hand, occur synchronously to program execution. An exception occurs based on the instruction about to be executed. The change of flow due to an exception occurs prior to the offending instruction actually being executed. Later in this chapter, we'll describe the most widely used exception handler in an embedded processor—the handler that manages pages describing memory attributes. Now, however, we will focus on interrupts rather than exceptions, because managing interrupts plays such a critical role in achieving peak performance.

1.2.1 System Interrupts

System level interrupts (those that are generated by peripherals) are handled in two stages— first in the system domain, and then in the core domain. Once the system interrupt controller (SIC) acknowledges an interrupt request from a peripheral, it compares the peripheral's assigned priority to all current activity from other peripherals to decide when to service this particular interrupt request. The most important peripherals in an application should be mapped to the highest priority levels. In general, the highest-bandwidth peripherals need the highest priority. One "exception" to this rule (pardon the pun!) is where an external processor or supervisory circuit uses a nonmaskable interrupt (NMI) to indicate the occurrence of an important event, such as powering down.

When the SIC is ready, it passes the interrupt request information to the core event controller (CEC), which handles all types of events, not just interrupts. Every interrupt from the SIC maps into a priority level at the CEC that regulates how to service interrupts with respect to one another, as Figure 1.1 shows. The CEC checks the "vector" assignment for the current interrupt request, to find the address of the appropriate interrupt service routine (ISR). Finally, it loads this address into the processor's execution pipeline to start executing the ISR.

There are two key interrupt-related questions you need to ask when building your system. The first is, "How long does the processor take to respond to an interrupt?" The second is, "How long can any given task afford to wait when an interrupt comes in?"

The answers to these questions will determine what your processor can actually perform within an interrupt or exception handler.

For the purposes of this discussion, we define interrupt response time as the number of cycles it takes from when the interrupt is generated at the source (including the time it takes for the current instruction to finish executing) to the time that the first instruction is executed in the

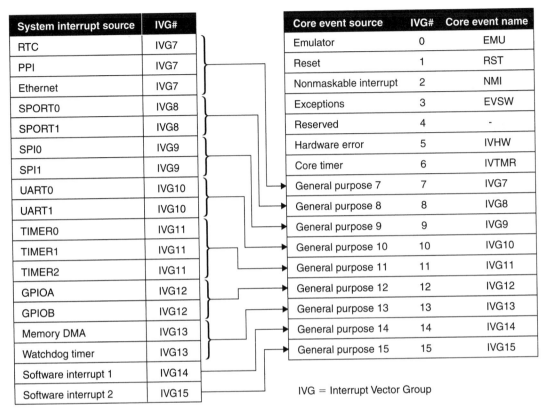

System interrupt source	IVG#
RTC	IVG7
PPI	IVG7
Ethernet	IVG7
SPORT0	IVG8
SPORT1	IVG8
SPI0	IVG9
SPI1	IVG9
UART0	IVG10
UART1	IVG10
TIMER0	IVG11
TIMER1	IVG11
TIMER2	IVG11
GPIOA	IVG12
GPIOB	IVG12
Memory DMA	IVG13
Watchdog timer	IVG13
Software interrupt 1	IVG14
Software interrupt 2	IVG15

Core event source	IVG#	Core event name
Emulator	0	EMU
Reset	1	RST
Nonmaskable interrupt	2	NMI
Exceptions	3	EVSW
Reserved	4	-
Hardware error	5	IVHW
Core timer	6	IVTMR
General purpose 7	7	IVG7
General purpose 8	8	IVG8
General purpose 9	9	IVG9
General purpose 10	10	IVG10
General purpose 11	11	IVG11
General purpose 12	12	IVG12
General purpose 13	13	IVG13
General purpose 14	14	IVG14
General purpose 15	15	IVG15

IVG = Interrupt Vector Group

Figure 1.1: Sample system-to-core interrupt mapping

interrupt service routine. In our experience, the most common method software engineers use to evaluate this interval for themselves is to set up a programmable flag to generate an interrupt when its pin is triggered by an externally generated pulse. The first instruction in the interrupt service routine then performs a write to a different flag pin. The resulting time difference is then measured on an oscilloscope. This method only provides a rough idea of the time taken to service interrupts, including the time required to latch an interrupt at the peripheral, propagate the interrupt through to the core, and then vector the core to the first instruction in the interrupt service routine. Thus, it is important to run a benchmark that more closely simulates the profile of your end application.

Once the processor is running code in an ISR, other higher priority interrupts are held off until the return address associated with the current interrupt is saved off to the stack. This is an important point, because even if you designate all other interrupt channels as higher priority than the currently serviced interrupt, these other channels will all be held off until you save the return address to the stack. The mechanism to re-enable interrupts kicks in automatically when you save the return address. When you program in C, any register the ISR uses will automatically be saved to the stack. Before exiting the ISR, the registers are restored from the stack. This also happens automatically, but depending on where your stack is located and how many registers are involved, saving and restoring data to the stack can take a significant amount of cycles.

Interrupt service routines often perform some type of processing. For example, when a line of video data arrives into its destination buffer, the ISR might run code to filter or downsample it. For this case, when the handler does the work, other interrupts are held off (provided that nesting is disabled) until the processor services the current interrupt.

When an operating system or kernel is used, however, the most common technique is to service the interrupt as soon as possible, release a semaphore, and perhaps make a call to a callback function, which then does the actual processing. The semaphore in this context provides a way to signal other tasks that it is okay to continue or to assume control over some resource.

For example, we can allocate a semaphore to a routine in shared memory. To prevent more than one task from accessing the routine, one task takes the semaphore while it is using the routine, and the other task has to wait until the semaphore has been relinquished before it can use the routine. A Callback Manager can optionally assist with this activity by allocating a callback function to each interrupt. This adds a protocol layer on top of the lowest layer of application code, but in turn it allows the processor to exit the ISR as soon as possible and return to a lower-priority task. Once the ISR is exited, the intended processing can occur without holding off new interrupts.

We already mentioned that a higher-priority interrupt can break into an existing ISR once you save the return address to the stack. However, some processors (like Blackfin) also support self-nesting of core interrupts, where an interrupt of one priority level can interrupt an ISR of the same level, once the return address is saved. This feature can be useful for building a simple scheduler or kernel that uses low-priority software-generated interrupts to preempt an ISR and allow the processing of ongoing tasks.

There are two additional performance-related issues to consider when you plan out your interrupt usage. The first is the placement of your ISR code. For interrupts that run most frequently, every attempt should be made to locate these in L1 instruction memory. On Blackfin processors, this strategy allows single-cycle access time. Moreover, if the processor were in the midst of a multi-cycle fetch from external memory, the fetch would be interrupted, and the processor would vector to the ISR code.

Keep in mind that before you re-enable higher priority interrupts, you have to save more than just the return address to the stack. Any register used inside the current ISR must also be saved. This is one reason why the stack should be located in the fastest available memory in your system. An L1 "scratchpad" memory bank, usually smaller in size than the other L1 data banks, can be used to hold the stack. This allows the fastest context switching when taking an interrupt.

1.3 Programming Methodology

It's nice not to have to be an expert in your chosen processor, but even if you program in a high-level language, it's important to understand certain things about the architecture for which you're writing code.

One mandatory task when undertaking a signal-processing-intensive project is deciding what kind of programming methodology to use. The choice is usually between assembly language and a high-level language (HLL) like C or C++. This decision revolves around many factors, so it's important to understand the benefits and drawbacks each approach entails.

The obvious benefits of C/C++ include modularity, portability and reusability. Not only do the majority of embedded programmers have experience with one of these high-level languages, but also a huge code base exists that can be ported from an existing processor domain to a new processor in a relatively straightforward manner. Because assembly language is architecture-specific, reuse is typically restricted to devices in the same processor family. Also, within a development team it is often desirable to have various teams coding different system modules, and an HLL allows these cross-functional teams to be processor-agnostic.

One reason assembly has been difficult to program is its focus on actual data flow between the processor register sets, computational units and memories. In C/C++, this manipulation occurs at a much more abstract level through the use of variables and function/procedure calls, making the code easier to follow and maintain.

The C/C++ compilers available today are quite resourceful, and they do a great job of compiling the HLL code into tight assembly code. One common mistake happens when programmers try to "outsmart" the compiler. In trying to make it easier for the compiler, they in fact make things more difficult! It's often best to just let the optimizing compiler do its job. However, the fact remains that compiler performance is tuned to a specific set of features that the tool developer considered most important. Therefore, it cannot exceed handcrafted assembly code performance in all situations.

The bottom line is that developers use assembly language only when it is necessary to optimize important processing-intensive code blocks for efficient execution. Compiler features can do a very good job, but nothing beats thoughtful, direct control of your application data flow and computation.

1.4 Architectural Features for Efficient Programming

In order to achieve high performance media processing capability, you must understand the types of core processor structures that can help optimize performance. These include the following capabilities:

- Multiple operations per cycle

- Hardware loop constructs

- Specialized addressing modes

- Interlocked instruction pipelines

These features can make an enormous difference in computational efficiency. Let's discuss each one in turn.

1.4.1 Multiple Operations per Cycle

Processors are often benchmarked by how many millions of instructions they can execute per second (MIPS). However, for today's processors, this can be misleading because of the confusion surrounding what actually constitutes an instruction. For example, multi-issue instructions, which were once reserved for use in higher-cost parallel processors, are now also available in low-cost, fixed-point processors. In addition to performing multiple ALU/MAC operations each core processor cycle, additional data loads and stores can be completed in the same cycle. This type of construct has obvious advantages in code density and execution time.

An example of a Blackfin multi-operation instruction is shown in Figure 1.2. In addition to two separate MAC operations, a data fetch and data store (or two data fetches) can also be accomplished in the same processor clock cycle. Correspondingly, each address can be updated in the same cycle that all of the other activities are occurring.

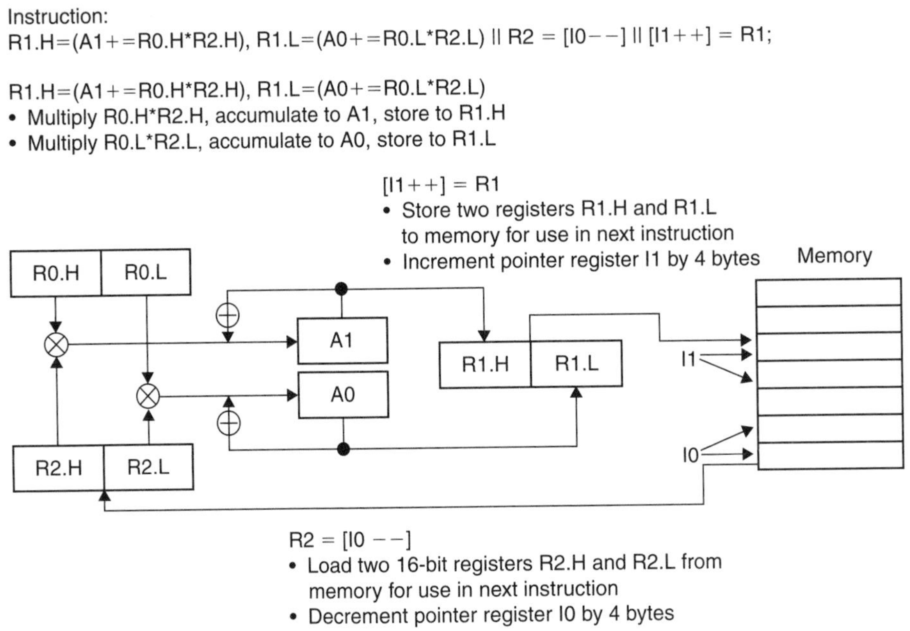

Instruction:
R1.H=(A1+=R0.H*R2.H), R1.L=(A0+=R0.L*R2.L) || R2 = [I0−−] || [I1++] = R1;

R1.H=(A1+=R0.H*R2.H), R1.L=(A0+=R0.L*R2.L)
- Multiply R0.H*R2.H, accumulate to A1, store to R1.H
- Multiply R0.L*R2.L, accumulate to A0, store to R1.L

[I1++] = R1
- Store two registers R1.H and R1.L
  to memory for use in next instruction
- Increment pointer register I1 by 4 bytes

R2 = [I0 −−]
- Load two 16-bit registers R2.H and R2.L from
  memory for use in next instruction
- Decrement pointer register I0 by 4 bytes

Figure 1.2: Example of single-cycle, multi-issue instruction

1.4.2 Hardware Loop Constructs

Looping is a critical feature in real-time processing algorithms. There are two key looping-related features that can improve performance on a wide variety of algorithms: *zero-overhead hardware loops* and *hardware loop buffers*.

Zero-overhead loops allow programmers to initialize loops simply by setting up a count value and defining the loop bounds. The processor will continue to execute this loop until the count has been reached. In contrast, a software implementation would add overhead that would cut into the real-time processing budget.

Many processors offer zero-overhead loops, but hardware loop buffers, which are less common, can really add increased performance in looping constructs. They act as a kind of cache for instructions being executed in the loop. For example, after the first time through a loop, the instructions can be kept in the loop buffer, eliminating the need to re-fetch the same code each time through the loop. This can produce a significant savings in cycles by keeping several loop instructions in a buffer where they can be accessed in a single cycle. The use of the hardware loop construct comes at no cost to the HLL programmer, since the compiler should automatically use hardware looping instead of conditional jumps.

Let's look at some examples to illustrate the concepts we've just discussed.

Example 1.1   Dot Product

The dot product, or scalar product, is an operation useful in measuring orthogonality of two vectors. It's also a fundamental operator in digital filter computations. Most C programmers should be familiar with the following implementation of a dot product:

```
short dot(const short a[], const short b[], int size) {
```

/* Note: It is important to declare the input buffer arrays as const, because this gives the compiler a guarantee that neither "a" nor "b" will be modified by the function. */

```
    int i;
    int output=0;

    for(i=0; i<size; i++){
       output += (a[i] * b[i]);
    }

    return output;
}
```

Below is the main portion of the equivalent assembly code:

```
/* P0 = Loop Count, P1 & I0 hold starting addresses of a & b
   arrays */
A1 = A0 = 0;              /* A0 & A1 are accumulators */
LSETUP (loop1, loop1) LC0 = P0 ;    /* Set up hardware loop
   starting and ending at label loop1 */
loop1: A1 += R1.H * R0.H, A0 += R1.L * R0.L || R1 = [ P1 ++ ]
   || R0 = [ I0 ++ ];
```

The following points illustrate how a processor's architectural features can facilitate this tight coding.

Hardware loop buffers and loop counters eliminate the need for a jump instruction at the end of each iteration. Since a dot product is a summation of products, it is

---

implemented in a loop. Some processors use a JUMP instruction at the end of each iteration in order to process the next iteration of the loop. This contrasts with the assembly program above, which shows the LSETUP instruction as the only instruction needed to implement a loop.

Multi-issue instructions allow computation and two data accesses with pointer updates in the same cycle. In each iteration, the values a[i] and b[i] must be read, then multiplied, and finally written back to the running summation in the variable output. On many microcontroller platforms, this effectively amounts to four instructions. The last line of the assembly code shows that all of these operations can be executed in one cycle.

Parallel ALU operations allow two 16-bit instructions to be executed simultaneously. The assembly code shows two accumulator units (A0 and A1) used in each iteration. This reduces the number of iterations by 50%, effectively halving the original execution time.

1.4.3 Specialized Addressing Modes

1.4.3.1 Byte Addressability

Allowing the processor to access multiple data words in a single cycle requires substantial flexibility in address generation. In addition to the more signal-processing-centric access sizes along 16-and 32-bit boundaries, byte addressing is required for the most efficient processing. This is important for multimedia processing because many video-based systems operate on 8-bit data. When memory accesses are restricted to a single boundary, the processor may spend extra cycles to mask off relevant bits.

1.4.3.2 Circular Buffering

Another beneficial addressing capability is *circular buffering*. We will look at it from the processor's perspective. For maximum efficiency, this feature must be supported directly by the processor, with no special management overhead. Circular buffering allows a programmer to define buffers in memory and stride through them automatically. Once the buffer is set up, no special software interaction is required to navigate through the data. The address generator handles nonunity strides and, more importantly, handles the "wrap-around" feature illustrated in Figure 1.3. Without this automated address generation, the programmer would have to manually keep track of buffer pointer positions, thus wasting valuable processing cycles.

Many optimizing compilers will automatically use hardware circular buffering when they encounter array addressing with a modulus operator.

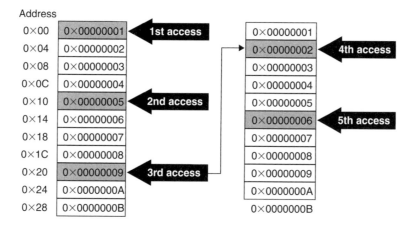

- Base address and starting index address = 0×0
- Index address register I0 points to address 0×0
- Buffer length L = 44 (11 data elements * 4 bytes/element)
- Modify register M0 = 16 (4 elements * 4 bytes/element)

Sample code:

```
R0 = [I0++M0];  // R0 = 1 and I0 points to 0×10 after execution
R1 = [I0++M0];  // R1 = 5 and I0 points to 0×20 after execution
R2 = [I0++M0];  // R2 = 9 and I0 points to 0×04 after execution
R3 = [I0++M0];  // R3 = 2 and I0 points to 0×14 after execution
R4 = [I0++M0];  // R4 = 6 and I0 points to 0×24 after execution
```

Figure 1.3: Circular buffer in hardware

Example 1.2   Single-Sample FIR

The finite impulse response filter is a very common filter structure equivalent to the convolution operator. A straightforward C implementation follows:

```
// sample the signal into a circular buffer
x[cur] = sampling_function();
cur = (cur+1)%TAPS; // advance cur pointer in circular fashion

// perform the multiply-addition
y = 0;
```

```
for (k=0; k<TAPS; k++) {
  y += h[k] * x[(cur+k)%TAPS];
}
```

The essential part of an FIR kernel written in assembly is shown below.

```
/* the samples are stored in the R0 register, while the
   coefficients are stored in the R1 register */
LSETUP (loop_begin, loop_end) LC0 = P0;    /* loop counter set to
   traverse the filter */
loop_begin: A1+=R0.H*R1.L, A0+=R0.L*R1.L || R0.L=[I0++]; /*
   perform MAC and fetch next data */
loop_end: A1+=R0.L*R1.H, A0+=R0.H*R1.H || R0.H=[I0++] || R1=
   [I1++];/* perform MAC and fetch next data */
```

In the C code snippet, the % (modulus) operator provides a mechanism for circular buffering. As shown in the assembly kernel, this modulus operator does not get translated into an additional instruction inside the loop. Instead, the Data Address Generator registers I0 and I1 are configured outside the loop to automatically wrap around to the beginning upon hitting the buffer boundary.

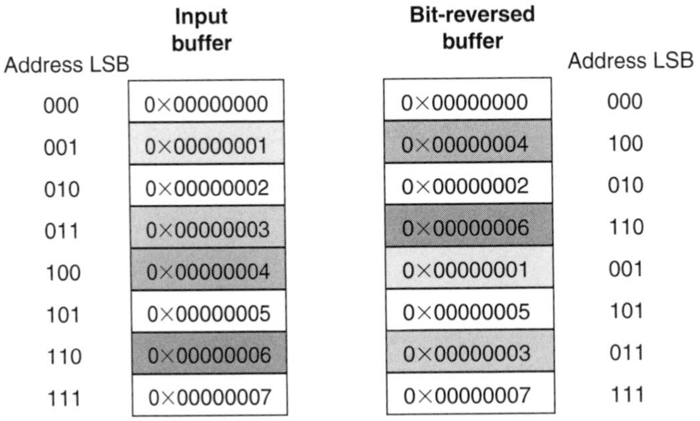

	Input buffer	Bit-reversed buffer	
Address LSB			Address LSB
000	0×00000000	0×00000000	000
001	0×00000001	0×00000004	100
010	0×00000002	0×00000002	010
011	0×00000003	0×00000006	110
100	0×00000004	0×00000001	001
101	0×00000005	0×00000005	101
110	0×00000006	0×00000003	011
111	0×00000007	0×00000007	111

Sample code:

```
LSETUP(start,end) LC0 = P0;          //Loop count = 8
start: R0 = [I0] || I0 += M0 (BREV);  // I0 points to input buffer, automatically incremented in
                                      //bit-reversed progression
end:[I2++] = R0;                      // I2 points to bit-reversed buffer
```

Figure 1.4: Bit reversal in hardware

1.4.3.3 Bit Reversal

An essential addressing mode for efficient signal-processing operations such as the FFT and DCT is bit reversal. Just as the name implies, bit reversal involves reversing the bits in a binary address. That is, the least significant bits are swapped in position with the most significant bits. The data ordering required by a radix-2 butterfly is in "bit-reversed" order, so bit-reversed indices are used to combine FFT stages. It is possible to calculate these bit-reversed indices in software, but this is very inefficient. An example of bit reversal address flow is shown in Figure 1.4.

Since bit reversal is very specific to algorithms like fast Fourier transforms and discrete Fourier transforms, it is difficult for any HLL compiler to employ hardware bit reversal. For this reason, comprehensive knowledge of the underlying architecture and assembly language are key to fully utilizing this addressing mode.

Example 1.3   FFT

A fast Fourier transform is an integral part of many signal-processing algorithms. One of its peculiarities is that if the input vector is in sequential time order, the output comes out in bit-reversed order. Most traditional general-purpose processors require the programmer to implement a separate routine to unscramble the bit-reversed output. On a media processor, bit reversal is often designed into the addressing engine.

Allowing the hardware to automatically bit-reverse the output of an FFT algorithm relieves the programmer from writing additional utilities, and thus improves performance.

1.4.4 Interlocked Instruction Pipelines

As processors increase in speed, it is necessary to add stages to the processing pipeline. For instances where a high-level language is the sole programming language, the compiler is responsible for dealing with instruction scheduling to maximize performance through the pipeline. That said, the following information is important to understand even if you're programming in C.

On older processor architectures, pipelines are usually not interlocked. On these architectures, executing certain combinations of neighboring instructions can yield incorrect results. Interlocked pipelines like the one in Figure 1.5, on the other hand, make assembly

IF1-3: Instruction fetch
DC: Decode
AC: Address calculation
EX1-4: Execution
WB: Writeback

		Pipeline stage								
	IF1	IF2	IF3	DC	AC	EX1	EX2	EX3	EX4	WB
1	Inst1									
2	Inst2	Inst1								
3	Inst3	Inst2	Inst1							
4	Inst4	Inst3	Inst2	Inst1						
5	Inst5	Inst4	Inst3	Inst2	Inst1					
6	Branch	Inst5	Inst4	Inst3	Inst2	Inst1				
7	Stall	Branch	Inst5	Inst4	Inst3	Inst2	Inst1			
8	Stall	Stall	Branch	Inst5	Inst4	Inst3	Inst2	Inst1		
9	Stall	Stall	Stall	Branch	Inst5	Inst4	Inst3	Inst2	Inst1	
10	Stall	Stall	Stall	Stall	Branch	Inst5	Inst4	Inst3	Inst2	Inst1

Time (vertical axis label on left)

Figure 1.5: Example of interlocked pipeline architecture with stalls inserted

programming (as well as the life of compiler engineers) easier by automatically inserting stalls when necessary. This prevents the assembly programmer from scheduling instructions in a way that will produce inaccurate results. It should be noted that, even if the pipeline is interlocked, instruction rearrangement can still yield optimization improvements by eliminating unnecessary stalls.

Let's take a look at stalls in more detail. Stalls will show up for one of four reasons:

1. The instruction in question may itself take more than one cycle to execute. When this is the case, there isn't anything you can do to eliminate the stall. For example, a 32-bit integer multiply might take three core-clock cycles to execute on a 16-bit processor. This will cause a "bubble" in two pipeline stages for a three-cycle instruction.

2. The second case involves the location of one instruction in the pipeline with respect to an instruction that follows it. For example, in some instructions, a stall may exist because the result of the first instruction is used as an operand of the following instruction. When

this happens and you are programming in assembly, it is often possible to move the instruction so that the stall is not in the critical path of execution.

Here are some simple examples on Blackfin processors that demonstrate these concepts.

Register Transfer/Multiply latencies (One stall, due to R0 being used in the multiply):

```
R0 = R4;  /* load R0 with contents of R4 */
<STALL>
R2.H = R1.L * R0.H;  /* R0 is used as an operand */
```

In this example, any instruction that does not change the value of the operands can be placed in-between the two instructions to hide the stall.

When we load a pointer register and try to use the content in the next instruction, there is a latency of three stalls:

```
P3 = [SP++];  /* Pointer register loaded from stack */

<STALL>
<STALL>
<STALL>

R0 = P3;  /* Use contents of P3 after it gets its value
    from earlier instruction */
```

---

3. The third case involves a change of flow. While a deeper pipeline allows increased clock speeds, any time a change of flow occurs, a portion of the pipeline is flushed, and this consumes core-clock cycles. The branching latency associated with a change of flow varies based on the pipeline depth. Blackfin's 10-stage pipeline yields the following latencies:

Instruction flow dependencies	(Static Prediction):
Correctly predicted branch	(4 stalls)
Incorrectly predicted branch	(8 stalls)
Unconditional branch	(8 stalls)
"Drop-through" conditional branch	(0 stalls)

The term "predicted" is used to describe what the sequencer does as instructions that will complete ten core-clock cycles later enter the pipeline. You can see that when the sequencer does not take a branch, and in effect "drops through" to the next instruction after the conditional one, there are no added cycles. When an unconditional branch occurs, the maximum number of stalls occurs (eight cycles). When the processor predicts that a branch occurs and it actually is taken, the number of stalls is four. In the case where it predicted no branch, but one is actually taken, it mirrors the case of an unconditional branch.

One more note here. The maximum number of stalls is eight, while the depth of the pipeline is ten. This shows that the branching logic in an architecture does not implicitly have to match the full size of the pipeline.

4. The last case involves a conflict when the processor is accessing the same memory space as another resource (or simply fetching data from memory other than L1). For instance, a core fetch from SDRAM will take multiple core-clock cycles. As another example, if the processor and a DMA channel are trying to access the same memory bank, stalls will occur until the resource is available to the lower-priority process.

1.5 Compiler Considerations for Efficient Programming

Since the compiler's foremost task is to create correct code, there are cases where the optimizer is too conservative. In these cases, providing the compiler with extra information (through pragmas, built-in keywords, or command-line switches) will help it create more optimized code.

In general, compilers can't make assumptions about what an application is doing. This is why pragmas exist—to let the compiler know it is okay to make certain assumptions. For example, a pragma can instruct the compiler that variables in a loop are aligned and that they are not referenced to the same memory location. This extra information allows the compiler to optimize more aggressively, because the programmer has made a guarantee dictated by the pragma.

In general, a four-step process can be used to optimize an application consisting primarily of HLL code:

1. Compile with an HLL-optimizing compiler.

2. Profile the resulting code to determine the "hot spots" that consume the most processing bandwidth.

3. Update HLL code with pragmas, built-in keywords, and compiler switches to speed up the "hot spots."

4. Replace HLL procedures/functions with assembly routines in places where the optimizer did not meet the timing budget.

For maximum efficiency, it is always a good idea to inspect the most frequently executed compiler-generated assembly code to make a judgment on whether the code could be more vectorized. Sometimes, the HLL program can be changed to help the compiler produce faster code through more use of multi-issue instructions. If this still fails to produce code that is fast enough, then it is up to the assembly programmer to fine-tune the code line-by-line to keep all available hardware resources from idling.

1.5.1 Choosing Data Types

It is important to remember how the standard data types available in C actually map to the architecture you are using. For Blackfin processors, each type is shown in Table 1.1.

The **float** (32-bit), **double** (32-bit), **long long** (64-bit) and **unsigned long long** (64-bit) formats are not supported natively by the processor, but these can be emulated.

1.5.2 Arrays versus Pointers

We are often asked whether it is better to use arrays to represent data buffers in C, or whether pointers are better. Compiler performance engineers always point out that arrays are easier to analyze. Consider the example:

Table 1.1: C data types and their mapping to Blackfin registers

C type	Blackfin equivalent
char	8-bit signed
unsigned char	8-bit unsigned
short	16-bit signed integer
unsigned short	16-bit unsigned integer
int	32-bit signed integer
unsigned int	32-bit unsigned integer
long	32-bit signed integer
unsigned long	32-bit unsigned integer

```
void array_example (int a[], int b[], int sum[], int n)
{
    int i;
    for (i = 0; i < n; ++i)
    sum[i] = a[i] + b[i];
}
```

Even though we chose a simple example, the point is that these constructs are very easy to follow.

Now let's look at the same function using pointers. With pointers, the code is "closer" to the processor's native language.

```
void pointer_example (int a[], int b[], int sum[], int n) {
    int i;
    for (i = 0; i < n; ++i)
        *out++ = *a++ + *b++ ;
}
```

Which produces the most efficient code? Actually, there is usually very little difference. It is best to start by using the array notation because it is easier to read. An array format can be better for "alias" analysis in helping to ensure there is no overlap between elements in a buffer. If performance is not adequate with arrays (for instance, in the case of tight inner loops), pointers may be more useful.

1.5.3 Division

Fixed-point processors often do not support division natively. Instead, they offer division primitives in the instruction set, and these help accelerate division.

The "cost" of division depends on the range of the inputs. There are two possibilities: You can use division primitives where the result and divisor each fit into 16 bits. On Blackfin processors, this results in an operation of $\sim$40 cycles. For more precise, bitwise 32-bit division, the result is $\sim$10$\times$ more cycles.

If possible, it is best to avoid division, because of the additional overhead it entails. Consider the example:

```
if ( X/Y > A/B )
```

This can easily be rewritten as:

```
if ( X * B > A * Y )
```

to eliminate the division.

Keep in mind that the compiler does not know anything about the data precision in your application. For example, in the context of the above equation rewrite, two 12-bit inputs are "safe," because the result of the multiplication will be 24 bits maximum. This quick check will indicate when you can take a shortcut, and when you have to use actual division.

1.5.4 Loops

We already discussed hardware looping constructs. Here we'll talk about software looping in C. We will attempt to summarize what you can do to ensure best performance for your application.

1. Try to keep loops short. Large loop bodies are usually more complex and difficult to optimize. Additionally, they may require register data to be stored in memory, decreasing code density and execution performance.

2. Avoid loop-carried dependencies. These occur when computations in the present iteration depend on values from previous iterations. Dependencies prevent the compiler from taking advantage of loop overlapping (i.e., nested loops).

3. Avoid manually unrolling loops. This confuses the compiler and cheats it out of a job at which it typically excels.

4. Don't execute loads and stores from a noncurrent iteration while doing computations in the current loop iteration. This introduces loop-carried dependencies. This means avoiding loop array writes of the form:

```
for (i = 0; i < n; ++i)
    a[i] = b[i] * a[c[i]]; /* has array dependency*/
```

5. Make sure that inner loops iterate more than outer loops, since most optimizers focus on inner loop performance.

6. Avoid conditional code in loops. Large control-flow latencies may occur if the compiler needs to generate conditional jumps.

As an example,

```
for {
    if { ….. } else {…..}
}
```

should be replaced, if possible, by:

```
if {
    for {…..}
} else {
    for    {…..}
}
```

7. Don't place function calls in loops. This prevents the compiler from using hardware loop constructs, as we described earlier in this chapter.

8. Try to avoid using variables to specify stride values. The compiler may need to use division to figure out the number of loop iterations required, and you now know why this is not desirable!

1.5.5 Data Buffers

It is important to think about how data is represented in your system. It's better to prearrange the data in anticipation of "wider" data fetches—that is, data fetches that optimize the amount of data accessed with each fetch. Let's look at an example that represents complex data.

One approach that may seem intuitive is:

```
short Real_Part[ N ];
short Imaginary_Part [ N ];
```

While this is perfectly adequate, data will be fetched in two separate 16-bit accesses. It is often better to arrange the array in one of the following ways:

```
short Complex [ N*2 ];
    or
long Complex [ N ];
```

Here, the data can be fetched via one 32-bit load and used whenever it's needed. This single fetch is faster than the previous approach.

On a related note, a common performance-degrading buffer layout involves constructing a 2D array with a column of pointers to **malloc**'d rows of data. While this allows complete flexibility in row and column size and storage, it may inhibit a compiler's ability to optimize, because the compiler no longer knows if one row follows another, and therefore it can see no constant offset between the rows.

1.5.6 Intrinsics and In-lining

It is difficult for compilers to solve all of your problems automatically and consistently. This is why you should, if possible, avail yourself of "in-line" assembly instructions and intrinsics.

In-lining allows you to insert an assembly instruction into your C code directly. Sometimes this is unavoidable, so you should probably learn how to in-line for the compiler you're using.

In addition to in-lining, most compilers support intrinsics, and their optimizers fully understand intrinsics and their effects. The Blackfin compiler supports a comprehensive array of 16-bit intrinsic functions, which must be programmed explicitly. Below is a simple example of an intrinsic that multiplies two 16-bit values.

```
#include <fract.h>
fract32 fdot(fract16 *x, fract16 *y, int n)
{
    fract32 sum = 0;
    int i;
    for (i = 0; i < n; i++)
        sum = add_fr1X32(sum, mult_fr1X32 (x[i], y[i]));
    return sum;
}
```

Here are some other operations that can be accomplished through intrinsics:

- Align operations
- Packing operations
- Disaligned loads
- Unpacking
- Quad 8-bit add/subtract
- Dual 16-bit add/clip
- Quad 8-bit average
- Accumulator extract with addition
- Subtract/absolute value/accumulate

The intrinsics that perform the above functions allow the compiler to take advantage of video-specific instructions that improve performance but that are difficult for a compiler to use natively.

When should you use in-lining, and when should you use intrinsics? Well, you really don't have to choose between the two. Rather, it is important to understand the results of using both, so that they become tools in your programming arsenal. With regard to in-lining of assembly instructions, look for an option where you can include in the in-lining construct the registers you will be "touching" in the assembly instruction. Without this information, the compiler will invariably spend more cycles, because it's limited in the assumptions it can make and therefore has to take steps that can result in lower performance. With intrinsics, the compiler can use its knowledge to improve the code it generates on both sides of the intrinsic code. In addition, the fact that the intrinsic exists means someone who knows the compiler and architecture very well has already translated a common function to an optimized code section.

1.5.7 Volatile Data

The **volatile** data type is essential for peripheral-related registers and interrupt-related data.

Some variables may be accessed by resources not visible to the compiler. For example, they may be accessed by interrupt routines, or they may be set or read by peripherals.

The **volatile** attribute forces all operations with that variable to occur exactly as written in the code. This means that a variable is read from memory each time it is needed, and it's written back to memory each time it's modified. The exact order of events is preserved. Missing a **volatile** qualifier is the largest single cause of trouble when engineers port from one C-based processor to another. Architectures that don't require **volatile** for hardware-related accesses probably treat all accesses as volatile by default and thus may perform at a lower performance level than those that require you to state this explicitly. When a C program works with optimization turned off but doesn't work with optimization on, a missing **volatile** qualifier is usually the culprit.

1.6 System and Core Synchronization

Earlier we discussed the importance of an interlocked pipeline, but we also need to discuss the implications of the pipeline on the different operating domains of a processor. On Blackfin devices, there are two synchronization instructions that help manage the relationship between when the core and the peripherals complete specific instructions or sequences. While these instructions are very straightforward, they are sometimes used more than necessary. The CSYNC instruction prevents any other instructions from entering the pipeline until all pending core activities have completed. The SSYNC behaves in a similar manner, except that it holds off new instructions until all pending system actions have completed. The performance impact from a CSYNC is measured in multiple CCLK cycles, while the impact of an SSYNC is measured in multiple SCLKs. When either of these instructions is used too often, performance will suffer needlessly.

So when do you need these instructions? We'll find out in a minute. But first we need to talk about memory transaction ordering.

1.6.1 Load/Store Synchronization

Many embedded processors support the concept of a Load/Store data access mechanism. What does this mean, and how does it impact your application? "Load/Store" refers to the characteristic in an architecture where memory operations (loads and stores) are intentionally separated from the arithmetic functions that use the results of fetches from memory operations. The separation is made because memory operations, especially instructions that access off-chip memory or I/O devices, take multiple cycles to complete and would normally halt the processor, preventing an instruction execution rate of one instruction per core-clock cycle. To avoid this situation, data is brought into a data register from a source memory location, and once it is in the register, it can be fed into a computation unit.

In write operations, the "store" instruction is considered complete as soon as it executes, even though many clock cycles may occur before the data is actually written to an external memory or I/O location. This arrangement allows the processor to execute one instruction per clock cycle, and it implies that the synchronization between when writes complete and when subsequent instructions execute is not guaranteed. This synchronization is considered unimportant in the context of most memory operations. With the presence of a write buffer that sits between the processor and external memory, multiple writes can, in fact, be made without stalling the processor.

For example, consider the case where we write a simple code sequence consisting of a single write to L3 memory surrounded by five NOP ("no operation") instructions. Measuring the cycle count of this sequence running from L1 memory shows that it takes six cycles to execute. Now let's add another write to L3 memory and measure the cycle count again. We will see the cycle count increase by one cycle each time, until we reach the limits of the write buffer, at which point it will increase substantially until the write buffer is drained.

1.6.2 Ordering

The relaxation of synchronization between memory accesses and their surrounding instructions is referred to as "weak ordering" of loads and stores. Weak ordering implies that the timing of the actual completion of the memory operations—even the order in which these events occur—may not align with how they appear in the sequence of a program's source code.

In a system with weak ordering, only the following items are guaranteed:

- Load operations will complete before a subsequent instruction uses the returned data.

- Load operations using previously written data will use the updated values, even if they haven't yet propagated out to memory.

- Store operations will eventually propagate to their ultimate destination.

Because of weak ordering, the memory system is allowed to prioritize reads over writes. In this case, a write that is queued anywhere in the pipeline, but not completed, may be deferred by a subsequent read operation, and the read is allowed to be completed before the write. Reads are prioritized over writes because the read operation has a dependent operation waiting on its completion, whereas the processor considers the write operation complete, and the write does not stall the pipeline if it takes more cycles to propagate the value out to memory.

For most applications, this behavior will greatly improve performance. Consider the case where we are writing to some variable in external memory. If the processor performs a write to one location followed by a read from a different location, we would prefer to have the read complete before the write.

This ordering provides significant performance advantages in the operation of most memory instructions. However, it can cause side effects—when writing to or reading from non-memory locations such as I/O device registers, the order of how read and write operations complete is often significant.

For example, a read of a status register may depend on a write to a control register. If the address in either case is the same, the read would return a value from the write buffer rather than from the actual I/O device register, and the order of the read and write at the register may be reversed. Both of these outcomes could cause undesirable side effects. To prevent these occurrences in code that requires precise (strong) ordering of load and store operations, synchronization instructions like CSYNC or SSYNC should be used.

The CSYNC instruction ensures all pending core operations have completed and the core buffer (between the processor core and the L1 memories) has been flushed before proceeding to the next instruction. Pending core operations may include any pending interrupts, speculative states (such as branch predictions) and exceptions. A CSYNC is typically required after writing to a control register that is in the core domain. It ensures that whatever action you wanted to happen by writing to the register takes place before you execute the next instruction.

The SSYNC instruction does everything the CSYNC does, and more. As with CSYNC, it ensures all pending operations have to be completed between the processor core and the L1 memories. SSYNC further ensures completion of all operations between the processor core, external memory and the system peripherals. There are many cases where this is important, but the best example is when an interrupt condition needs to be cleared at a peripheral before an interrupt service routine (ISR) is exited. Somewhere in the ISR, a write is made to a peripheral register to "clear" and, in effect, acknowledge the interrupt. Because of differing clock domains between the core and system portions of the processor, the SSYNC ensures the peripheral clears the interrupt before exiting the ISR. If the ISR were exited before the interrupt was cleared, the processor might jump right back into the ISR.

Load operations from memory do not change the state of the memory value itself. Consequently, issuing a speculative memory-read operation for a subsequent load instruction usually has no undesirable side effect. In some code sequences, such as a conditional branch instruction followed by a load, performance may be improved by speculatively issuing the read request to the memory system before the conditional branch is resolved.

For example,

```
IF CC JUMP away_from_here
R0 = [P2];
...
away_from_here:
```

If the branch is taken, then the load is flushed from the pipeline, and any results that are in the process of being returned can be ignored. Conversely, if the branch is not taken, the memory will have returned the correct value earlier than if the operation were stalled until the branch condition was resolved.

However, this could cause an undesirable side effect for a peripheral that returns sequential data from a FIFO or from a register that changes value based on the number of reads that are requested. To avoid this effect, use an SSYNC instruction to guarantee the correct behavior between read operations.

Store operations never access memory speculatively, because this could cause modification of a memory value before it is determined whether the instruction should have executed.

1.6.3 Atomic Operations

We have already introduced several ways to use semaphores in a system. While there are many ways to implement a semaphore, using atomic operations is preferable, because they provide noninterruptible memory operations in support of semaphores between tasks.

The Blackfin processor provides a single atomic operation: TESTSET. The TESTSET instruction loads an indirectly addressed memory word, tests whether the low byte is zero, and then sets the most significant bit of the low memory byte without affecting any other bits. If the byte is originally zero, the instruction sets a status bit. If the byte is originally nonzero, the instruction clears the status bit. The sequence of this memory transaction is atomic—hardware bus locking insures that no other memory operation can occur between the test and set portions of this instruction. The TESTSET instruction can be interrupted by the core. If this happens, the TESTSET instruction is executed again upon return from the interrupt. Without something like this TESTSET facility, it is difficult to ensure true protection when more than one entity (for example, two cores in a dual-core device) vies for a shared resource.

1.7 Memory Architecture—The Need for Management

In this section, we will discuss how to best use memory in your application.

1.7.1 Memory Access Trade-offs

Embedded media processors usually have a small amount of fast, on-chip memory, whereas microcontrollers usually have access to large external memories. A hierarchical memory architecture combines the best of both approaches, providing several tiers of memory with different performance levels. For applications that require the most determinism, on-chip SRAM can be accessed in a single core-clock cycle. Systems with larger code sizes can utilize bigger, higher-latency on-chip and off-chip memories.

Most complex programs today are large enough to require external memory, and this would dictate an unacceptably slow execution speed. As a result, programmers would be forced to manually move key code in and out of internal SRAM. However, by adding data and instruction caches into the architecture, external memory becomes much more manageable. The cache reduces the manual movement of instructions and data into the processor core, thus greatly simplifying the programming model.

Figure 1.6 demonstrates a typical memory configuration where instructions are brought in from external memory as they are needed. Instruction cache usually operates with some type

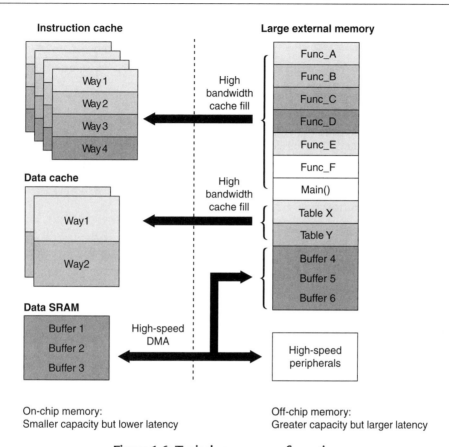

Figure 1.6: Typical memory configuration

of least recently used (LRU) algorithm, insuring that instructions that run more often get replaced less often. The figure also illustrates that having the ability to configure some on-chip data memory as cache and some as SRAM can optimize performance. DMA controllers can feed the core directly, while data from tables can be brought into the data cache as they are needed.

When porting existing applications to a new processor, "out-of-the-box" performance is important. As we saw earlier, there are many features compilers exploit that require minimal developer involvement. Yet, there are many other techniques that, with a little extra effort by the programmer, can have a big impact on system performance.

Proper memory configuration and data placement always pays big dividends in improving system performance. On high-performance media processors, there are typically three paths into a memory bank. This allows the core to make multiple accesses in a single clock cycle (e.g., a load and store, or two loads). By laying out an intelligent data flow, a developer can avoid conflicts created when the core processor and DMA vie for access to the same memory bank.

1.7.2 Instruction Memory Management—To Cache or To DMA?

Maximum performance is only realized when code runs from internal L1 memory. Of course, the ideal embedded processor would have an unlimited amount of L1 memory, but this is not practical. Therefore, programmers must consider several alternatives to take advantage of the L1 memory that exists in the processor, while optimizing memory and data flows for their particular system. Let's examine some of these scenarios.

The first, and most straightforward, situation is when the target application code fits entirely into L1 instruction memory. For this case, there are no special actions required, other than for the programmer to map the application code directly to this memory space. It thus becomes intuitive that media processors must excel in code density at the architectural level.

In the second scenario, a caching mechanism is used to allow programmers access to larger, less expensive external memories. The cache serves as a way to automatically bring code into L1 instruction memory as needed. The key advantage of this process is that the programmer does not have to manage the movement of code into and out of the cache. This method is best when the code being executed is somewhat linear in nature. For nonlinear code, cache lines may be replaced too often to allow any real performance improvement.

The instruction cache really performs two roles. For one, it helps pre-fetch instructions from external memory in a more efficient manner. That is, when a cache miss occurs, a cache-line fill will fetch the desired instruction, along with the other instructions contained within the cache line. This ensures that, by the time the first instruction in the line has been executed, the instructions that immediately follow have also been fetched. In addition, since caches usually operate with an LRU algorithm, instructions that run most often tend to be retained in cache.

Some strict real-time programmers tend not to trust cache to obtain the best system performance. Their argument is that if a set of instructions is not in cache when needed for execution, performance will degrade. Taking advantage of cache-locking mechanisms can offset this issue. Once the critical instructions are loaded into cache, the cache lines can be locked, and thus not replaced. This gives programmers the ability to keep what they need in cache and to let the caching mechanism manage less-critical instructions.

In a final scenario, code can be moved into and out of L1 memory using a DMA channel that is independent of the processor core. While the core is operating on one section of memory, the DMA is bringing in the section to be executed next. This scheme is commonly referred to as an overlay technique.

While overlaying code into L1 instruction memory via DMA provides more determinism than caching it, the trade-off comes in the form of increased programmer involvement. In other words, the programmer needs to map out an overlay strategy and configure the DMA channels appropriately. Still, the performance payoff for a well-planned approach can be well worth the extra effort.

1.7.3 Data Memory Management

The data memory architecture of an embedded media processor is just as important to the overall system performance as the instruction clock speed. Because multiple data transfers take place simultaneously in a multimedia application, the bus structure must support both core and DMA accesses to all areas of internal and external memory. It is critical that arbitration between the DMA controller and the processor core be handled automatically, or performance will be greatly reduced. Core-to-DMA interaction should only be required to set up the DMA controller, and then again to respond to interrupts when data is ready to be processed.

A processor performs data fetches as part of its basic functionality. While this is typically the least efficient mechanism for transferring data to or from off-chip memory, it provides the simplest programming model. A small, fast scratchpad memory is sometimes available as part of L1 data memory, but for larger, off-chip buffers, access time will suffer if the core must fetch everything from external memory. Not only will it take multiple cycles to fetch the data, but the core will also be busy doing the fetches. It is important to consider how the core processor handles reads and writes. As we detailed above, Blackfin processors possess a multi-slot write buffer that can allow the core to proceed with subsequent instructions before all posted writes have completed. For example, in the following code sample, if the pointer register P0 points to an address in external memory and P1 points to an address in internal memory, line 50 will be executed before R0 (from line 46) is written to external memory:

```
...
Line 45: R0 = R1+R2;
Line 46: [P0] = R0; /* Write the value contained in R0 to slower
    external memory */
Line 47: R3 = 0×0 (z);
Line 48: R4 = 0×0 (z);
Line 49: R5 = 0×0 (z);
Line 50: [P1] = R0; /* Write the value contained in R0 to faster
    internal memory */
```

In applications where large data stores constantly move into and out of external DRAM, relying on core accesses creates a difficult situation. While core fetches are inevitably needed at times, DMA should be used for large data transfers, in order to preserve performance.

1.7.3.1 What About Data Cache?

The flexibility of the DMA controller is a double-edged sword. When a large C/C++ application is ported between processors, a programmer is sometimes hesitant to integrate DMA functionality into already-working code. This is where data cache can be very useful, bringing data into L1 memory for the fastest processing. The data cache is attractive because it acts like a mini-DMA, but with minimal interaction on the programmer's part.

Because of the nature of cache-line fills, data cache is most useful when the processor operates on consecutive data locations in external memory. This is because the cache doesn't just store the immediate data currently being processed; instead, it prefetches data in a region contiguous to the current data. In other words, the cache mechanism assumes there's a good chance that the current data word is part of a block of neighboring data about to be processed. For multimedia streams, this is a reasonable conjecture.

Since data buffers usually originate from external peripherals, operating with data cache is not always as easy as with instruction cache. This is due to the fact that coherency must be managed manually in "non-snooping" caches. "Non-snooping" means that the cache is not aware of when data changes in source memory unless it makes the change directly. For these caches, the data buffer must be invalidated before making any attempt to access the new data. In the context of a C-based application, this type of data is "volatile." This situation is shown in Figure 1.7.

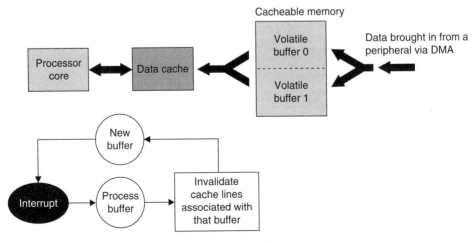

Figure 1.7: Data cache and DMA coherency

In the general case, when the value of a variable stored in cache is different from its value in the source memory, this can mean that the cache line is "dirty" and still needs to be written back to memory. This concept does not apply for volatile data. Rather, in this case the cache line may be "clean," but the source memory may have changed without the knowledge of the core processor. In this scenario, before the core can safely access a volatile variable in data cache, it must invalidate (but not flush!) the affected cache line.

This can be performed in one of two ways. The cache tag associated with the cache line can be directly written, *or* a "Cache Invalidate" instruction can be executed to invalidate the target memory address. Both techniques can be used interchangeably, but the direct method is usually a better option when a large data buffer is present (e.g., one greater in size than the data cache size). The Invalidate instruction is always preferable when the buffer size is smaller than the size of the cache. This is true even when a loop is required, since the Invalidate instruction usually increments by the size of each cache line instead of by the more typical 1-, 2- or 4-byte increment of normal addressing modes.

From a performance perspective, this use of data cache cuts down on improvement gains, in that data has to be brought into cache each time a new buffer arrives. In this case, the benefit of caching is derived solely from the pre-fetch nature of a cache-line fill. Recall that the prime benefit of cache is that the data is present the second time through the loop.

One more important point about volatile variables, regardless of whether or not they are cached—if they are shared by both the core processor and the DMA controller, the programmer must implement some type of semaphore for safe operation. In sum, it is best to keep volatiles out of data cache altogether.

1.7.4 System Guidelines for Choosing Between DMA and Cache

Let's consider three widely used system configurations to shed some light on which approach works best for different system classifications.

1.7.4.1 Instruction Cache, Data DMA

This is perhaps the most popular system model, because media processors are often architected with this usage profile in mind. Caching the code alleviates complex instruction flow management, assuming the application can afford this luxury. This works well when the system has no hard real-time constraints, so that a cache miss would not wreak havoc on the timing of tightly coupled events (for example, video refresh or audio/video synchronization).

Also, in cases where processor performance far outstrips processing demand, caching instructions is often a safe path to follow, since cache misses are then less likely to cause bottlenecks. Although it might seem unusual to consider that an "oversized" processor would ever be used in practice, consider the case of a portable media player that can decode and play both compressed video and audio. In its audio-only mode, its performance requirements will be only a fraction of its needs during video playback. Therefore, the instruction/data management mechanism could be different in each mode.

Managing data through DMA is the natural choice for most multimedia applications, because these usually involve manipulating large buffers of compressed and uncompressed video, graphics and audio. Except in cases where the data is quasi-static (for instance, a graphics icon constantly displayed on a screen), caching these buffers makes little sense, since the data changes rapidly and constantly. Furthermore, as discussed above, there are usually multiple data buffers moving around the chip at one time—unprocessed blocks headed for conditioning, partly conditioned sections headed for temporary storage, and completely processed segments destined for external display or storage. DMA is the logical management tool for these buffers, since it allows the core to operate on them without having to worry about how to move them around.

1.7.4.2 Instruction Cache, Data DMA/Cache

This approach is similar to the one we just described, except in this case part of L1 data memory is partitioned as cache, and the rest is left as SRAM for DMA access. This structure is very useful for handling algorithms that involve a lot of static coefficients or lookup tables. For example, storing a sine/cosine table in data cache facilitates quick computation of FFTs. Or, quantization tables could be cached to expedite JPEG encoding or decoding.

Keep in mind that this approach involves an inherent tradeoff. While the application gains single-cycle access to commonly used constants and tables, it relinquishes the equivalent amount of L1 data SRAM, thus limiting the buffer size available for single-cycle access to data. A useful way to evaluate this trade-off is to try alternate scenarios (Data DMA/Cache versus only DMA) in a Statistical Profiler (offered in many development tools suites) to determine the percentage of time spent in code blocks under each circumstance.

1.7.4.3 Instruction DMA, Data DMA

In this scenario, data and code dependencies are so tightly intertwined that the developer must manually schedule when instruction and data segments move through the chip. In such hard real-time systems, determinism is mandatory, and thus cache isn't ideal.

Although this approach requires more planning, the reward is a deterministic system where code is always present before the data needed to execute it, and no data blocks are lost via buffer overruns. Because DMA processes can link together without core involvement, the start of a new process guarantees that the last one has finished, so that the data or code movement is verified to have happened. This is the most efficient way to synchronize data and instruction blocks.

The Instruction/Data DMA combination is also noteworthy for another reason. It provides a convenient way to test code and data flows in a system during emulation and debug. The programmer can then make adjustments or highlight "trouble spots" in the system configuration.

An example of a system that might require DMA for both instructions and data is a video encoder/decoder. Certainly, video and its associated audio need to be deterministic for a satisfactory user experience. If the DMA signaled an interrupt to the core after each complete buffer transfer, this could introduce significant latency into the system, since the interrupt would need to compete in priority with other events. What's more, the context switch at the beginning and end of an interrupt service routine would consume several core processor cycles. All of these factors interfere with the primary objective of keeping the system deterministic.

Figures 1.8 and 1.9 provide guidance in choosing between cache and DMA for instructions and data, as well as how to navigate the trade-off between using cache and using SRAM, based on the guidelines we discussed previously.

As a real-world illustration of these flowchart choices, Tables 1.2 and 1.3 provide actual benchmarks for G.729 and GSM AMR algorithms running on a Blackfin processor under various cache and DMA scenarios. You can see that the best performance can be obtained when a balance is achieved between cache and SRAM.

In short, there is no single answer as to whether cache or DMA should be the mechanism of choice for code and data movement in a given multimedia system. However, once developers are aware of the trade-offs involved, they should settle into the "middle ground," the perfect optimization point for their system.

1.7.5 Memory Management Unit (MMU)

An MMU in a processor controls the way memory is set up and accessed in a system. The most basic capabilities of an MMU provides for memory protection, and when cache is used, it also determines whether or not a memory page is cacheable. Explicitly using the MMU is usually optional, because you can default to the standard memory properties on your processor.

On Blackfin processors, the MMU contains a set of registers that can define the properties of a given memory space. Using something *called cacheability protection look-aside buffers*

Instruction cache vs code overlay decision flow

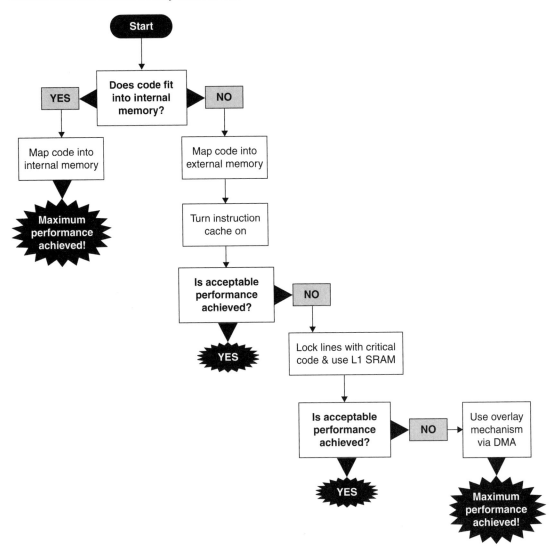

Figure 1.8: Checklist for choosing between instruction cache and DMA

(CPLBs), you can define parameters such as whether or not a memory page is cacheable, and whether or not a memory space can be accessed. Because the 32-bit-addressable external memory space is so large, it is likely that CPLBs will have to be swapped in and out of the MMU registers.

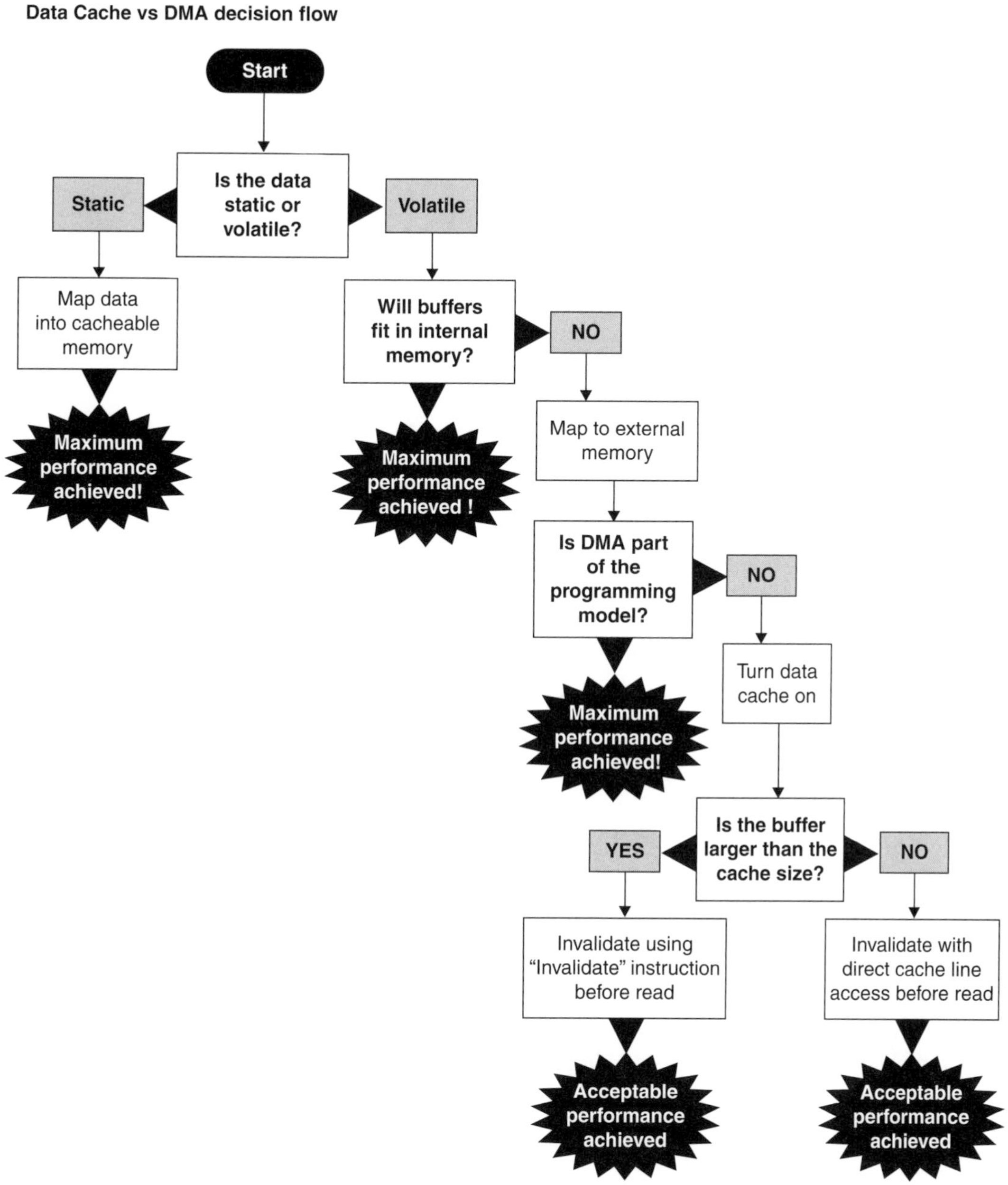

Data Cache vs DMA decision flow

Figure 1.9: Checklist for choosing between data cache and DMA

Table 1.2: Benchmarks (relative cycles per frame) for G.729a algorithm with cache enabled

| | L1 banks configured as SRAM | | L1 banks configured as cache | | | Cache + SRAM |
	All L2	L1	Code only	Code + DataA	Code + DataB	DataA cache, DataB SRAM
Coder	1.00	0.24	0.70	0.21	0.21	0.21
Decoder	1.00	0.19	0.80	0.20	0.19	0.19

Table 1.3: Benchmarks (relative cycles per frame) for GSM AMR algorithm with cache enabled

| | L1 banks configured as SRAM | | L1 banks configured as cache | | | Cache + SRAM |
	All L2	L1	Code	Code + DataA	Code + DataB	DataA cache, DataB SRAM
Coder	1.00	0.34	0.74	0.20	0.20	0.20
Decoder	1.00	0.42	0.75	0.23	0.23	0.23

1.7.5.1 CPLB Management

Because the amount of memory in an application can greatly exceed the number of available CPLBs, it may be necessary to use a CPLB manager. If so, it's important to tackle some issues that could otherwise lead to performance degradation. First, whenever CPLBs are enabled, any access to a location without a valid CPLB will result in an exception being executed prior to the instruction completing. In the exception handler, the code must free up a CPLB and re-allocate it to the location about to be accessed. When the processor returns from the exception handler, the instruction that generated the exception then executes.

If you take this exception too often, it will impact performance, because every time you take an exception, you have to save off the resources used in your exception handler. The processor then has to execute code to re-program the CPLB. One way to alleviate this problem is to profile the code and data access patterns. Since the CPLBs can be "locked," you can protect the most frequently used CPLBs from repeated page swaps.

Another performance consideration involves the search method for finding new page information. For example, a "nonexistent CPLB" exception handler only knows the address where an access was attempted. This information must be used to find the corresponding address "range" that needs to be swapped into a valid page. By locking the most frequently used pages and setting up a sensible search based on your memory access usage (for instructions and/ or data), exception-handling cycles can be amortized across thousands of accesses.

1.7.5.2 Memory Translation

A given MMU may also provide memory translation capabilities, enabling what's known as *virtual memory*. This feature is controlled in a manner that is analogous to memory protection. Instead of CPLBs, *translation look-aside buffers* (TLBs) are used to describe physical memory space. There are two main ways in which memory translation is used in an application. As a holdover from older systems that had limited memory resources, operating systems would have to swap code in and out of a memory space from which execution could take place.

A more common use on today's embedded systems still relates to operating system support. In this case, all software applications run thinking they are at the same physical memory space, when, of course, they are not. On processors that support memory translation, operating systems can use this feature to have the MMU translate the actual physical memory address to the same virtual address based on which specific task is running. This translation is done transparently, without the software application getting involved.

1.8 Physics of Data Movement

So far, we've seen that the compiler and assembler provide a bunch of ways to maximize performance on code segments in your system. Using of cache and DMA provide the next level for potential optimization. We will now review the third tier of optimization in your system—it's a matter of physics.

Understanding the "physics" of data movement in a system is a required step at the start of any project. Determining if the desired throughput is even possible for an application can yield big performance savings without much initial investment.

For multimedia applications, on-chip memory is almost always insufficient for storing entire video frames. Therefore, the system must usually rely on L3 DRAM to support relatively fast access to large buffers. The processor interface to off-chip memory constitutes a major factor in designing efficient media frameworks, because access patterns to external memory must be well planned in order to guarantee optimal data throughput. There are several high-level steps that can ensure that data flows smoothly through memory in any system. Some of these are discussed below and play a key role in the design of system frameworks.

1. Grouping Like Transfers to Minimize Memory Bus Turnarounds

Accesses to external memory are most efficient when they are made in the same direction (e.g., consecutive reads or consecutive writes). For example, when accessing off-chip synchronous memory, 16 reads followed by 16 writes is always completed sooner than 16 individual read/write sequences. This is because a write followed by a read incurs latency. Random accesses to external memory generate a high probability of bus turnarounds. This added latency can easily halve available bandwidth. Therefore, it is important to take advantage of the ability to control the number of transfers in a given direction. This can be done either automatically (as we'll see here) or by manually scheduling your data movements.

A DMA channel garners access according to its priority, signified on Blackfin processors by its channel number. Higher priority channels are granted access to the DMA bus(es) first. Because of this, you should always assign higher priority DMA channels to peripherals with the highest data rates or with requirements for lowest latency.

To this end, MemDMA streams are always lower in priority than peripheral DMA activity. This is due to the fact that with Memory DMA, no external devices will be held off or starved of data. Since a Memory DMA channel requests access to the DMA bus as long as the channel is active, efficient use of any time slots unused by a peripheral DMA are applied to MemDMA transfers. By default, when more than one MemDMA stream is enabled and ready, only the highest priority MemDMA stream is granted.

When it is desirable for the MemDMA streams to share the available DMA bus bandwidth, however, the DMA controller can be programmed to select each stream in turn for a fixed number of transfers.

This "Direction Control" facility is an important consideration in optimizing use of DMA resources on each DMA bus. By grouping same-direction transfers together, it provides a way to manage how frequently the transfer direction changes on the DMA buses. This is a handy way to perform a first level of optimization without real-time processor intervention. More importantly, there's no need to manually schedule bursts into the DMA streams.

When direction control features are used, the DMA controller preferentially grants data transfers on the DMA or memory buses that are going in the same read/write direction as in the previous transfer, until either the direction control counter times out, or until traffic stops or changes direction on its own. When the direction counter reaches zero, the DMA controller changes its preference to the opposite flow direction.

In this case, reversing direction wastes no bus cycles other than any physical bus turnaround delay time. This type of traffic control represents a trade-off of increased latency for improved utilization (efficiency). Higher block transfer values might increase the length of time each request waits for its grant, but they can dramatically improve the maximum attainable bandwidth in congested systems, often to above 90%.

Here's an example that puts these concepts into some perspective:

Example 1.4

First, we set up a memory DMA from L1 to L3 memory, using 16-bit transfers, that takes about 1100 system clock (`SCLK`) cycles to move 1024 16-bit words.

We then begin a transfer from a different bank of external memory to the video port (PPI). Using 16-bit unpacking in the PPI, we continuously feed an NTSC video encoder with 8-bit data. Since the PPI sends out an 8-bit quantity at a 27 MHz rate, the DMA bus bandwidth required for the PPI transfer is roughly 13.5 M transfers/second.

When we measure the time it takes to complete the same 1024-word MemDMA transfer with the PPI transferring simultaneously, it now takes three times as long.

Why is this? It's because the PPI DMA activity takes priority over the MemDMA channel transactions. Every time the PPI is ready for its next sample, the bus effectively reverses direction. This translates into cycles that are lost both at the external memory interface and on the various internal DMA buses.

When we enable Direction Control, the performance increases because there are fewer bus turn-arounds.

As a rule of thumb, it is best to maximize same-direction contiguous transfers during moderate system activity. For the most taxing system flows, however, it is best to select a block transfer value in the middle of the range to ensure no one peripheral gets locked out of accesses to external memory. This is especially crucial when at least two high-bandwidth peripherals (like PPIs) are used in the system.

In addition to using direction control, transfers among MDMA streams can be alternated in a "round-robin" fashion on the bus as the application requires. With this type of arbitration, the first DMA process is granted access to the DMA bus for some number of cycles, followed by the second DMA process, and then back to the first. The channels alternate in this pattern until all of the data is transferred. This capability is most useful on dual-core processors (for example, when both core processors have tasks that are awaiting a data stream transfer). Without this "round-robin" feature, the first set of DMA transfers will occur, and the second DMA process will be held off until the first one completes. Round-robin prioritization can help insure that both transfer streams will complete back-to-back.

Another thing to note: using DMA and/or cache will always help performance because these types of transactions transfer large data blocks in the same direction. For example, a DMA transfer typically moves a large data buffer from one location to another. Similarly, a cache-line fill moves a set of consecutive memory locations into the device, by utilizing block transfers in the same direction.

Buffering data bound for L3 in on-chip memory serves many important roles. For one, the processor core can access on-chip buffers for pre-processing functions with much lower latency than it can by going off-chip for the same accesses. This leads to a direct increase in system performance. Moreover, buffering this data in on-chip memory allows more efficient peripheral DMA access to this data. For instance, transferring a video frame on-the-fly through a video port and into L3 memory creates a situation where other peripherals might be locked out from accessing the data they need, because the video transfer is a high-priority process. However, by transferring lines incrementally from the video port into L1 or L2 memory, a Memory DMA stream can be initiated that will quietly transfer this data into L3 as a low-priority process, allowing system peripherals access to the needed data.

2. Understanding Core and DMA SDRAM Accesses

Consider that on a Blackfin processor, core reads from L1 memory take one *core*-clock cycle, whereas core reads from SDRAM consume eight *system* clock cycles. Based on typical CCLK/SCLK ratios, this could mean that eight SCLK cycles equate to 40 CCLKs. Incidentally, these eight SCLKs reduce to only one SCLK by using a DMA controller in a burst mode instead of direct core accesses.

There is another point to make on this topic. For processors that have multiple data fetch units, it is better to use a dual-fetch instruction instead of back-to-back fetches. On Blackfin processors with a 32-bit external bus, a dual-fetch instruction with two 32-bit fetches takes nine SCLKs (eight for the first fetch and one for the second). Back-to-back fetches in separate instructions take 16 SCLKs (eight for each). The difference is that, in the first case, the request for the second fetch in the single instruction is pipelined, so it has a head start.

Similarly, when the external bus is 16 bits in width, it is better to use a 32-bit access rather than two 16-bit fetches. For example, when the data is in consecutive locations, the 32-bit fetch takes nine SCLKs (eight for the first 16 bits and one for the second). Two 16-bit fetches take 16 SCLKs (eight for each).

3. Keeping SDRAM Rows Open and Performing Multiple Passes on Data

Each access to SDRAM can take several SCLK cycles, especially if the required SDRAM row has not yet been activated. Once a row is active, it is possible to read data from an entire row without reopening that row on every access. In other words, it is possible to access any location in memory on every SCLK cycle, as long as those locations are within the same row in SDRAM. Multiple SDRAM clock cycles are needed to close a row, and therefore constant row closures can severely restrict SDRAM throughput. Just to put this into perspective, an SDRAM page miss can take 20–50 CCLK cycles, depending on the SDRAM type.

Applications should take advantage of open SDRAM banks by placing data buffers appropriately and managing accesses whenever possible. Blackfin processors, as an example, keep track of up to four open SDRAM rows at a time, so as to reduce the setup time—and thus increase throughput—for subsequent accesses to the same row within an open bank. For example, in a system with one row open, row activation latency would greatly reduce the overall performance. With four rows open at one time, on the other hand, row activation latency can be amortized over hundreds of accesses. Let's look at an example that illustrates the impact this SDRAM row management can have on memory access bandwidth:

Figure 1.10 shows two different scenarios of data and code mapped to a single *external* SDRAM bank. In the first case, all of the code and data buffers in external memory fit in a single bank, but because the access patterns of each code and data line are random, almost

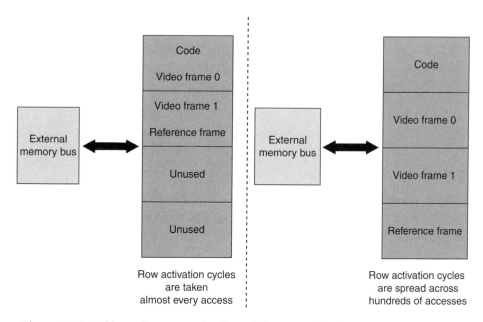

Figure 1.10: Taking advantage of code and data partitioning in external memory

every access involves the activation of a new row. In the second case, even though the access patterns are randomly interspersed between code and data accesses, each set of accesses has a high probability of being within the same row. For example, even when an instruction fetch occurs immediately before and after a data access, two rows are kept open and no additional row activation cycles are incurred.

When we ran an MPEG-4 encoder from external memory (with both code and data in SDRAM), we gained a 6.5% performance improvement by properly spreading out the code and data in external memory.

4. Optimizing the System Clock Settings and Ensuring Refresh Rates Are Tuned for the Speed at Which SDRAM Runs

External DRAM requires periodic refreshes to ensure that the data stored in memory retains its proper value. Accesses by the core processor or DMA engine are held off until an in-process refresh cycle has completed. If the refresh occurs too frequently, the processor can't access SDRAM as often, and throughput to SDRAM decreases as a result.

On the Blackfin processor, the SDRAM Refresh Rate Control register provides a flexible mechanism for specifying the Auto-Refresh timing. Since the clock frequency supplied to the SDRAM can vary, this register implements a programmable refresh counter. This counter coordinates the supplied clock rate with the SDRAM device's required refresh rate.

Once the desired delay (in number of SDRAM clock cycles) between consecutive refresh counter time-outs is specified, a subsequent refresh counter time-out triggers an Auto-Refresh command to all external SDRAM devices.

Not only should you take care not to refresh SDRAM too often, but also be sure you're refreshing it often enough. Otherwise, stored data will start to decay because the SDRAM controller will not be able to keep corresponding memory cells refreshed.

Table 1.4 shows the impact of running with the best clock values and optimal refresh rates. Just in case you were wondering, RGB, CYMK and YIQ are imaging/video formats. Conversion between the formats involves basic linear transformation that is common in video-based systems. Table 1.4 illustrates that the performance degradation can be significant with a non-optimal refresh rate, depending on your actual access patterns. In this example, CCLK is reduced to run with an increased SCLK to illustrate this point. Doing this improves performance for this algorithm because the code fits into L1 memory and the data is partially in L3 memory. By increasing the SCLK rate, data can be fetched faster. What's more, by setting the optimal refresh rate, performance increases a bit more.

Table 1.4: Using the optimal refresh rate

	Sub-optimal SDRAM refresh rate		Optimal SDRAM refresh rate	
CCLK	594 MHz	526 MHz	526 MHz	
SCLK	119 MHz	132 MHz	132 MHz	
RGB to CMYK Conversion (iterations per second)	226	244	250	
RGB to YIQ Conversion (iterations per second)	266	276	282	Total
Cumulative Improvement		5%	2%	7%

5. Exploiting Priority and Arbitration Schemes Between System Resources

Another important consideration is the priority and arbitration schemes that regulate how processor subsystems behave with respect to one another. For instance, on Blackfin processors, the core has priority over DMA accesses, by default, for transactions involving L3 memory that arrive at the same time. This means that if a core read from L3 occurs at the same time a DMA controller requests a read from L3, the core will win, and its read will be completed first.

Let's look at a scenario that can cause trouble in a real-time system. When the processor has priority over the DMA controller on accesses to a shared resource like L3 memory, it can lock out a DMA channel that also may be trying to access the memory. Consider the case where the processor executes a tight loop that involves fetching data from external memory. DMA activity will be held off until the processor loop has completed. It's not only a loop with a read embedded inside that can cause trouble. Activities like cache line fills or nonlinear code execution from L3 memory can also cause problems because they can result in a series of uninterruptible accesses.

There is always a temptation to rely on core accesses (instead of DMA) at early stages in a project, for a number of reasons. The first is that this mimics the way data is accessed on a typical prototype system. The second is that you don't always want to dig into the internal workings of DMA functionality and performance. However, with the core and DMA arbitration flexibility, using the memory DMA controller to bring data into and out of internal memory gives you more control of your destiny early on in a project.

Bibliography

Analog Devices, Inc. (September 2004). *ADSP-BF533 Blackfin Processor Hardware Reference, Rev 3.0.*

Analog Devices, Inc. *VisualDSP++ 4.0 C/C++ Compiler and Library Manual for Blackfin Processors.*

Steve, Heath. (2003). *Embedded Systems Design* (second edition). Newnes: Elsevier.

CHAPTER 2

Low Power Design Techniques, Design Methodology, and Tools

Findlay Shearer

Every design involves trade-offs, but few are more difficult than performance vs. power in portable designs. The management of energy consumption is the greatest challenge facing designers of high-performance, feature-rich handsets that work from a tiny battery. Fortunately a lot of smart people have been working on the problem over the last few years in particular. In the next two chapters Findlay Shearer nicely distills and presents the results of those efforts for his readers.

"Low Power Design Techniques, Design Methodology and Tools" focuses primarily on SoC design, though many of the techniques also apply at the overall system level. Shearer goes into some detail about a wide range of techniques to help reduce power consumption, including voltage and power gating, clock gating, multi-Vt optimization, multiple voltage islands, substrate biasing, dynamic voltage/frequency scaling and power shutoff. Often a combination of these techniques must be implemented to accommodate power and performance constraints.

Since dynamic power is proportional to V_{DD}^2, simply lowering the core voltage on a chip significantly reduces power consumption. Thus the use of voltage islands and power gating has become common, though lowering the voltage also slows logic gates, and you then get to play "whack-a-mole" with issues of performance, coordinated wake-up sequences and in-rush current. Shearer spends some time on clock gating and asynchronous techniques; since a third to a half of an SoC's dynamic power is consumed by the clock distribution system, this is time well spent.

After a brief recap of the SoC design flow—from system design to tapeout—Shearer looks to see what help leading EDA vendors have to offer the hapless designer. Here there are two competing camps: Cadence, with its Common Power Format (CPF) and Synopsys, Mentor, Magma, et al., with the Unified Power Format (UPF). Shearer runs through both the CPF and UPF low-power design flows, from concept to physical implementation, showing where each of Cadence's and Synopsys's tools come into play. He gives a good explanation of the benefits of both power formats, though some of them still remain to be implemented at this time.

Since one SoC design in four fails because of power-related issues, this chapter is highly recommended reading before your next design start.

—**John Donovan**

2.1 Low Power Design Techniques

Many design techniques have been developed to reduce power and by the judicious application of these techniques, systems are tuned for the best power/performance trade-offs.

2.1.1 Dynamic Process Temperature Compensation

A common engineering philosophy when designing a system-on-a-chip (SoC) is to insure that they perform under "worst-case" conditions. Worst case in semiconductor manufacturing applies to very high temperatures and variations in the manufacturing process; transistor performance varies in a predefined range of parameters. Thus, some SoCs from the same wafer lot are capable of supporting higher operating frequencies (best case—fast process) or lower frequencies at the bottom of the predefined performance window (worst case—slow process) at the given voltage (Figure 2.1).

Dynamic process temperature compensation (DPTC) mechanism measures the frequency of a reference circuit that is dependent on the process speed and temperature. This reference circuit captures the product's speed dependency on the process technology and existing operating temperature and lowers the voltage to the minimum level needed to support the existing required operating frequency (Figure 2.2).

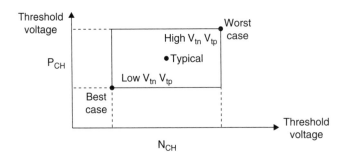

Figure 2.1: Process variations defining varying SoC performance

A mobile device containing a fast-process SoC operating in a moderate climate condition can be expected to work at the worst-case calculated voltage to support the required frequency. This is less than an optimum energy savings.

The DPTC concept allows the supply voltage to be adjusted to match the process corner and SoC temperature. If the process corner is "best case" a lower supply voltage can be applied to support the required performance of the SoC. Similarly, the temperature of the part can be used to adjust the supply voltage.

The available performance is monitored by different types of reference circuits comprised of free-running ring oscillators. The inputs from reference "sense" circuits are processed by internal control and compare logic and written to software readable registers. If there is a significant (predefined) change in the reference circuit delay values, an interrupt is triggered. The relevant software interrupt routine calculates the new required voltage and re-programs the Power Management IC (PMIC) to supply the new voltage to the SoC. A new voltage is applied, based on the reference circuit delay, values change, providing feedback and closing the loop of the DPTC mechanism. This insures that the system stabilizes at the proper voltage level. Software control permits fast and simple changes.

DPTC can result in an approximate power savings of 35%, significantly improving the battery life.

2.1.2 Static Process Compensation

Static process compensation (SPC) follows a similar path to DPTC but without the temperature compensation aspect. SoCs are designed at the worst-case process corner (see Figure 2.1). However, production wafer lots are typically manufactured close to a typical point in the "box" and as a result can run at a lower voltage and still meet performance requirements.

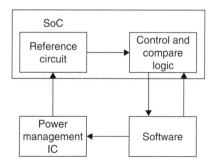

Figure 2.2: DPTC mechanism

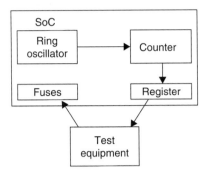

Figure 2.3: SPC basic circuits

SPC is a technique of identifying minimum operating voltage for each SoC at the production line and programming the fuses with the information. Software reads the fuses to set the operating voltage for the SoC.

The basic circuits required to support SPC are similar to those employed in DPTC and integrated into the SoC. They include a ring oscillator, support register, and fuses (Figure 2.3).

The frequency of the ring oscillator correlates with the process corner. The support register captures the oscillator frequency and the fuses are programmed to define the counter frequency and voltage for the SoC.

2.1.2.1 Compared to DPTC

SPC does not compensate voltage for temperature and the operating voltage is not changed dynamically in SPC. In addition, the SoC manufacturer can test the SoC to the SPC defined voltage. Given that SPC is a subset of DPTC, it has been demonstrated that the temperature compensation aspect of DPTC provides marginal benefit to energy conservation.

2.1.3 Power Gating

Like voltage gating, power gating involves temporarily shutting down blocks in a design when the blocks are not in use. And, like voltage gating, the technique is complex. With power gating, the designer has to worry about it at the SoC design phase, specifically at the Register Transfer Level (RTL). The engineer has to design a power controller that is going to control what blocks need to shut down at a particular time and has to think about what voltage to run different blocks (Figure 2.4).

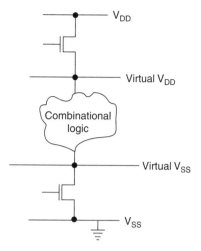

Figure 2.4: Power gating

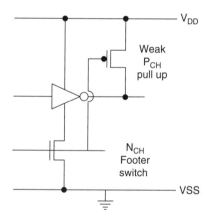

Figure 2.5: Fine power gating

Traditionally, two methods for power gating are fine-grained and coarse-grained. In fine-grained power gating, designers place a switch transistor between ground and each gate. This approach allows designers to shut off the connection to ground whenever a series of functions is not in use. This technique is done with every cell in the library. There are trade-offs with fine-grained power gating because it is fairly easy to do power characterization of each cell. However, the problem is the area hit is very significant: two to four times larger (Figure 2.5).

In order to keep the area overhead to a minimum, fine-grained power gates are implemented as footer switches to ground as NMOS transistors. The timing impact of the IR drop across the switch and the behavior of the clamp are easy to characterize. It is still possible to use a traditional design flow to deploy fine-grained power gating.

Designers can also mix and match cells having some power gated and others not. Cells with high threshold voltage need not use power gating. For the most part, the power penalty is just too large, and many design groups are instead using coarse-grained power gating, in which designers create a power switch network. This is essentially, a group of switch transistors that in parallel turn entire blocks on and off. The technique does not have the area hit of the fine-grained technique because for a given block of logic the switching activity will be less than the 100%. However, due to the propagation delay through the cells, the switching activity will be distributed in time. In addition, it is harder to characterize on a cell-by-cell basis (Figure 2.6).

Unlike fine-grained power gating, when the power is switched in coarse-grained power gating, the power is disconnected from all logic, including the registers, resulting in the loss of all states. If the state is to be preserved when the power is disconnected then it must be stored somewhere, where it is not power gated. Most commonly this is done locally to the registers by swapping in special "retention" registers which have an extra storage node that is separately powered. There are a number of retention register designs which trade-off performance against area. Some use the existing slave latch as the storage node whilst others add an additional "balloon" latch storage node. However, they all require one or more extra control signals to save and restore the state.

The key advantage of retention registers is that they are simple to use and are very quick to save and restore state. This means that they have a relatively low energy cost of entering and leaving standby mode and so are often used to implement "light sleep." However, in order to minimize the leakage power of these retention registers during standby, it is important that the

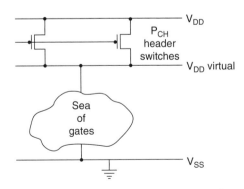

Figure 2.6: Coarse-grained power gating

storage node and associated control signal buffering are implemented using high threshold low leakage transistors.

If very low standby leakage is required then it is possible to store the state in main memory and cut the power to all logic including the retention registers. However, this technique is more complex to implement and also takes much longer to save and restore state. This means that it has a higher energy cost of entering and leaving standby mode and so is more likely to be used to implement "deep sleep."

A key challenge in power gating is managing the in-rush current when the power is reconnected. This in-rush current must be carefully controlled in order to avoid excessive IR drop in the power network as this could result in the collapse of the main power supply and loss of the retained state.

2.1.4 State-Retention Power Gating

The major motivation of this technique is to significantly reduce the leakage power for the SoC when in the inactive mode. State-retention power gating (SRPG) is a technique that allows the voltage supply to be reduced to zero for the majority of a block's logic gates while maintaining the supply for the state elements of that block. The state of the SoC is always saved in the sequential components. Combinational elements propagate the state of the flip-flops. Using the SRPG technique, when in the inactive mode, power to the combinational logic is turned off and the sequential stays on. SRPG can thereby greatly reduce power consumption when the application is in stop mode, yet it still accommodates fast wake-up times.

Reducing the supply to zero in the stop mode allows both the dynamic and static power to be removed. Retaining the supply on the state elements allows a quick continuation of processing when exiting the stop mode.

Since the state of the digital logic is stored in the flip-flops, if the flip-flops are kept on a constantly powered voltage grid, the intermediate logic can be put onto a voltage grid that can be power gated. When the voltage is reapplied to the intermediate logic, the state of the flip-flops will be re-propagated through the logic and the system can start where it has left off as illustrated in Figure 2.7.

In a full SRPG implementation the entire target platform is entered into state retention and all (100%) flip-flops retain the state during power down. There is a specific power up sequence required and the power down sequence is also predefined. Power up/down time is dependent

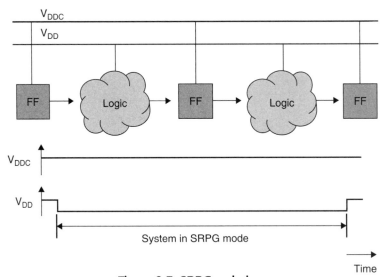

Figure 2.7: SRPG technique

Source: http://www.freescale.com

on number of flops. Expected "wake-up" latency is less than 1 ns and "sleep" is less than 500 ns.

2.1.4.1 Partial SRPG

Partial SRPG further reduces the leakage from a full SRPG implementation.

In this case only a few flip-flops are made capable of state retention and all other flip-flops are turned off.

After power up, the SRPG flip-flops are restored to original state and the others are restored to reset state of that flip-flop. System software understands the non-saved registers are lost and should either re-program those registers or ensure the reset state meets the software requirements.

2.2 Low Power Architectural and Subsystem Techniques

2.2.1 Clock Gating

A tried-and-true technique for reducing power is clock gating. One-third to one-half of an IC design's dynamic power is in the SoC's clock-distribution system. The concept is simple, if you do not need a clock running, shut it down (Figure 2.8).

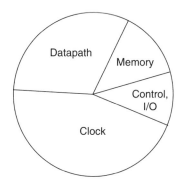

Figure 2.8: Power distribution in a high performance processor[2]

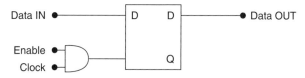

Figure 2.9: Local clock gating

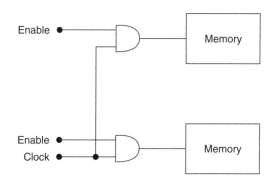

Figure 2.10: Global clock gating

Today, the two popular methods of clock gating are local and global. If you feed old data to the output of a flip-flop back into its input through a multiplexer, you typically need not clock again. Therefore, you can replace each feedback multiplexer with a clock gating cell that clocks the signal off. You would then use the enable signal that controls the multiplexer to control the clock cell to clock the signal off (Figure 2.9).

The other popular approach of clock gating, global clock gating, is to simply turn off the clock to the whole block, typically from a central-clock-generator module. This method functionally shuts down the block, unlike local clock gating, but even further reduces dynamic power because it shuts down the entire clock tree (Figure 2.10).

2.2.2 Asynchronous Techniques: GALS

The increasing demands of mobile devices create several problems for system design and integration. Challenges like the integration of complex systems, timing closure including clock generation and control, system noise characteristic, and power consumption for mobile applications, in particular wireless devices.

Most digital systems designed today operate synchronously. One significant challenge is the generation of the clock tree. The clock tree in complex digital systems includes clock gating supporting circuitry, clock dividers for different clock domains, phase locked loops (PLL), and complex clock-phasing blocks.

Mobile communication devices have one very critical constraint, power consumption. The limited capacity of the batteries creates firm limits for the system power consumption. In addition, the power demands of complex systems are usually high and hence, power consumption must be controlled and minimized. Partly, this can be achieved by using known methods for minimization and localization of switching power like clock gating, asynchronous design, or voltage scaling. However, clock gating makes the design of the clock tree even harder.

Clockless logic employs a protocol of local handshaking rather than a global clock to define transaction timing. Handshakes are implemented by means of simple request and acknowledge signals that mark the validity of the data. Only those parts of the system actively involved in the task execution draw power.

Globally asynchronous locally synchronous (GALS) techniques have the potential to solve some of the most challenging design issues of SoC integration of mobile devices. The basic tenet behind GALS is that system blocks can internally operate synchronously and communicate asynchronously[1].

Figure 2.11 illustrates the two scenarios. The first is a normal synchronous system with a master clock, while the second is a GALS system in which two blocks talk to each other with a handshake interface. Although each block has its own local clock, the overall system works without a global clock. In SoC designs, the GALS architecture (Figure 2.12) helps to solve the increasingly difficult problem of integrating multiple clock domains into a single SoC and reducing the overall power consumption[2].

A GALS system consists of a number of locally synchronous modules each surrounded with an asynchronous wrapper. Communication between synchronous blocks is performed indirectly via asynchronous wrappers (Figure 2.13).

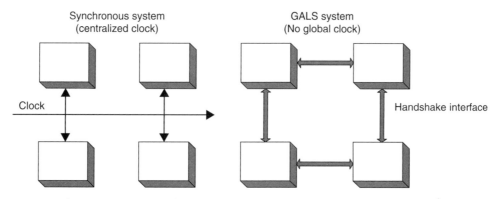

Figure 2.11: Comparison of a synchronous system and GALS system[1]

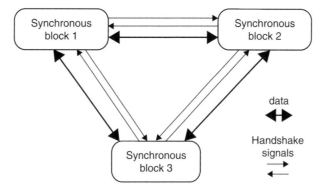

Figure 2.12: The GALS architecture[2]

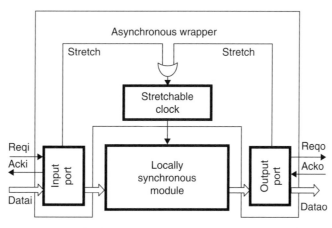

Figure 2.13: Communications are performed with asynchronous wrappers[2]

The purpose of the asynchronous wrapper is to perform all operations for safe data transfer between locally synchronous modules. Accordingly, a locally synchronous module should just acquire the input data delivered from the asynchronous wrapper, process it and provide the processed data to the output stage (port) of the asynchronous wrapper. Every locally synchronous module can operate independently, minimizing the problem of clock skew and clock tree generation.

A GALS design methodology should include power saving mechanisms. The goal is to completely integrate power saving mechanisms into the asynchronous wrappers. Implementing GALS would automatically introduce a certain power reduction. However, the power saving in GALS is based on the same assumptions as clock gating in the synchronous design. The main idea is identical, lowering of dynamic power consumption by disabling the clock signal.

Locally synchronous modules are usually surrounded by asynchronous wrappers. Local clocks drive those synchronous circuit blocks. Stoppable ring oscillators are frequently used to generate the local clocks. Data transfer between different blocks requires stopping of the local clocks during data transfer in order to avoid meta-stability problems. The asynchronous wrappers should perform all necessary activities for safe data transfer between the blocks. Locally synchronous modules do not play any role in providing the prerequisites for block-to-block data transfer.

The power savings of a GALS design has been investigated. Most of these investigations are based on the application of GALS to high-speed processor implementations. However, these results show some general trend. In Figure 2.14, the power distribution in a high performance processor is given.

The clock signal is the dominant source of power consumption in such a high performance processor. Within GALS design, the clock network is split into several smaller sub-networks

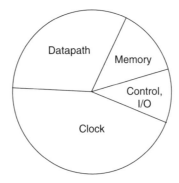

Figure 2.14: Power distribution in a high performance processor[2]

with lower power consumption. First estimations, according to[11], showed that about 30% of power savings could be expected in the clock net due to the application of GALS techniques.

In addition, GALS design techniques offer independent setting of frequency and voltage levels for each locally synchronous module. When using dynamic voltage scaling (DVS), an average energy reduction of up to 30% can be reached, yet associated with a performance drop of 10%, as reported in[12]. The power saving techniques that are immanent with the use of GALS have similar limits as clock gating. Further energy reduction is possible only with the application of DVS in conjunction with GALS.

The ARM996HS processor is an example of a commercially available processor that uses Handshake Solutions clock less technology[3].

2.2.3 Power Saving Modes

SoCs offer many different power saving modes providing the mobile device developer the ability to trade-off between power consumption in standby and recovery times as shown in Figure 2.15:

Run: This is the normal, functional operating mode of the SoC. Core frequency and operating voltage can be dynamically changed within a range.

Wait: In this mode, the processor core clock is gated. Operation resumes on interrupt. This mode is useful for running low MIPS applications that primarily involve peripheral activity, such as a viewfinder.

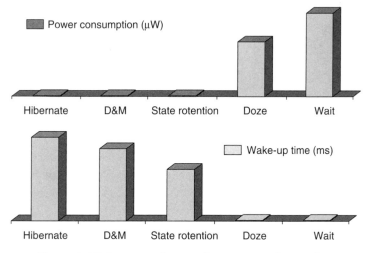

Figure 2.15: Power saving modes versus wake-up time

Source: http://www.freescale.com

Doze: In this mode, both the processor core and MAX clocks are gated. Clocks for specific peripherals can be switched off automatically in Doze mode by preprogramming the Clock Controller module. This mode is useful for processes that require quick reactivation. Normal operation resumes on interrupt.

State retention: In this mode, all clocks are switched off and the PLL is disabled. External memory is put in low power (self-refresh) mode. MCU and peripheral clocks are gated. Supply voltage can be dropped to a minimum. State-retention uses less power and has a longer wake-up time than Doze mode, but there is no need to recover any data after the wake-up.

Deep Sleep: In this mode the clocks are gated. MCU core platform power supply is turned off. Any relevant register data should be saved before entering Deep Sleep mode. Normal operation resumes on interrupt.

Hibernate: The power supply of the entire SoC is shut down. System is completely dead. Operation resume is equivalent to a cold boot. All internal data should be saved to external memory prior to hibernate.

The power saving modes can be extended to board-level applications. For example, in a memory hold mode, everything can be powered down except the PMIC and memory, which is kept in a state-retention mode. This is a very low power state, and in many applications different components, such as an application or baseband processor, will have to reboot at wake-up, creating some latency issues.

However, if the core processor is kept in a low leakage standby mode, wake-up is much quicker because the baseband reboot will not be necessary. The application requirements will often dictate which memories hold state is used for best performance/energy efficiency optimization.

2.3 Low Power SoC Design Methodology, Tools, and Standards

2.3.1 Introduction

There has been increasing concern in system design houses and semiconductor suppliers that power consumption of ICs is beginning to impact design complexity, design speed, design area, and manufacturing costs.

Historically power consumption has not been an issue or a high priority in the semiconductor industry. The industry was driven by the competitive market forces of simultaneously increasing integration and reducing SoC die area. In the high-volume processor market segment, integration was driven by the need to increase the performance. Higher performance was achieved through the use of new architectures (e.g. super scalar and super pipelined); integrating more and more system functions on the SoC increasing the processor's speed and increasing the processor's ability to perform computations. The trend in the large, and cost sensitive, embedded processor market segment is toward higher frequencies.

Today one in four IC designs fail due to power-related issues. Power management techniques are considered a critical part of the SoC design flow.

In rising to the challenge to reduce power the semiconductor industry has adopted a multifaceted approach, attacking the problem on three fronts:

1. Reducing chip capacitance through process scaling: This approach to reduce the power is very expensive and has a very low return on investment (ROI).

2. Reducing voltage: While this approach provides the maximum relative reduction in power it is complex, difficult to achieve and requires moving the DRAM and systems industries to a new voltage standard. The ROI is low.

3. Employing better architectural and circuit design techniques: This approach promises to be the most successful because the investment to reduce power by design is relatively small in comparison to the other two approaches.

The need to reduce power consumption has become more critical as larger, faster ICs move into portable applications. As a result techniques for managing power throughout the design flow are evolving to assure that all parts of the product receive power properly and efficiently and that the product is reliable. However, there is one significant drawback, namely the design problem has essentially changed from an optimization of the design in two dimensions, i.e., performance and area, to optimizing a three-dimensional problem, i.e., performance, area, and power (see Figure 2.16).

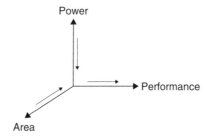

Figure 2.16: Performance, area, and power trade-off

As stated earlier, the best power management decisions are made at the system level during system design. At this point in the development cycle, system designers make initial decisions about the end-product power requirements and distribution. That entails determining the overall power budget, making critical hardware/software trade-offs, and working with designers to select the proper process technology for semiconductor fabrication, such as gate length and multi-voltage/multi-threshold. Electronic Design Automation (EDA) gives system designers the design and verification environment they need to achieve power/performance estimation at the system level. Such system-level decisions drive all subsequent software and semiconductor requirements.

Just as system-level decisions drive the semiconductor requirements, ICs together with process technology drive EDA tool and flow requirements. IC designers select EDA tools and flows for power management based on the following criteria:

- Process and application support.

- Design optimization capabilities—needed to ensure operations stay within power budget.

- Design verification and signoff capabilities—needed to ensure device reliability and power integrity.

EDA tools address all these requirements, offering power management throughout the design flow. The most effective EDA power management solutions are characterized by the ability to deliver comprehensive power estimation, optimization, analysis, and reliability signoff within a power-aware synthesis and signoff environment.

Unlike timing, which is localized to critical paths within the IC, power is distributed across the device. Power requirements are also dependent on the particular application software that is running and the operating mode. In general, three areas of power need to be managed in the flow to ensure the design remains within the required power budget and reliability targets:

- Dynamic power: The designer must manage how much power is used when a circuit switches or is active (changes from 1 to 0 or 0 to 1) due to charging/discharging capacitance.

- Leakage (static) power: The designer must manage how much power "leaks" from transistors or is wasted when the circuit is static or unchanged.

- Peak power: The designer must manage peak-load periods.

A variety of techniques have been developed to address the various aspects of the power problem and to meet aggressive power specifications. Examples of these include the use of clock gating, multi-switching threshold transistors, multi-supply multi-voltage, substrate biasing, dynamic voltage and frequency scaling, and power shut-off (PSO). Table 2.1 illustrates the power, timing, and area trade-offs among the various power management techniques.

The use of more advanced approaches reduces power consumption, but at the same time increase the complexity associated with design, verification, and implementation methodologies. Although using a single technique in isolation could be relatively simple, often a combination of these techniques must be used to meet the required timing and power targets. Using multiple techniques concurrently could result in an extremely complex design flow. A key requirement is consistency throughout the flow such that the use of one technique preserves any gains from other techniques. This also requires a good understanding of the various low power techniques and their respective trade-offs, as highlighted in Table 2.1.

Table 2.1: Cadence low power solution architecting, designing, implementing, and verifying low power digital SoCs (www.cadence.com)

Power Reduction Technique	Power Benefit	Timing Penalty	Area Penalty	Methodology Impact			
				Architecture	Design	Verification	Implementation
Multi-Vt optimization	Medium	Little	Little	Low	Low	None	Low
Clock gating	Medium	Little	Little	Low	Low	None	Low
Multi-supply voltage	Large	Some	Little	High	Medium	Low	Medium
Power shut-off	Huge	Some	Some	High	High	High	High
Dynamic and adaptive voltage frequency scaling	Large	Some	Some	High	High	High	High
Substrate biasing	Large	Some	Some	Medium	None	None	High

Furthermore, "low power" is not just something that is "bolted" on at the end of the development process. To meet aggressive design schedules, it is no longer sufficient to consider power only in the implementation phase of the design. The size and complexity of today's ICs makes it imperative to consider power throughout the design process, from the SoC/system architectural phase; through the implementation architecture phase; through design (including micro-architecture decisions); and all the way to implementation with power-aware synthesis, placement, and routing. Similarly, to prevent functional issues from surfacing in the final silicon, power-aware verification must be performed throughout the development process.

2.3.2 Low Power Design Process

2.3.2.1 SoC Design Flow

To address these issues, power management techniques should be applied during every step of the flow, from design planning through reliability signoff analysis (see Figure 2.17).

2.3.2.2 System Design

For a new product, a well-defined product specification is the first critical step to achieving a successful product. The design specification typically is driven by a product requirements document (PRD), or user specification, typically generated by the technical marketing organization, which contains a synthesis of customer requirements in the form of features, cost, and schedule.

The design or architectural specification is derived from the PRD. The architectural specification can be hierarchical in nature with the top level describing an abstract picture of the SoC under development. The subsequent layers divide the top-level blocks into finer detail and at the lowest layer the SoC designer is left with a description of all the IP blocks that comprise the SoC.

2.3.2.3 RTL Design

In a SoC design, RTL description is a way of describing the operation of a digital circuit. In RTL design, a circuit's behavior is defined in terms of the flow of signals (or transfer of data) between registers, and the logical operations performed on those signals.

This step converts the user specification (what the user wants the SoC to do) into a RTL description. The RTL specifies, in painstaking detail, exactly what every bit of the SoC should do on every clock cycle.

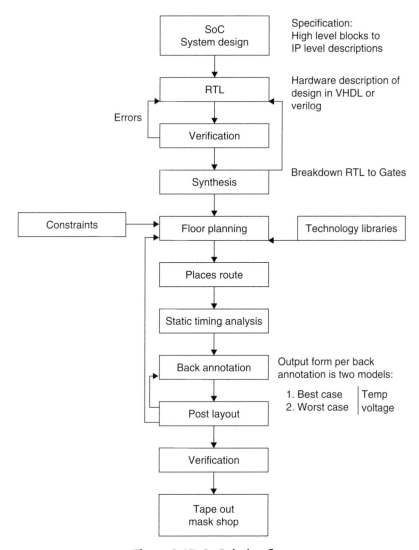

Figure 2.17: SoC design flow

RTL abstraction is used in hardware description languages (HDLs) like Verilog and VHDL to create high-level representations of a circuit, from which lower-level representations and ultimately actual wiring can be derived.

When designing digital ICs with a HDL, the designs are usually engineered at a higher level of abstraction than transistor or gate level. In HDLs, the designer declares the registers (which roughly corresponds to variables in computer programming languages), and describes the combination logic by using constructs that are familiar from programming languages such as if-then-else and arithmetic operations. This level is called Register Transfer Level. The term refers to the fact that RTL focuses on describing the flow of signals between registers.

2.3.2.4 Verification

Verification is the act of proving or disproving the correctness of intended algorithms underlying a system with respect to a certain formal specification. It answers the question "are we building the product correctly?"

2.3.2.5 Synthesis

Using an EDA tool for synthesis, the RTL description can be directly translated to equivalent hardware implementation files, typically in terms of logic gates, for an SoC. Logic gates can take the form of AND, OR, NAND, NOR, and other logical functions.

During synthesis, EDA tools help the designer complete RTL and gate-level power optimization together with peak and average power debug and analysis. Here, power management is needed to ensure the design's dynamic and leakage power remains within the power budget and reliability targets. Power optimization and peak power analysis solutions include support for multi-voltage, multi-threshold, clock gating, data, and power-gating; RTL estimation analysis at full-SoC/block level; and gate-level peak and average power analysis. The synthesis tool also performs logic optimization.

2.3.2.6 Floorplanning

During Floorplanning, the RTL of the SoC is assigned to gross regions of the SoC, input/ output (I/O) pins are assigned and large objects (arrays, cores, etc.) are placed.

From a low power perspective EDA tools help designers create the power grid and perform power network analysis to assure the power remains within the power budget and reliability targets. Power-aware floorplanning solutions include support for power grid planning, multi-voltage region planning, clock tree planning, and estimation of voltage (IR) drop and electromigration effects.

2.3.2.7 Place and Route: Physical Implementation

During Place and Route, a placer takes a given synthesized circuit, an output from the Floorplanning, together with a technology library and produces a valid placement layout. The layout is optimized according to the constraints and ready for cell resizing and buffering, considered an essential step for timing and signal integrity satisfaction. Clock tree synthesis (CTS) and routing follow, completing the physical design process.

Routing is the process of creating all the wires needed to properly connect all the placed components, while obeying the design rules of the process. Place and Route are often lumped together as necessary physical operations.

During Place and Route, EDA helps designers implement power saving features to manage both dynamic and leakage power in the design. Power implementation solutions today include support for multi-voltage and multi-threshold designs.

2.3.2.8 Tape-out and Mask Making

In SoC design, tape-out is the final stage of the design phase of an SoC such as a microprocessor. It is the point at which the description of a circuit is sent for manufacture, initially to a mask shop and then to a wafer fabrication facility.

2.3.2.9 Reliability Signoff Analysis

During reliability signoff analysis, EDA provides post-route analysis of power grids to detect power anomalies such as voltage drop and electromigration which, if left uncorrected, can lead to design failure.

Power management starts at the system level and is driven by the application power budget requirements throughout the semiconductor value chain. IC designers collaborate closely with their EDA tool providers to ensure power management is an integral part of their design flow.

EDA offers designers the much-needed infrastructure and tools for enabling power management throughout the flow. There are future opportunities for EDA to contribute to more power savings during system definition. The design for low power problem cannot be achieved without good tools. The following sections will cover the design process, some key EDA vendor approaches and standardization.

2.3.3 Key EDA Vendors Approach to Low Power Design

2.3.3.1 Cadence Low Power Flow

Cadence has developed a complete solution for the design, verification, and implementation of low power SoCs. The low power solution combines a number of technologies from several Cadence platforms. Each of these advanced products also leverages the Common Power Format (CPF) (Figure 2.18).

2.3.3.2 Cadence Low Power Methodology Kit

Low Power Methodology Kit, offered by Cadence, provides users with a complete end-to-end methodology for low power implementation and verification[4] (Figure 2.19).

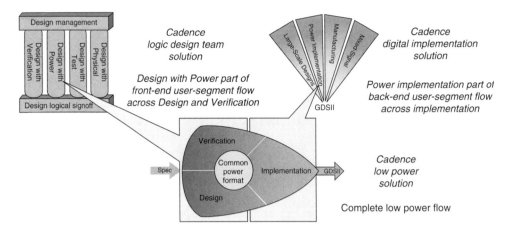

Figure 2.18: Low power flow

Source: http://www.cadence.com

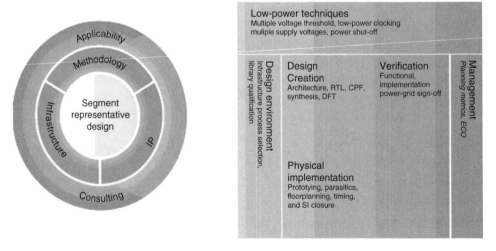

Figure 2.19: Low power methodology kit

Source: http://www.cadence.com

The kit is modularized to allow for incremental adoption and to allow teams to focus on what is most critical for the design (Figure 2.20).

- The Cadence Low Power Methodology Kit includes the following:

- Wireless segment representative design (SRD), including all required views for RTL design, physical implementation, and verification (including test bench)

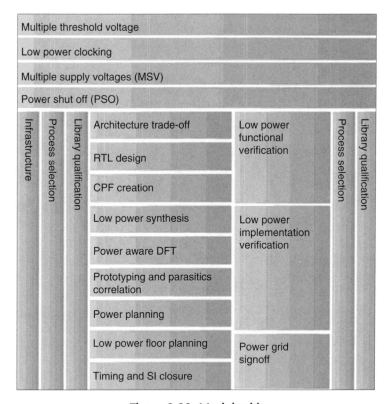

Figure 2.20: Modular kit

Source: http://www.cadence.com

- Detailed low power methodology guide, covering all aspects of low power implementation

- Reference flow implementations with step-by-step walkthroughs

- Detailed documentation of the SRD and reference flow

- Detailed flow checklists and trade-off analysis

- Expert consulting designed to map the verified and demonstrated methodologies to a specific customer design

The Low Power Methodology Kit utilizes and integrates with the following technologies (Figure 2.21):

- Cadence Logic Design Team Solution

- Incisive Plan-To-Closure Methodology

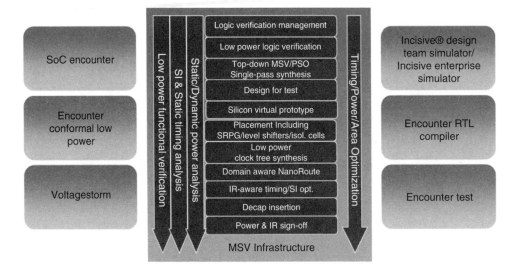

Figure 2.21: Low power design flow

Source: http://www.cadence.com

- Incisive Enterprise Family

- Incisive Formal Verifier

- Encounter RTL Compiler

- Encounter Test

- First Encounter

- SoC Encounter

- Encounter Conformal EC

- Encounter Conformal LP

- Voltage Storm

2.3.3.3 Incisive Design Team and Enterprise Simulators

The Incisive simulators help to capture functional failures due to inserted power management structures and identify bugs that could lead to excessive power consumption. In addition, the incisive simulators help engineers explore different power architectures early in the design flow without having to wait for gate-level netlists or needing to develop resource consuming modeling techniques.

2.3.3.4 Encounter RTL Compiler: Low Power Synthesis

It performs multi-objective optimization that simultaneously considers timing, power, and area to create logic structures that meet all these goals in a single pass. Encounter RTL Compiler synthesis techniques address advanced SoC design needs, such as decreased power (both dynamic and static leakage) without sacrificing performance.

Encounter RTL Compiler global synthesis allows designers to use the CPF to explore the trade-offs of employing different power management techniques. It then performs top-down power-domain-aware synthesis, optimizing for timing, power, and area simultaneously.

2.3.3.5 SoC Encounter

The SoC Encounter system combines RTL synthesis, silicon virtual prototyping, and full-SoC implementation in a single system. It enables engineers to synthesize to a flat virtual prototype implementation at the beginning of the design cycle. With the SoC Encounter system, engineers have a view of whether the design will meet its targets and be physically realizable. Designers can then choose to either complete the final implementation or to revisit the RTL design phase. It also supports advanced timing closure and routing, as well as signoff analysis engines for final implementation.

2.3.3.6 Encounter Conformal LP

During development, a low power design undergoes numerous iterations prior to final layout, and each step in this process has the potential to introduce logical bugs. Encounter Conformal Low Power checks the functional equivalence of different versions of a low power design at these various stages and enables the identification and correction of errors as soon as they are introduced. It supports advanced dynamic and static power synthesis optimizations such as clock gating and signal gating, Multi-Vt libraries, as well as de-cloning and re-cloning of gated clocks during CTS and optimization.

In addition, it supports the CPF specification language. It uses CPF as guidance to independently insert and connect low power cells (level shifters, isolation, and state-retention registers) into an RTL design, thus enabling true low power RTL-to-gate equivalence checking.

2.3.3.7 Encounter Test

Based on the power intent read through the CPF, Encounter Test automatically creates distinct test modes for the required power domain and PSO combinations. It inserts special purpose design-for-test (DFT) structures to enable control of PSO during test. The power-aware automatic test pattern generation (ATPG) engine targets low power structures such as level shifters and isolation cells, and generates low power scan vectors that significantly reduce power consumption during test. This results in the highest quality of test and the lowest power consumption during test for low power devices.

2.3.3.8 VoltageStorm

VoltageStorm hierarchical solution gives design teams the confidence that IR (voltage) drop and power rail electromigration are managed effectively. VoltageStorm technology has evolved to become an integral component of design creation, which requires early and up-front power rail analysis to help create robust power networks during power planning. Employing parasitic extraction that is manufacturing aware, and using static and dynamic algorithms, VoltageStorm technology delivers power estimation and power rail analysis functionality and automation to both analyze and optimize power networks throughout the design flow.

2.3.3.9 Synopsis

Synopsis provides an automated, end-to-end power management solution that delivers the low power implementations. It ensures consistent correlation from RTL to silicon, enabling design teams to benefit from reduced iterations and improved productivity. The Synopsys' solution utilizes industry standards such as Unified Power Format (UPF) and is complemented by a strong ecosystem of IP, modeling techniques, and libraries (Figure 2.22).

2.3.3.10 System Design

Early insight into power consumption at the system level enables the most significant power savings. Synopsys' **DesignWare® Virtual Platforms**[5] enable power modeling at the system level, providing a relative measure of power consumption on an application-by-application basis using the real applications' software and target OS.

Virtual platforms effectively contribute to lowering the risk and decreasing the time-to-market, through early software development and hardware/software integration. The benefits of virtual platforms include:

- Enabling concurrent hardware/software development, by using Virtual Platform technology and models, system and application software development can begin months earlier in the product development cycle, and concurrent with hardware development.

- Reducing the SoC-based product release cycle by 6–9 months.

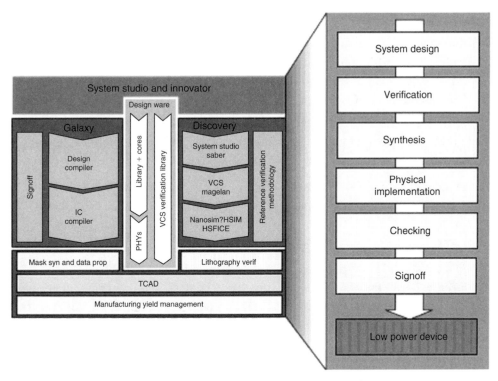

Figure 2.22: The synopsis low power solution
Source: http://www.synopsis.com

- Improving software development productivity, typically by a factor of 2-5x, by allowing unlimited observability and controllability of the target hardware, and by predictable and repeatable execution of debug scenarios.

- Promoting continuous hardware/software integration, rather than software and hardware meeting near the end of silicon development which is no longer practical in today's fast-paced markets.

The **Innovator** integrated development environment (IDE) helps designers develop, run and debug Virtual Platforms. Innovator allows designers to rapidly build Virtual Platforms and instantiate, configure, and connect low power components for early and accurate exploration of low power architecture.

System Studio is a system-level design creation, simulation, and analysis tool that supports transaction-level modeling. System Studio allows design teams to make early hardware–software architectural trade-offs, including modeling of power management strategies.

2.3.3.11 Verification

VCS® solution is the most comprehensive RTL verification solution in a single product, providing advanced bug-finding technologies, a built-in debug and visualization environment and support for all popular design and verification languages including Verilog, VHDL, System Verilog, and SystemC. The VCS solution understands and acts on the power intent definitions, enabling correct simulation of power domains in all modes of operation.

Verification tools use the definition of power domains and how they are controlled and shut down during RTL simulation to verify the functional behavior. Similarly, simulation must be able to model the use of isolation and retention cells so that designers can model the functionality correctly in different power states.

2.3.3.12 RTL Synthesis

The design team should know as early as possible if there is any risk that a design is not going to meet its power budget. In addition, the RTL synthesis should exploit all available power optimization techniques concurrently with timing optimization and all of the other optimization parameters.

Power cannot be optimized in isolation, it must be done in the context of area, speed, yield, manufacturability, reliability, signal integrity, as well as for worst-case process corners/ variability and low power/voltage scenarios.

Design Compiler® Ultra with topographical technology delivers accurate correlation to post-layout power, timing, test, and area, without wire load models by using the same placement as layout engine. It is designed for RTL designers and requires no physical design expertise or changes to the synthesis use model. The accurate prediction of layout power, timing, area, and testability in Synopsys' Design Compiler Ultra enables RTL designers to identify and fix real design issues while still in synthesis and generate a better start point for physical design. The clear design benefit is reduced iterations and faster turnaround time.

Power Compiler automatically minimizes dynamic power consumption at the RTL and gate level, using both clock gating and simultaneous optimization for power, timing, and area. Power Compiler also supports the use of multi-threshold libraries for automatic leakage power optimization.

Power Compiler performs automatic clock gating at the RTL without requiring any changes to the RTL source. This enables fast and easy trade-off analysis and maintains technology-independent RTL source.

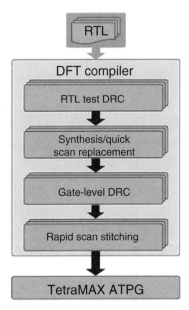

Figure 2.23: DFT compiler flow
Source: http://www.synopsis.com

DFT Compiler enables designers to conduct in-depth testability analysis of RTL, to implement the most effective test structures at the hierarchical block level, and, if necessary, to automatically repair test design rule checking violations at the gate level (Figure 2.23).

DFT Compiler transfers all information about the Adaptive Scan architecture to TetraMAX to automatically generate compressed test patterns with the highest test coverage.

TetraMAX ATPG automatically generates high-quality manufacturing test vectors. It is the only ATPG solution optimized for a wide range of test methodologies that is integrated with Synopsys' test synthesis tool. TetraMAX allows RTL designers to quickly create efficient, compact tests for even the most complex designs.

2.3.3.13 Physical Implementation

The Place and Route solution, **IC Compiler**, performs power-aware placement. By shortening high-activity nets and clustering registers close to the clock source. The clock tree has a critical part to play in dynamic power dissipation. Power-aware CTS can take advantage of merging and splitting integrated clock gating cells (ICGs) in the placed design. This capability allows more efficient use of ICGs, which saves significant dynamic power in the clock tree.

IC Compiler is a complete physical implementation solution with hierarchical design planning with automated power network synthesis and analysis. IC Compiler includes Extended Physical Synthesis (XPS) technology which provides a single convergent flow from netlist to silicon with support for multi-voltage designs, multi-threshold leakage, low power placement and CTS, and state-retention power gating.

JupiterXT™ design planning solution enables fast feasibility analysis for a preview of implementation results, and provides detailed floorplanning capabilities for flat or hierarchical physical design implementation styles. In addition, it provides fast, automatic placement, power network synthesis, and in-place optimization to allow designers to quickly generate prototype floorplans.

2.3.3.14 Checking

In multi-voltage designs, voltage-level shifting cells, and power domain isolation cells affect signal functionality; therefore, the design team needs to specify power intent in the RTL code to support automated RTL-to-gate equivalence checking. To ensure design integrity, designers also need to check special power-related structures against a whole battery of design rules.

Formality is used to make sure the RTL is equivalent to the gate-level implementation of the multi-voltage design. **Leda** checker is a programmable design and coding guideline checker that can be used to validate an entire SoC against a full range of low power checks. Leda provides over 50 low power checks, including insertion/location of level shifters and isolation cells and clock gating used to turn off power regions.

2.3.3.15 Signoff

Signoff must include power. Power management decrees whether an SoC will meet specifications. Power grid IR drop and multi-voltage design techniques directly affect SoC-level timing, so timing signoff must include the impact of voltage and temperature variation.

PrimeRail delivers full-SoC dynamic voltage-drop and electromigration analysis for power network signoff. By ensuring accurate correlation with design planning designers can be confident that the important decisions they take during the design planning stages with respect to power network planning will not have to be revisited at the end of the design cycle.

PrimeTime® PX enables full-SoC, concurrent timing, signal integrity, and power analysis in a single, correlated environment. It provides high accuracy dynamic and leakage power analysis concurrently with timing and signal integrity analysis. This capability offers major accuracy and productivity benefits over the use of separate or standalone timing and power analysis tools.

HSIMplus™ **PWRA** provides power net reliability analysis of IR drop and electromigration in power and ground buses. Concurrently **HSIMplus**™ **PWRA** accounts for the effects of interconnect resistance on dynamic circuit performance. In addition, it overcomes the limitations of simulation with millions of extracted parasitic resistors by incorporating built-in compression and reduction algorithms to maintain accuracy, capacity, and performance.

2.3.4 Low Power Format Standards

Limitations in the existing design infrastructure prevent power-related aspects of design intent from being specified across the design chain. The industry lacks automated power-control techniques capable of achieving both functional and structural verification of designs prior to incurring extensive manufacturing costs. Threatened by potentially high costs coupled with missed time-to-market opportunities, companies will remain reluctant to adopt advanced process geometries and effective low power methodologies with advanced process technologies.

The reluctance to adopt new process geometries has a negative impact across all business sectors in the electronics industry. Innovation by design teams is constrained due to the risk of low yields and costly re-spins. Library and IP vendors are unable to leverage the differentiated value of modeling the new processes, and tool providers are unable to sell new capabilities based on evolving process requirements. As a result, consumers are offered products that suffer from shorter battery life, higher heat dissipation, and other shortcomings that lead directly to lower sales—negatively affecting the profitability of businesses across the industry.

A key enabler of a modern power-aware design flow is the ability to capture and preserve the intent of the SoC architects and designers throughout the design flow. This requires a common specification that can be used and shared across the entire design chain, from architectural specification to verification. Power management concept is not solved for EDA tools and silicon libraries exhaustively.

Immature and proprietary power management tools have been developed by many EDA companies. Historically, there has not been any dialog in the industry to develop a power management SoC tool flow. There are point solutions but no complete solution (Figure 2.24).

Currently, EDA tools debugging capabilities are poor due to complexity issues and weak and misleading reports. In addition, functional correctness is difficult to verify. Another concern for the SoC designer is that most power management tools are at the gate level rather than at RTL level to meet the accuracy requirements.

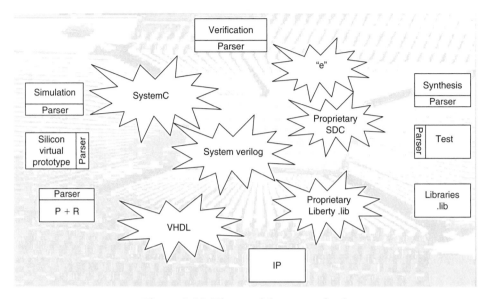

Figure 2.24: The need for a standard
Source: http://www.cadence.com

Power management problems concern all elements in the development chain, from system houses, silicon, IP providers, and tool providers that provides a 360 degree point of view.

Standardization is a key in building an efficient development flow and methodology. It insures a common language, a set of general methodologies developed around the standard, interoperable tools, and an opportunity for invention and creativity.

Power management tool requirements from SoC developers include:

Usability: Fills in existing tool flows minimizing learning curves, set up and deployment time.

Portability: Nonconflicting IP- and SoC-level activities. Insuring IP deliverables include information on how to implement and verify Power Management for IP when integrated into the SoC.

Reusability: Required for derivative designs.

Interoperability between tools: No scripting required to have tools work together.

Automatic verification: First time verification of the power management implementation accomplished automatically.

Standards should be speedily authored and employ open standards for low power design sourced by a broad inclusive team from across the SoC design chain.

Speed: Start simple, address immediate need, and evolve over time. Fill in the gaps by making use of existing standards and technologies.

Open: Everything on the table. No IP licensing or confidentiality required.

Inclusive: Invite everyone who wants an industry-wide solution to participate.

2.3.4.1 Single File Format

Currently, there are two major efforts in developing a common power description format that could be read by tools throughout the design flow[6]:

1. Common Power Format

2. Unified Power Format

2.3.4.2 Common Power Format

An example of such a specification is the CPF[7], which is managed by the **Silicon Integration Initiative** (Si2) consortium's **Low Power Coalition**. CPF is a design language that addresses limitations in the design automation tool flow. It provides a mechanism to capture architects' and designers' intent for power management and it enables the automation of advanced low power design techniques. CPF allows all design, implementation, verification, and technology-related power objectives to be captured in a single file and then applies that data across the design flow, providing a consistent reference point for design development and production (Figure 2.25).

There are three major benefits of using CPF to drive the design, verification, and implementation steps of the development flow:

1. It helps designers achieve the required SoC specs by driving the implementation tools to achieve superior trade-off among timing, power, and area.

2. By integrating and automating the design flow, it increases designer productivity and improves the cycle time.

3. By eliminating the need for manual intervention and replacing ad hoc verification methodologies, it reduces the risk of silicon failure due to inadequate functional or structural verification.

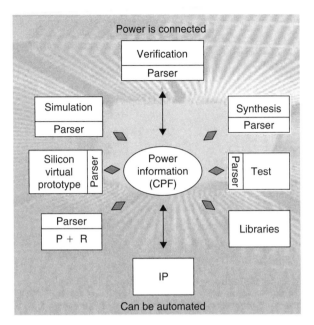

Figure 2.25: Standards-based solution

Source: http://www.cadence.com

2.3.4.3 *Power Forward Initiative*

The Power Forward Initiative's primary goal is to remove the barriers to automation of advanced low power design, and to provide a pathway to the development of a standards-based solution. By reducing the risks and costs associated with the design, verification, and implementation of designs using multiple/variable supply voltages and PSO design techniques, particularly at 90 nm and below, the door to a new era of innovation be opened.

Building a solution that fills the current void in the industry's infrastructure requires a departure from current approaches that have failed to create a holistic solution. In an effort to create a reset, the Power Forward Initiative addresses the entire scope of the industry – from the initial specification of design intent through to manufacturing and test (Figure 2.26).

However, this reset must not be disruptive to the portions of the existing infrastructure that are clearly working. What is required is a solution that serves as an overlay instead of a radical departure—one that does not require re-engineering of libraries, IP blocks, verification suites, or designs that can be reused. By definition, there are additive elements to the design environment and design process, but these new elements must not create a snowballing change throughout the ecosystem.

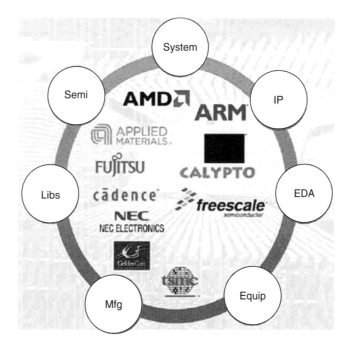

Figure 2.26: Ecosystem

Source: http://www.cadence.com

The Power Forward Initiative's initially focus is on the issues related to the use of complex power domains including multiple/variable supply voltages and power shut-off (PSO) techniques.

To achieve these goals, ongoing work on the **Common Power Format** is being considered within the larger industry frame of reference. Looking to satisfy the needs of the broad constituency of SoC design teams and providers of tools, equipment, IP, silicon, manufacturing, test, and services, the CPF has been created to deliver a comprehensive solution to the challenges posed by today's advanced power requirements. The CPF is being architected for future support of new design techniques and materials breakthroughs, including architecture, hardware and software system modeling, as well as analog and mixed-signal design.

Benefits of the CPF-based solution include:

- Functional verification of PSO using the same design description used for implementation

- Reduced turnaround time for physical implementation

- Enhanced optimization through simplified design exploration

- More accurate power utilization estimates

- Equivalence checking between functional description and implementation

Due to the fact that the new Power Forward approach is additive to the existing environment infrastructure and design flows, no changes to existing/legacy designs will be required. The CPF shows strong indications of viability for delivering productivity gains and improved Quality-of-Service (QoS) across the design chain. Automation is a key aspect to solve the industry's power management design challenges, and support for the CPF's extendable infrastructure will foster the adoption of automated low power design and verification solutions.

2.3.4.4 Unified Power Format

The UPF standard is a convergence of proven technology donations from a number of vendors[8]. EDA vendor contributions are derived from several years of successful use of their products on taped out low power designs. End customers contribute their internally developed optimization and analysis technologies which deal with application-specific power issues, especially for wireless and handheld devices. Strong collaborative participation by Accellera members and other dedicated companies results in an open standard.

When power consumption is a key consideration, describing low power design intent with the UPF improves the way complex ICs can be designed, verified, and implemented. The open standard permits all EDA tool providers to implement advanced tool features that enable the design of modern low power ICs. Starting at the RTL and progressing into the detailed levels of implementation and verification, UPF facilitates an interoperable, multi-vendor tool flow and ensures consistency throughout the design process.

A UPF specification defines how to create a supply network to supply power to each design element, how the individual supply nets behave with respect to one another, and how the logic functionality is extended to support dynamic power switching to these logic design elements. By controlling the operating voltages of each supply net and whether the supply nets (and their connected design elements) are turned on or off, the supply network only provides power at the level the functional areas of the SoC need to complete the computational task in a timely manner.

2.3.4.5 Accellera

Accellera, similar to Si2, provides design and verification standards for quick availability and use in the electronics industry. The organization and its members cooperatively deliver much-needed EDA standards that lower the cost of designing commercial IC and EDA products[9].

2.3.4.6 IEEE

In the near future SoC designers will have to cope with two standard power formats. Technically CPF and UPF are similar up to 90%. Both UPF and CPF use commands allow user to establish and manage separate power domains, set up level shifters, specify isolation and retention, and define power-related rules and constraints.

However, there is a silver lining, the IEEE has a working group currently studying both the CPF and UPF standards[10].

Clearly, this is one of the biggest challenges the EDA standards community has faced. An aggressive timetable, multiple standards organizations, and a highly competitive landscape make the task appear daunting.

2.4 Summary

Low power technologies and techniques for energy efficiency are distinctive combinations of advanced architectural techniques and circuit design techniques with the latest design methodology. These technologies and techniques provide the highest possible performance levels within a restricted power budget. They can be applied to any application, benefiting the user by extending battery life without impacting performance as well as keeping energy costs lower.

Various low power circuit and architectural techniques, for mitigating leakage power, have been described in this chapter. These include power gating, dynamic process temperature control, static process compensation, state-retention power gating, clock gating, low power modes, and asynchronous circuits.

Power consumption of mobile devices has moved to the forefront of SoC development concerns. Designing for low power can have a significant impact on the power budget of a mobile device. In addition, identifying and resolving power problems late in the flow requires time consuming and expensive iterations. However, designing for low power adds an additional degree of complexity to an already complex problem that has historically targeted performance and area. Optimizing the three variable equation requires a new class of EDA tools that are able to address the power problem from the start to the end of the SoC design flow.

In order to handle complex interrelationships between diverse effects, it is necessary to have a design system whose tools are fully integrated with each other and also with other analysis tools. There are a number of EDA tool vendors that have made, and continue to make, significant contributions in the area of low power design. Two major players in the EDA industry are Cadence and Synopsis. Each has low power flows that provide low power end-to-end solutions from RTL to GDSII. A diverse variety of low power design techniques have been developed to address the various aspects of the power problem.

In addition, comprehensive power reduction also demand a strong ecosystem that includes low power hardware IP blocks, low power libraries, modeling techniques, and silicon wafer fabrication flows that support the low power design methodology.

Two competitive standards for low power SoC design are available. The UPF has been released by Accellera. The CPF is managed by Silicon Integration Initiative's Low Power Coalition. Can two standards survive in this highly competitive market? Possibly yes, however, one will dominate defined by market forces.

Future instantiations of the low power EDA tools will have to address the unique and higher-level power issues associated with analog functionality, packaging, and systems-level challenges.

References

1. Arora M. GALS technique fells massive clock trees. *Chip Design magazine*. http://www.chipdesignmag.com; October/November 2006.

2. Krstic M. *Request-driven GALS technique for datapath architectures*. http://www.deposit.ddb.de/cgibin/dokserv?idn=978773063&dok_var=dl&dok_ext=pdf&.filename=978773036.pdf; February 2006, pp. 12–18.

3. Bink A, de Clercq M, York R. *ARM996HS Processor, ARM White Paper*. February 2006.

4. *Cadence Low-Power Methodology Kit, Datasheet*. http://www.cadence.com; 2007.

5. *Synopsis Low-Power Solution, White Paper*. http://www.synopsis.com; June 2007.

6. Peterman D. *Unifying the Industry behind a Power Format*. http://www.TI.com; October 2006.

7. *Using a Low-Power Kit to Improve Productivity, Reduce Risk, and Increase Quality, White Paper*. http://www.cadence.com; 2007.

8. Bailey S. *Low Power Design Specification from RTL through GDSII*. http://www.edadesignline.com; July 2007.

9. Brophy D. *Accellera Update and Roadmap*. http://www.accellera.org; April 2007.

10. Goering R. *Dueling Power Standard Camps Stall Specification Merger*. http://www.eetimes.com; March 2007.

11. Hemani A, Meincke T, Kumar S, Postula A, Olson T, Nilsson P, Öberg J, Ellervee P, Lindqvist D. Lowering power consumption in clock by using globally asynchronous locally synchronous design style. *Proceedings of AC/WIEEE Design Automation Conference* 1999.

12. Talpes E, Marculescu D. Toward a multiple clock: Voltage island design style for power-aware processors. *IEEE Transactions on Very Large Scale Integrations (VLSI) Systems* 2005; **13**(5):591–603.

CHAPTER 3

System-Level Approach to Energy Conservation

Findlay Shearer

"System Level Approach to Energy Conservation" tells how the Big Boys do it—including Freescale, ARM, National, TI, Intel and Transmeta—and how you can, too, once you go out and buy a Power Management IC (PMIC), figure out how the System Power Management (SPM) Architectural Framework of the Mobile Industry Processor Interface (MIPI) Alliance works, and then build your portable design around it. Reading this chapter first, of course, is recommended.

Here Shearer puts the various power management techniques that he introduced earlier into a system-level perspective. To implement dynamic voltage and frequency scaling (DVFS), for example, you need a hardware-aware software algorithm that drives it and an operating system that supports it. Read this section and then read the documentation that came with your RTOS to see if it's up to the task. It's no wonder that semiconductor companies are now hiring more software engineers than hardware designers.

Power management problems in portables have become so complex that even when equipped with the same extensive bag of tricks, different manufacturers take quite different approaches:

Freescale's eXtreme Energy Conservation (XEC) software creates a generic software/hardware framework, application programming interface (API) and hardware interface that is portable across different Freescale platforms. The software makes performance predictions and then invokes a range of power-saving techniques to control the various systems on both the chip and board.

ARM's Intelligent Energy Manager (IEM) also makes predictions based on both a policy database and historical data to scale the processor to an optimal level of performance, using adaptive voltage/frequency scaling (AVFS) to control numerous voltage islands.

National's PowerWise technology embeds an Advanced Power Controller (APC) into the target SoC, using AVFS to control the SoC and external PMICs via a two-wire interface for the rest of the system. Threshold scaling addresses the static power problem that looms at smaller geometries.

TI's SmartReflex uses every trick mentioned so far, plus a combination of silicon IP that targets static leakage power and system software that supports Symbian and Linux.

Intel's SpeedStep includes three low power states (standby, deep idle, deep sleep); dynamic frequency and voltage scaling; and a Power Manager Software Architecture that enables detailed software via the OS.

There are a lot of considerations in choosing a processor for a portable design, but power management has to be at the top of the list. Read this chapter before you commit.

—**John Donovan**

3.1 Introduction

The challenging energy demands of new portable devices are forcing radical approaches to solve complex power management problems. It means expanding the definition of power management beyond power delivery, to include power distribution and power consumption. It means interfacing the power-delivery system to the power consumption system, to permit the systems to communicate with each other to vastly improve power conservation.

To achieve these quantum leaps manufacturers need a new industry model. Historically power management and processor integrated circuit (IC) suppliers have developed their technologies independently. However, power efficiencies are now reaching levels where only minimal gains can be achieved through this conventional-isolated approach. No longer are piecemeal short-term solutions addressing the power efficiency of individual components. Rather the entire system has to be considered as a whole and the opportunities for system components to work together have to be leveraged to obtain power performance level required by next generation devices.

Effective power management requires more than simply turning devices on or off. A flexible power architecture is needed. For example, many processors and peripherals support multiple power modes such as Standby and Sleep to power down unused functions except when they are in active use. Multiple power zones are required to support the various supply requirements and specialized operational modes to extend battery life for portable products. Each zone is supplied by an independent linear or switching regulator, giving developers a high degree of control over power.

Further optimization can be achieved through techniques such as coordinated wake-up sequences; instead of turning components on all at once, power is conserved by firing up various circuits only when they are needed. For example, in a cell phone, there is no point in powering analog amplifiers before the radio frequency (RF) synthesizer has stabilized, nor turning on the digital signal processor (DSP) before a signal has been received that needs processing. Using tuned and coordinated wake-up sequences helps to manage the supply perturbations due to current surges flowing through the source impedance of the battery. By spreading out startup surges and reducing battery dips, the end-of-life cutoff for minimum battery level can be extended with a resulting improvement in effective battery lifetime between charge cycles.

With a highly integrated approach, power management integrated circuits (PMICs) enable developers to optimize power consumption at the system level, for specific applications, while significantly reducing the design complexity required to achieve these gains. For this reason, early consideration of power management details is fundamental for designing power efficient hardware and software that can be relied upon to provide optimal battery life.

Solutions require a systems approach where portions of the power management circuitry exist in separate ICs and portions are integrated into the system chips. Companies such as AMD, Intel, Texas Instruments, Freescale, ARM, National Semiconductors, and Transmeta have obtained good results using this type of frequency–voltage management in addition to clock gating.

3.2 Low Power System Framework

Basic Energy Management System: An Energy Management System (EMS) employs techniques to minimize the drain from the battery or other limited power source. The diagram shows a basic EMS solution comprising three parts:

1. The Platform Hardware with various power managed components (PMC).

2. Performance Estimator employing dynamic voltage and frequency scaling (DVFS) and other power saving algorithms.

3. Performance Setter that maps the estimations to specific platform operating points (OP), i.e., combinations of PMC settings.

The best operating settings for the PMCs depends on the changing workload demanded by the processor, peripherals, and the software. This is application specific, and typically several application programs run concurrently.

One approach is to characterize the workload for each application beforehand and cumulate for all known use cases. However, this is suboptimal and an efficient solution would use prediction and other techniques to compute the required performance in real time (Figure 3.1).

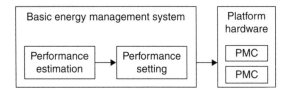

Figure 3.1: Basic energy management system

3.2.1 Advanced Energy Management Solution

The EMS dynamically adjusts the performance-power settings of the device hardware to levels that achieve the appropriate level of performance to handle the software workload within its real-time deadlines.

The Performance Estimator uses additional performance predictors and system event monitors to improve its performance estimation. Incrementally, it could have an interface to "power-aware" application programs and middleware software capable of specifying their required performance levels dynamically.

The Performance Estimator computes the actual workload from moment to moment. However, the performance setting-function does not change the hardware settings for each new estimation due to the cost, measured in time and power, in changing from one operating point to another. An advanced EMS performs a cost–benefit analysis in real time to decide whether a change in OP makes the energy saving better or worse. For example, processor low power idle modes, like Sleep and Doze, each have a "break-even" time. Unless the device spends at least the break-even time in the low power mode, you would be better off, in energy-saving terms, staying in the Run mode. If the device has reliable event-driven information to determine how long it can be idle for, the EMS can make intelligent selections of which low power mode it can employ as required (Figure 3.2).

An advanced EMS systematically encodes hardware design data so that its run-time choices are optimized for the chipset on which it is running. These are called System Cost Rules.

3.2.2 Software for Self-Optimizing Systems

The EMS algorithms are fed with relevant real-time events to make minor corrections for changing scenarios. These minor corrections include adapting the optimal device settings in response to product user inputs via user policies.

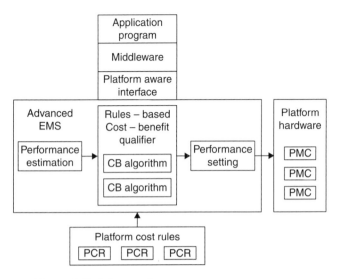

Figure 3.2: Advanced energy management system

A "user-friendly" advanced EMS should allow the device developer to set approximately optimal settings in the EMS for a broad set of software use cases and applications. The adaptive nature of the EMS manages the run-time variations. This is not just software optimization but software for self-optimizing systems. The concepts developed in EMS are the foundation of a number of available power conservation techniques.

3.3 Low Power System/Software Techniques

Power-saving technologies have been developed to address sources of power waste. Many are hardware solutions such as smaller transistor geometries and active well biasing. However, other hardware technologies require software and different software techniques must be used to exploit each of these different hardware technologies effectively. Dynamic power management (DPM) describes a system that sets the power states of its hardware modules in real time to minimize power waste, and still meets performance needs. DPM includes techniques such as dynamic voltage and frequency scaling (DVFS), dynamic process and temperature compensation (DPTC), and idle time prediction for controlling low power idle modes (Doze, Sleep, Hibernate, etc.)

The software techniques shown in Figure 3.3 include Dynamic Voltage Frequency Scaling and Dynamic Process Temperature Compensation. Application programs are monitored

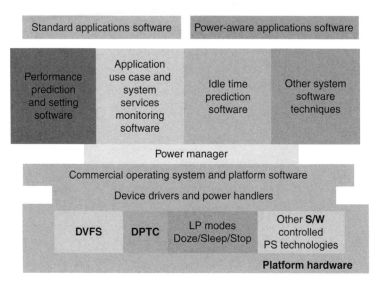

Figure 3.3: Power-saving software techniques
Source: www.freescale.com

during execution by one or more of the software techniques illustrated. Some of these applications know their performance-power needs ("power-aware" software) and others do not. These techniques are employed to control a power manager that drives the hardware power-saving mechanisms using software drivers and power handlers in the operating system.

3.3.1 Dynamic Frequency Scaling

System-on-chip (SoC) dynamic power consumption is proportional to operating frequency. It makes sense to lower the clock frequency of a processor to the lowest value that still meets the required processing performance. Although the software runs more slowly, it still meets its real-time deadlines with acceptable margins. This is performed dynamically and requires special power management software to compute which frequency setting is acceptable.

It may seem intuitive that although this lowers the instantaneous power consumption, it might not reduce the overall energy requirement, but instead spreads it over a longer period. In fact, some memory systems, such as integrated level 2 caches and SDRAMs, will incur fewer accesses at the reduced operation frequency, so their circuits incur reduced switching and therefore reduced power consumption. However, the benefits of frequency scaling alone to total energy management are marginal.

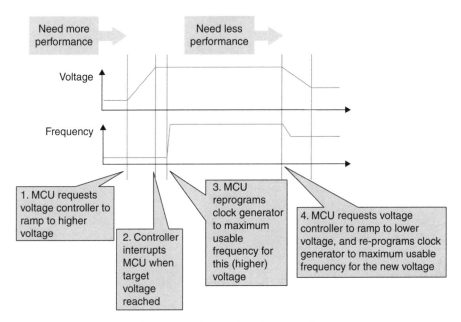

Figure 3.4: Dynamic voltage scaling

Source: www.freescale.com

3.3.2 Dynamic Voltage Scaling

Since power varies with the square of voltage, square-law power savings are possible with voltage scaling. If voltage scaling and frequency scaling are both used, the combination, called DVFS, yields power savings roughly proportional to the cube of operating voltage. These square-law and cube-law power savings depend not only on the configuration and efficiency of the voltage control circuits, but also on the efficiency of the prediction software used to set the voltage/frequency settings.

For a given SoC design, the operating voltage determines the maximum usable operating frequency. The voltage, and hence frequency, are scaled to trade required performance against minimal power waste. However, when scaling the voltage the operating frequency must be scaled with the voltage to meet the design constraints.

Figure 3.4 shows the required sequence for changing:

- First, to a higher voltage and frequency, and then

- Later, back to a lower voltage and frequency.

At point 1 a request is made to the voltage controller to ramp the processor voltage up to the new higher value. The voltage will increase at a defined slew rate, to avoid excessive power surges, and the controller will notify the requester, via a processor interrupt, when the target voltage has been reached at point 2. The processor must continue at the lower frequency until then, at which point software then signals its clock generator, typically a programmable phase lock loop, to change the processor frequency to a higher one, usually the highest usable frequency for the new voltage, which occurs at point 3. The frequency change is quick compared to the slow slew rate that must be adopted.

When going to lower voltages/frequencies, the processor changes its frequency and requests the voltage change at point 4. The processor need not wait until the voltage ramp down has completed before it continues normal software execution, since the new lower frequency is already within the operating range of the voltage profile.

3.3.3 Dynamic Process and Temperature Compensation

The operating voltage setting chosen with DVFS to achieve a specific operating frequency is a worst-case value. It includes a voltage margin for variations in process and temperature. This margin represents power waste because the SoC is operating at a slightly higher voltage than it needs for the operating frequency. By monitoring the SoC using an on-chip process and temperature-dependent structure, it is possible to calculate a lower operating voltage that is very close to process limits, which thereby minimizes power waste. If the SoC is manufactured in two or more silicon processes that information could also be fed into the calculations.

3.3.4 Handling Idle Modes

DVFS technology addresses varying but continuous software workloads. However, what happens when the processor or other power-hungry device has periods of no activity? When the processor has periods of inactivity, the hardware devices have low power idle modes that software can exploit to save power. However, some devices have more than one idle mode. A processor core could have Doze, Sleep, and Stop modes. The deeper modes yield more power savings, but usually come with a greater "cost" penalty in time and transition power when entering and exiting the mode. For example, an idle mode based on power gating requires the hardware's internal memory state to be saved and restored, either by hardware or software.

In a real-time system, the device's run state must be re-entered in time-to-service events that otherwise would degrade the user's quality of service. Smart software must either predict or have advanced knowledge of an event and the time taken to restore the hardware from idle to run mode. Based on the predicted start and end of an idle period, it controls the shut-down and wake-up mechanisms associated with the idle modes. The prediction software has to determine which of the idle modes will yield a net power saving for each idle period it encounters.

3.4 Software Techniques and Intelligent Algorithms

A simple DVFS software driver attached to the operating system is used to interface to the integrated DVFS control logic to increase or decrease the processor's operating frequency and voltage. Controlling and monitoring the DVFS mechanism is the easy part.

Knowing in real time which frequency/voltage setting to use, and when, is more challenging. Intelligent software is needed to compute dynamically how much processor performance the application programs and other system software require at any given time.

How intelligent that software is will determine how close you can get to the theoretical hardware limit on power saving. Higher performance software is more complex and more costly to develop. There is a performance versus complexity/cost trade-off which will vary for different applications and products. The trick is to develop intelligent energy management technology that is both flexible and scalable to give improved performance at acceptable cost across a wide range of devices. That includes the ability to support one or more algorithms for many hardware power-saving technologies.

3.4.1 Operating System

Typically, each microprocessor and DSP in wireless applications uses operating system software to manage the large number of hardware and software resources. A DPM approach should treat the OS as a state machine in which each state may require its own power management techniques. For example, DVFS could be used using normal software execution of tasks and operating system, idle mode predictors used during software idling, and full-speed execution during interrupt handling.

Figure 3.5 shows a possible architecture for the power management software for a modern applications domain operating system. The hardware modules may each have several power states including a fully powered, and one or more low power idle states. A DVFS-controlled processor will have multiple active states and may also have several idle states.

The instantaneous values of the collective power states could be related to the phone use modes in many ways. The main criterion for judging an algorithm is the average power saving

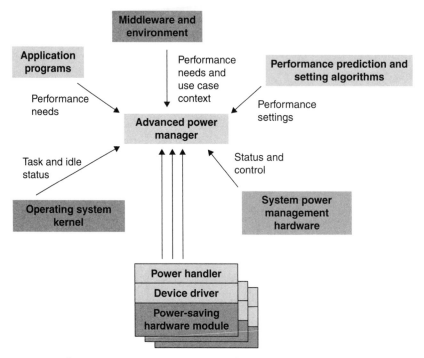

Figure 3.5: Power management in the operating system

Source: www.freescale.com

obtained over a range of typical use cases compared to a system which has no software power minimization.

Performance-prediction and performance-setting algorithms are used to control the performance-power states of the system hardware dynamically. These algorithms are very sophisticated and may need to be adapted or tuned to suit different system designs.

3.4.2 Typical DVFS Algorithm

Many algorithms exist for use with DVFS-based processors that set the processor's operating frequency and voltage based on predicting the short-term software workload on the processor.

An example algorithm in this class tracks the recent software workload history of each task running in the OS. This technique assumes a reasonable correlation between the recent past workload of a task and that of the near future. The task status information must be supplied by the OS kernel.

The algorithm maintains estimates of workload and unused idling time to predict the aggregate workload for all tasks. This normalized MCU processing level is translated by associated software into the relevant frequency and voltage settings required for the specific DVFS mechanism used. The algorithm continuously re-calculates and supplies new predictions in response to changing software workloads.

Ideally, the algorithm correctly predicts the required processor performance that meets individual deadlines for each OS task. The algorithm works well for OS tasks whose workloads do not change very rapidly.

3.4.3 Scope Within Wireless Mobile Applications

Some may question whether such complex prediction techniques are really necessary in a mobile application. After all, most of the "engine" software that runs on the cellular modem DSP, the Bluetooth® and WLAN processors, although it involves complex protocols, is fairly deterministic and can be well characterized and optimized for energy conservation at design time. The main and largely unexploited opportunity for large power savings exists in the applications domain. In the application domain the processor and multimedia applications, such as video and audio, run in real-time.

3.5 Freescale's XEC: Technology-Specific Intelligent Algorithms

For power-managed platforms, Freescale takes an architectural approach, and creates a generic software/hardware XEC framework with applications programming interfaces (APIs) and hardware interfaces. Its purpose is to help to maximize the portability of XEC technology across Freescale platforms.

Prediction-based software control of power-saving techniques is an immature technology and is just starting to emerge within the mobile wireless device industry. As yet there are no existing, widely accepted power management standards among mobile device manufacturers, operating system vendors, semiconductor companies, and others.

Algorithmic software, known as performance predictors, is used to predict runtime workloads of specific modules or power-saving technologies (e.g., DVFS). A predictor module can be included or removed from the software build depending on whether or not its corresponding hardware technology is present.

An advanced energy-saving solution such as XEC uses two or more performance predictors for certain technologies such as DVFS, to ensure high performance under a wide range of operating conditions (Figure 3.6).

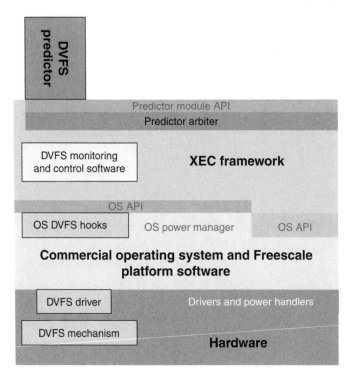

Figure 3.6: DVFS predictor

Source: www.freescale.com

Typically, a performance predictor operates abstractly. For example, a DVFS predictor only tries to predict required processing power as a normalized fraction of the maximum possible. It does not deal with the processor frequencies, operating voltages, MIPS, etc.

These device specific details of the DVFS hardware technology are handled by the XEC framework and OS driver software. Ideally, performance predictors are designed to be independent of the specific OS and hardware, to enable easy porting from one platform to another.

Depending on the software activity, the processor and other hardware modules will have periods of idling where they could be put into low power idle modes. The XEC technology uses an idle time predictor. It contains prediction software with connections to the OS and other event-monitoring software. The monitoring and control software specific to the platform's idle mode hardware is located in the XEC framework and in the OS device drivers for the hardware (Figure 3.7).

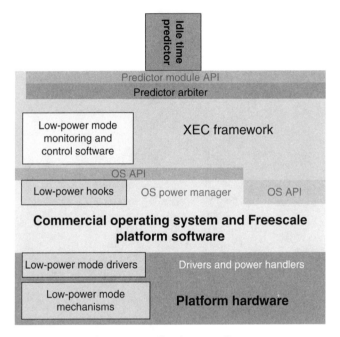

Figure 3.7: Idle time predictor

Source: www.freescale.com

In principle, any hardware module with software-settable power-saving modes can be managed by the idle time policy software for minimum power waste (Figure 3.8).

3.5.1 XEC Framework

The XEC framework isolates the algorithmic components, namely the performance predictors, from the details of the specific platform OS and hardware. It contains platform-specific mechanisms for controlling and monitoring hardware power-saving technologies such as DVFS. In addition, it also provides common APIs for power-aware application programs and replaceable XEC modules like the performance predictors.

Because performance predictors may operate concurrently, a predictor arbiter arbitrates between multiple predictors to resolve which of their performance-power recommendations to select at any given time. The arbiter is programmable, so that the priority of any one performance predictor may be changed during normal operation depending on the runtime circumstances. Another framework component called the policy manager, acting as an agent

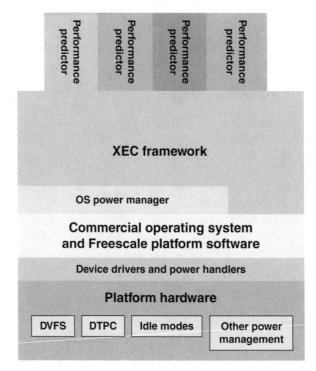

Figure 3.8: Performance predictors

Source: www.Freescale.com

for the product user, dynamically selects a policy, from a group of policies that sets the criteria for trading off performance, power, and energy in general or for specific situations or programs.

The XEC framework uses platform-specific power-cost rules to determine when and if to transition a particular hardware module from its current power state. Finally, the framework has an OS adaptation layer containing OS-specific code. Its purpose is to minimize the porting effort and configuration management problems to migrate the XEC software to the operating system.

Based on advanced runtime performance algorithms, XEC uses a standardized software framework that supports multiple concurrent predictors, policies, power-cost rules, etc. It runs as system software with commercial operating systems. The XEC solution targets more potential areas for energy savings such as SoC-specific low power modes and LCD panel hardware.

XEC is transparent to application programs and middleware, unless that software is optimized as a "power-aware" application; in that case, the application could communicate its power needs to XEC.

Using advanced techniques, XEC dynamically discovers each piece of software's required performance, sets hardware power management features like DVFS, multiple low power idle modes, etc., for just-enough performance and minimal power waste, and achieves much more energy savings than traditional techniques. XEC has performance predictors, real-time cost–benefit analyzers, policies and a policy manager, and many other advanced algorithms.

3.6 ARM's Intelligent Energy Manager

Completing a task before its deadline, and then idling, is significantly less energy efficient than running the task more slowly so that the deadline is met exactly. The goal of ARM's Intelligent Energy Manager (IEM) is to reduce the performance level of the processor without allowing applications to miss their deadlines. The central issue is how the right level of performance can be predicted for the application.

ARM has developed a system level solution to energy management. The IEM[1] framework provides a hardware and software mechanism for achieving these goals: it standardizes the interface for setting the processor's performance level, specifies counters for measuring the amount of work that is being accomplished, and includes operating system and application-level algorithms for predicting future behavior.

ARM Intelligent Energy Manager (IEM) technology implements advanced algorithms to optimally balance processor workload and energy consumption, while maximizing system responsiveness to meet end-user performance expectations. The IEM technology works with the operating system and applications running on the mobile device to dynamically adjust the required processor performance level through a standard programmer's model (Figure 3.9).

The IEM software component uses information from the OS to build up a historical view of the execution of the application software running on the system. A number of different software algorithms are applied to classify the types of activity and to analyze their processor utilization patterns. The results of each analysis are combined to make a global prediction about the future performance requirement for the system. This performance setting is communicated to the hardware component so that the SoC-specific scaling hardware can be controlled to bring the processor to that level of performance.

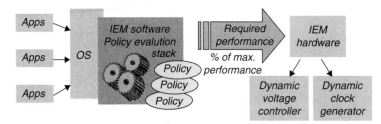

Figure 3.9: IEM technology controlling voltage and frequency[2]

3.6.1 IEM Policies and Operating System Events

IEM Policies are short, modifiable software routines that control what to do when an OS event occurs. An IEM kernel employs policies to determine a response to an event like increase or decrease the voltage and frequency to the CPU. Some software tasks are predictable in their behavior and can track past behavior to predict future performance requirements. IEM policies and their relationship to the IEM software architecture are illustrated in Figure 3.10.

An event occurs in the operating system: for example, a new task starts or battery low is detected. The operating system informs IEM and policies are checked to determine the appropriate action. A new frequency requirement is determined and sent to the performance controller.

3.6.2 Types of Policy

Figure 3.11 indicates a conceptual view of IEM policies[3].

A *Step Policy* monitors the time spent by the operating system in idle task. If the time is large it reduces the frequency and if small it increases the frequency.

A *Perspective Policy* monitors the performance from the perspective of the task. The policy minimizes idle time available between each time the task is run. In addition it removes the effect of the task being preempted.

A *Mean Policy* generates the average workload for tasks recently run. If more time is available, the policy reduces frequency the next time the task is run. If the task did not complete in time, the policy increases the frequency the next time.

Customer Policies are created by customers for applications such as gaming or playing music or events like low battery.

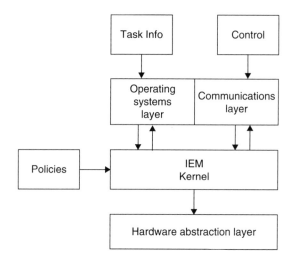

Figure 3.10: IEM software architecture

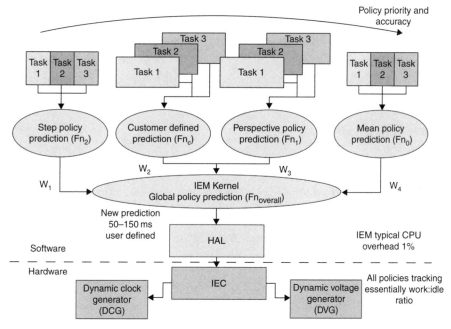

Figure 3.11: Types of policies[3]

Policies are split into standard part and fast part. The standard part is run at a scheduled interval and may be preempted by higher priority tasks. However, the fast part is run during every OS event including context switch, task creating and deletion. It is preemptible only if the OS supports preemption of the kernel.

The fast part of the policy is used to capture time-critical information that will not be available when the standard part runs. In addition it reacts quickly to events like real time applications being scheduled to run and to capture information that is not accessible to the standard part.

The IEM software layer has the ability to combine the results of multiple algorithms and arrive at a single global decision. The policy stack supports multiple independent performance-setting policies in a unified manner. The primary reason for having multiple policies is to allow the specialization of performance-setting algorithms to specific situations, instead of having to make a single algorithm perform well under all conditions. The policy stack keeps track of commands and performance-level requests from each policy and uses this information to combine them into a single global performance-level decision when needed.

Using this system, performance requests can be submitted any time and a new result computed without explicitly having to invoke all the performance-setting policies. While policies can be triggered by any event in the system and they may submit a new performance request at any time, there are sets of common events of interest to all. On these events, instead of re-computing the global performance level each time a policy modifies its request, the performance level is computed only once after all interested policies' event handlers have been invoked.

3.6.3 Generic IEM Solution

The IEM system components, illustrated in Figure 3.12,[4] include:

- *Intelligent Energy Manager (IEM)*: A software component that measures the workload and predicts performance level needed to achieve it.

- *Intelligent Energy Controller (IEC)*: A hardware block that assists the measurement of the workload and transforms requirement to a percentage frequency value.

- *Dynamic Voltage Controller (DVC)*: A hardware module that transforms desired performance level to voltage and frequency via an open and closed-loop approach. Adaptive Power Controller (APC) from National Semiconductors is an example of a DVC.

- *Power Supply Unit (PSU)*. A highly controllable and responsive power supply.

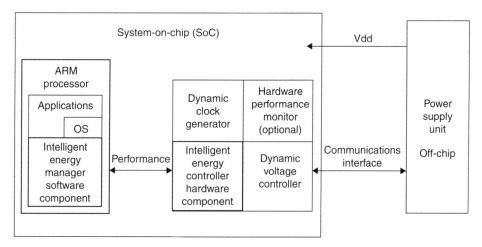

Figure 3.12: IEM block diagram[4]

The IEM software and hardware monitor the system workload to generate a performance request. The DVC can then set the correct operating voltage in either open-loop or closed-loop mode without processor intervention. The DVC will transparently provide the fastest possible response while assuring that the processor will always receive the minimum safe operating voltage for any given clock frequency. The DVC would also coordinate all clocks switching including the verification of stable supply voltage. The IEM provides a uniform software interface to simplify implementation and reuse. The DVC provides an open-standard interface to the external power supply.

3.6.4 Intelligent Energy Controller

The IEC is an advanced microcontroller bus architecture (AMBA) bus-compliant peripheral which provides a number of support functions to the IEM software. The IEC provides an abstracted view of the SoC-specific performance scaling hardware. It is responsible for translating the performance prediction made by the IEM software (0–100% of maximum performance) into an appropriate performance point at which the system will run and then controlling the scaling hardware to achieve operation at that point.

The IEC also measures the work done in the system to ensure that the software deadlines are not going to be missed. Additionally, the IEC supports a "maximum performance" hardware request feature.

The three major parts of the IEC, illustrated in Figure 3.13, are:

1. An AMBA Advanced Peripheral Bus (APB) interface which provides the IEM
    software with a standard API interface and allows it to communicate with the dynamic
    performance controller (DPC) and dynamic performance monitor (DPM) blocks.
    It also provides the IEM software status information about the support provided by the
    system.

2. The DPC interfaces to the DCG and the DVC to set the target performance level to be
    achieved and requested by the IEM software.

3. The Dynamic Performance Monitor provides the IEM software with some of the system
    metrics it needs to decide the performance level required.

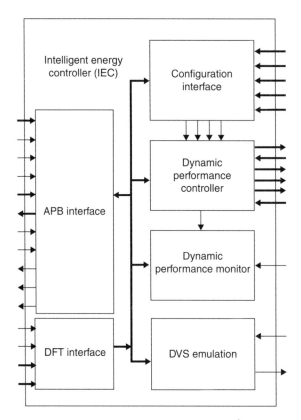

Figure 3.13: IEC block diagram[4]

3.6.5 Voltage Islands

In conventional multi-voltage approaches, a design is partitioned into voltage islands that use the lowest supply voltage that allows the maximum performance required by the island. Specifically, one or more voltage islands operate with lower supply voltages and run at correspondingly lower clock frequencies. Each island's voltage and frequency are fixed. In some dynamic voltage/frequency-scaling designs, an island's supply level and clock frequency can change, but only over a few fixed choices that designers set ahead of time.

An adaptive voltage/frequency-scaling approach, such as ARM IEM technology, offers much more flexibility in scaling voltages and frequencies on multiple islands. For this reason, an IEM system is better able to scale performance to meet task requirements. For example, power (and hence energy) is wasted when a processor executes a task quickly but then idles while waiting for the next task. To be truly efficient, the processor should never run faster than is necessary to meet application software deadlines, especially if the time saved on one process is spent idling/waiting for the next process. To help to ensure power and energy efficiency, IEM software policies determine the appropriate task durations for an application and set percentages of full voltage and clock frequency accordingly, to reduce idle states.

IEM technology includes both hardware and software for adaptive adjustment of voltage and frequency to suit processor workload. IEM software collects data directly from the OS kernel, eliminating the need for any special applications encoding to handle this task. IEM software uses various policies to categorize application workload and to profile performance requirements on the fly. IEM software can then set suitable OP for voltage and frequency that enable the processor to meet performance requirements while minimizing idle time. To implement IEM technology, an advanced design methodology and low power physical IP supporting multi-voltage and multi-threshold design are required. The physical IP must be characterized at multiple voltage points, include level shifters and isolation cells, and be available with multiple threshold versions to support leakage power optimization.

3.7 National Semiconductor's PowerWise® Technology

3.7.1 PowerWise Technology

National Semiconductor's PowerWise technology is a system-level approach that reduces the power consumption of SoC solutions used in portable devices. PowerWise technology is a three-part solution: embedded intelligence in digital processors such as the baseband or applications processors of a mobile phone; an open standard power management interface; and companion PMICs.

SoCs use multiple processor and hardware accelerators to provide the processing power required by the applications running on the system. Each of the separate processing engines requires dedicated power management and control to optimize its power consumption. PowerWise technology is suited for managing multiple independent processing engines inside a SoC either when fully operational, or when functions are idling, dormant or completely turned off. The technology creates closed-loop systems where the power-consuming SoC's and power-delivery systems operate in close cooperation, minimizing demands on the power source while providing peak energy efficiency.

3.7.2 Adaptive Power Controller

In addition the technology embeds a synthesizable AMBA-compliant core, the Advanced Power Controller (APC) into the target SoC, as shown in Figure 3.14. The APC ensures that power-consuming digital logic and power-delivery systems operate in close cooperation, minimizing demands on the power source while providing peak energy efficiency. The APC uses a variety of advanced power management algorithms to achieve this, including adaptive and DVFS. The APC is responsible for monitoring and adjusting the supply voltage of the SoC so that the supply voltage is always optimized for the current operating frequency.

PowerWise technology further addresses the needs of powering microprocessor cores with the ability to implement threshold scaling. With deep submicro technology, the static power dissipation due to leakage current becomes significant. Threshold scaling reduces static leakage current by offsetting the N- and P-well bulk biases so that transistors are more

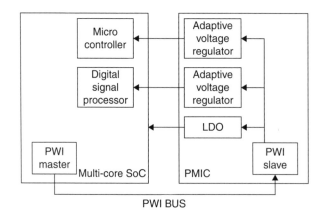

Figure 3.14: Multi-core SoC and PWI compliant PMIC

effectively driven "off." The APC interfaces to the rest of the system using three interfaces: The AMBA compliant host interface, the CMU interface and the open standard PowerWise Interface (PWI). The host interface is used to control and configure the APC2 while the CMU interface is used to coordinate voltage and frequency changes.

The PWI is used to communicate power management information to external power management ICs to adjust supply voltages. The APC enables the system to implement either dynamic voltage scaling (DVS) or fully adaptive voltage scaling (AVS) on the target system. The use of standard interfaces enables the APC to be easily embedded into any logic circuit and interfaced with other parts of the system. This enables system designers to very quickly develop platforms with next generation power efficiency, minimizing both time-to-market and risk (Figure 3.15).

PWI facilitates adoption of advanced power management techniques that can be used to monitor performance and control various processor voltages such as supply voltages, threshold voltages, etc. The PWI specification defines a communication interface between a processor and one or more external PMIC. PWI does not include OS interfaces for performance monitoring and control.

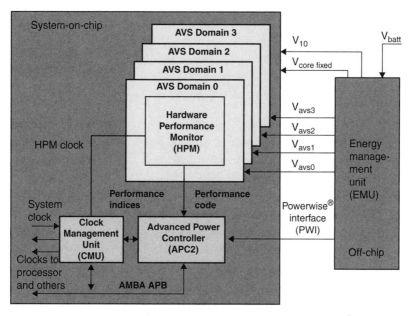

Figure 3.15: PowerWise Advanced Power Controller[5]

3.7.3 The PWI Specification

The demand for advanced power management techniques is driven by the increased functionality in portable devices. To provide longer battery life a number of power management techniques are available:

- Dynamic voltage and frequency scaling

- Adaptive voltage and frequency scaling

- Back-biasing for static leakage reduction

These power management techniques require a link between the SoC and the PMIC that permits fast data transfer and rapid PMIC response. In addition, this interface must be easy-to-use and add little cost.

In 2003 ARM, National Semiconductor and six other companies released the PWI 1.0 standard to support the use of advanced power management techniques in SoCs used in mobile battery-operated devices.

The PWI specification defines a two-wire serial interface connecting the integrated power controller of an SoC processor system with a PMIC voltage regulation system that allows system designers to dynamically adjust the supply and threshold voltages on digital processors.

The PWI specification defines the required functionality in the PWI-slave; the operating states, the physical interface, the register set, the command set, and the data communication protocol for messaging between the PWI-master and the PWI-slave. The PWI command set includes PMIC operating state control; register read, register write, and voltage adjust commands.

The PWI standard, PWI 2.0, address new and growing power management issues of complex SoC with multiple processors. The scope of PWI 2.0 covers:

- Advanced power management techniques needed to reduce system power consumption; both dynamic and static leakage power are important

- Multi-processor and multi-core systems

- Advanced hardware and software power management techniques
    - Accurately monitor and control performance level for a given workload
    - Accurately control various supply voltages based on performance level
    - Enable only those functions needed at any given moment
    - Implement system level energy control in real time

- Rapid deployment of advanced power management techniques at the system level requires standardization in hardware and OS interfaces

- PWI 2.0 is a hardware interface standard for advanced system-level power management

The PWI 2.0 specification defines a two-wire serial interface dedicated to deployment of advanced power management technologies. In addition it connects one or two SoC processor systems with external PMIC. The PWI 2.0 specification defines the physical interface layer, the register set, the command set, and the data communication protocol for messaging between PWI-masters and PWI-slaves.

Its command set includes PMIC operating state control, register read, register write, and voltage adjust commands as well as master-to-master communication protocol. Also the specification additionally provides provision for user-defined registers in the PWI-slaves and masters.

3.7.4 PowerWise PMU/EMU: Power/Energy Management Unit

A PWI™ 2.0 compliant Energy Management Unit (EMU) reduces power consumption of low power handheld applications such as dual-core processors and DSPs. A typical EMU (same as a PMIC) contains two advanced, digitally controlled switching regulators for supplying variable voltages to a SoC or processor. The device also incorporates five programmable low-dropout, low-noise linear regulators for powering I/O, peripheral logic blocks, auxiliary system functions, and maintaining memory retention (dual-domains) in shutdown mode.

The device is controlled via the high speed serial PWI 2.0 open-standard interface. It operates cooperatively with PowerWise® technology compatible processors to optimize supply voltages adaptively (adaptive voltage scaling (AVS)) over process and temperature variations. It also supports dynamic voltage (DVS) using frequency/voltage pairs from pre-characterized look-up tables (Figure 3.16).

3.7.5 Dynamic Voltage Scaling

DVS exploits the fact that the peak frequency of a processor implemented in CMOS is proportional to the supply voltage, while the amount of dynamic energy required for a given workload is proportional to the square of the processor's supply voltage[6]. Reducing the supply voltage, while slowing the processor's clock frequency, yields a quadratic reduction in energy consumption, at the cost of increased run time.

Completing a task before its deadline is an inefficient use of energy[7]. Employing performance-setting algorithms to reduce the processor's clock frequency only when it is

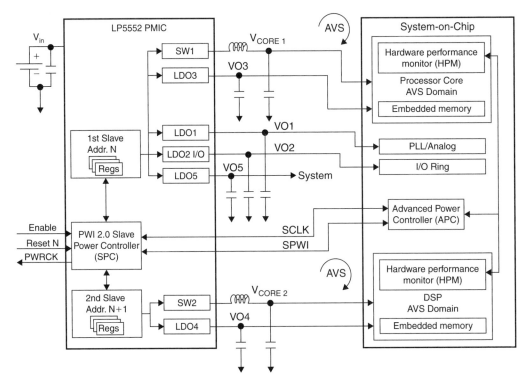

Figure 3.16: LP 5552 EMU/PMU block diagram

Source: www.nsc.com

not critical to meet the application's deadlines is critical for energy conservation. There is a significantly lower total energy consumption using DVS compared with traditional gated-clock power management, for the same workload. Note that with DVS, the lower supply voltage reduces static power even when the clock is gated off.

DVS solutions offer improved performance by reducing the supply voltage as the clock frequency is reduced. Open-loop DVS, as shown in Figure 3.17, allows the processor to set the supply voltage based on a table of frequency/voltage pairs. This table must be determined by characterization to assure sufficient margin for all operating conditions and process corners.

The processor must determine the desired operating frequency, request a new voltage, wait for the voltage to stabilize, and then switch itself to the new frequency. The switch may be made immediately when changing from a higher frequency to a lower frequency. When switching from a lower frequency to a higher frequency, the power supply voltage must be high enough to support the new frequency prior to changing the clock.

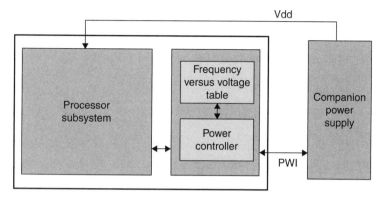

Figure 3.17: Open loop DVS[8]

Power supply stability can be assured either by a time delay or by an analog measurement. Use of a time delay is risky, since there will always be a desire to implement the minimum possible delay for enhanced processor response time.

Open-loop operation is simplified by creating an APC module to off-load the voltage scaling and clock management from the processor. The APC approach supports a common software API allowing the DVS function to be easily accessed by applications or the operating system. Providing a standard interface to the external power supply also simplifies system design and facilitates second-source options for the power supply component. An architecture using an APC is shown in Figure 3.17.

3.7.6 Adaptive Voltage Scaling (AVS)

AVS is a technique for determining the minimum supply voltage for a digital processor during operation. This technology significantly reduces the supply voltage of a digital processor by removing voltage margins associated with the effects of process and die temperature variation inside the SoC, as well as IR-drop and regulator tolerances in the system.

Performance-setting algorithms optimize power consumption based on workload variations. However, significant power efficiency can also be gained if the SoC does not have to operate under worst-case assumptions but can tune its operating parameters to temporal environmental conditions[9]. SoCs are designed to operate reliably over a wide range of temperature levels and variations of the silicon substrate. Increased voltage levels must be used to assure the large safe operating range at the cost of reduced power efficiency. By monitoring the margin between expected and actual operating conditions, the voltage level of the processor can be reduced without sacrificing operational stability. This closed-loop monitoring of system margin will be referred to as AVS.

Closed-loop or AVS offers improved performance and ease of implementation compared to open-loop DVS. In the closed-loop system shown in Figure 3.18, the voltage is set automatically by monitoring the system's performance margin and adjusting the supply voltage adaptively.

Using a performance request, an APC can set the correct operating voltage and interface with the clock-management unit to enable transitions to new frequencies. The APC receives commands from the Hardware Performance Monitor (HPM) when new, higher frequencies are to be deployed and enables the new clock frequencies on the CPU core and, if applicable, in on-chip cache, memory, and peripherals of the SoC. The HPM can require new voltage levels and fine-tune them by communicating to the external power supply.

While the core is still operating under its previous stable frequency, the next higher test frequency is sent to the HPM, its results checked again and again. Voltage is increased in steps until the HPM reports that the test frequency yielded a correct result by issuing a vdd_ok signal to the APC.

The test frequency yields a go/no-go answer which must be augmented by additional HPM logic that can deliver voltage adjustments as the chip's temperature rises and IR drops change, owing to changing demands in supply current. Feedback to the external power supply is based on a local spot on the chip only and may require several sensors for the tightest voltage performance. Downward frequency shifts are less problematic, since the lower frequencies

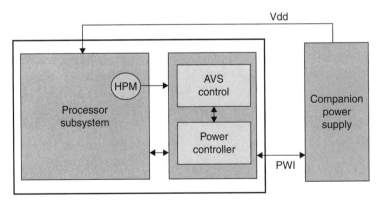

Figure 3.18: Adaptive voltage scaling[8]

are supported by higher voltages. The combined function of HPM and APC circumvents differences in process, fabrication, and temperature condition.

Since the system is closed-loop in nature, a much finer degree of control over the voltage is possible when compared to the discrete table values in an open-loop system. Response time of the AVS system can be much faster, since it is limited only by the external power supply. The performance measuring circuitry can be used to verify the power supply stability to offer the fastest possible switching from one clock frequency to the next.

Closed-loop AVS is distinguished from open loop, table-based voltage scaling techniques in that it regulates the propagation delay margin in logic cells. In this system, the power supply voltage is a variable that increases or decreases, and the delay margin is a fixed parameter that is regulated over parts, temperature, and clock frequency.

Many advantages arise from this methodology. Closed-loop AVS relaxes the characterization process. There is no need for characterizing voltage/frequency tables because a delay margin is maintained by the AVS feedback loop. Another incentive is that less demand is placed on power supply regulation. The AVS loop adjusts the supply voltage as necessary, compensating for the ±5% tolerance typically allocated to power supply regulation.

By far the most beneficial advantage is that the minimum operating voltage is realized for all conditions, and can dynamically change as conditions change. Below are the results comparing DVS with AVS at varying temperatures[10] that indicate significant energy savings with AVS (Figure 3.19).

3.8 Energy Conservation Partnership

ARM and National Semiconductor have developed energy conservation solutions in order to help mobile device manufacturers to maximize the battery life of their handheld, battery-powered devices. The modularized nature of the total solution means that the technology can be adapted to suit the underlying performance scaling hardware, including DVS and AVS.

The IEM prediction software determines the lowest performance level that the processor can run at while ensuring, with the aid of the IEC, that software deadlines are never missed.

The APC works with the external EMU using the performance prediction to bring the processor to the absolute lowest possible voltage and frequency that still operates the application software correctly (Figure 3.20).

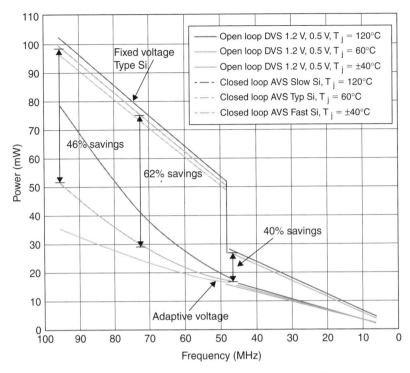

Figure 3.19: 130 nm bulk CMOS at 96 MHz[10]

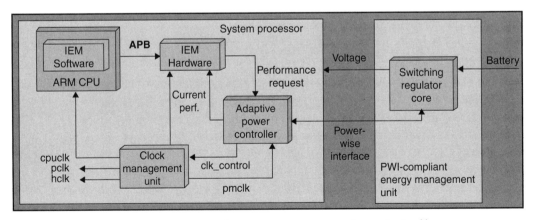

Figure 3.20: ARM and NSC energy conservation solution[11]

This complete solution reduces the energy consumed by the processor to the lowest possible given the constraints of the clock generator and power supply dynamics and the headroom available in the mix of application software.

The ARM IEM technology can be used to reduce the energy requirement of an embedded processor by up to 75%. National's PowerWise technology can reduce safety margins and provide additional energy savings of about 45% using AVS at room temperature as compared with an open loop voltage control solution.

Open-loop and closed-loop voltage scaling are only one area in the project ARM and NSC have undertaken. Clock gating, new cell libraries to minimize leakage current and power domains to turn it off are additional areas where the two companies are collaborating to reduce consumed energy.

3.9 Texas Instruments: SmartReflex

TI's SmartReflex[12] technologies take a system-wide perspective on the interrelated issues of power and performance to address both dynamic and static leakage power. SmartReflex technologies are comprised of three facets:

1. Silicon intellectual property (IP).

2. Techniques that can be applied at the SOC design level.

3. System software that manages many of the hardware-enabled SmartReflex technologies, which interface to other power management techniques, based in operating systems (OS) or third-party software subsystems.

3.9.1 Silicon IP

At the silicon level, TI has a track record as an industry pioneer in sophisticated power and performance capabilities, many of which have transitioned into SmartReflex technologies. One major emerging challenge addressed by SmartReflex technologies is static leakage power, which becomes a significantly greater part of a device's total power at smaller process nodes. Several SmartReflex technologies can be applied to drastically limit leakage from a device. For example, the static power leakage in the OMAP2420 processor is reduced by a factor of 40 with SmartReflex technologies. Today, many of TI's 90-nm wireless components already implement SmartReflex technologies to reduce leakage power. In the future, all new devices at the 90 nm, 65 nm and smaller process nodes will incorporate these breakthrough technologies.

Another SmartReflex technology at the silicon level is a library of power management cells that enable power switching, isolation, and voltage shifting to facilitate a granular approach to partitioning a device's power domains. By structuring the device with multiple power domains, functional blocks can be powered down or put into a standby power mode where they are not active, thus reducing power consumption while ensuring optimal performance. To simplify chip-level integration, SmartReflex technologies are supported by an easy-to-use, non-intrusive design flow:

- *Retention SRAM and logic*: SRAM and logic retention cells support dynamic power switching (DPS) without state loss, lowering voltage, and reducing leakage.

- *Dual-threshold voltages*: Higher threshold for lower leakage and lower threshold for higher performance.

- *Power management cell library*: Switching, isolation, and level shifters support multiple domains in SOC implementations.

- *Process and temperature sensor*: Adapts voltage dynamically in response to silicon processes and temperature variations.

- *Design flow support*: Complete, nonintrusive support for easily integrating SmartReflex technologies.

3.9.2 System-on-Chip

SmartReflex technologies include techniques at the architectural level of SoC design to address static leakage power. These include the following:

- *Adaptive voltage scaling (AVS)*: Maintains high performance while minimizing voltage based on silicon process and temperature.

- *Dynamic power switching (DPS)*: Dynamically switches between power modes based on system activity to reduce leakage power.

- *Dynamic voltage and frequency scaling (DVFS)*: Dynamically adjusts voltage and frequency to adapt to the performance required.

- *Multiple domains (voltage/power/clock)*: Enables distinct physical domains for granular power and performance management by software.

- *Static leakage management (SLM)*: Maintains lowest static power mode compatible with required system responsiveness to reduce leakage power.

3.9.3 System Software

At the level of system software, SmartReflex technologies include host processor power management, which features several capabilities that are deployed at the system level, such as a workload monitor, workload predictor, resource manager, and device driver power management software. Additionally, the SmartReflex framework features DSP/BIOS™ power management software.

SmartReflex technologies support multiple cores, hardware accelerators, functional blocks, peripherals, and other system components. In addition, their system-level technologies are open to OS-based and third-party power management software so that a collaborative and cooperative environment with regards to power and performance can be developed.

- *OS support*: Provides an open environment for blending with operating systems. Supports Symbian, and Linux.

- *Software power management framework*: Intelligent control for power and performance management. Transparent to application programs and legacy code. Monitors system activity, not just processor activity.

- *Workload monitoring and prediction*: Determines system performance needs used to make intelligent power and performance management decisions.

- *Policy and domain managers*: Dynamically controls the system, providing the performance needed at the lowest power.

- DSP/BIOS power management.

3.10 Intel SpeedStep

Intel's SpeedStep® Technology provides the ability to dynamically adjust the voltage and frequency of the processor based on CPU demand[13]. Wireless Intel SpeedStep Technology incorporates three low power states deep idle, standby, and deep sleep. Employing the Wireless Intel SpeedStep Power Manager Software, to intelligently manage the power and performance needs for the end user.

The Power Manager software can be used to manage power consumption and optimize system standby time and talk time in smart phones and other mobile devices. The software provides device programming interfaces (DPIs) to device drivers and applications programming interfaces (APIs) to applications. The platform-specific layer of the Power Manager software is used to adapt the Power Manager software to a given platform.

Device drivers must be set up as a client of the Power Manager software so that the device receives notifications on power policy changes and power states via the DPIs. Optionally, applications can use the APIs to further enhance power savings.

3.10.1 Usage Modes

The Power Manager software has the following usage modes:

- Standby mode (standby time)

- Voice communications (talk time)

- Data communications

- Multimedia (audio, video, and camera)

- Multimedia and data communications (video conferencing)

For each of these usage modes, the Power Manager software provides:

- Optimal power policy for dynamic scaling of power and performance

- Optimal operating frequency and voltage

- Usage of low power modes for the entire system including all of the devices

- State transition and power management for its devices

The Power Manager software changes the operating profile of a generic system as shown in Figure 3.21. DVM and DFM are used to dynamically scale the "Run" frequency and voltage

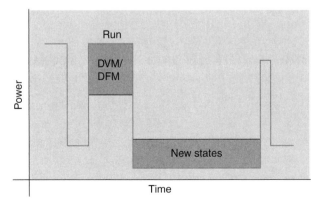

Figure 3.21: Operating profile with DVM, DFM, and new states[13]

to meet immediate performance requirements with minimum power consumption. New power states are used to minimize "Idle" power consumption.

3.10.2 Power Manager Architecture

The power manager architecture consists of the following five software components:

1. The *Policy Manager* is responsible for determining a system's power policy. The Policy Manager uses dynamic scaling of frequencies and voltages to help to provide the lowest power consumption under all types of workloads. The Policy Manager assesses information supplied by the Idle Profiler, Performance Profiler, User Settings, DPIs, and APIs. The Policy Manager defines power states that run the processor at the frequencies and voltages that are consistent with the lowest power consumption.

2. The *Idle Profiler* monitors parameters that are available in the operating system's Idle thread. The Idle Profiler then provides status information to the Policy Manager.

3. The *Performance Profiler* uses the Performance Monitoring Unit (PMU) within the processor to determine if the workload is CPU bound, memory bound, or both CPU and memory bound. The Performance Profiler then provides status information to the Policy Manager.

4. The *User Settings* allow the user to specify the parameters that are used by the Policy Manager to determine a system's power policy.

5. The *Operating System Mapping* allows the Power Manager software to be ported across multiple operating systems.

The PM software architecture with the five components is shown in Figure 3.22.

3.10.3 Speedstep DFM

The processor's core and peripheral clocks are derived from phase locked loops. The processor implements Intel DFM by allowing the core clock to be configured dynamically by software.

The core clock frequency can be changed in several ways:

*   Selecting the 13-MHz clock source

*   Changing the core PLL frequency

*   Enabling or disabling turbo mode or half turbo mode

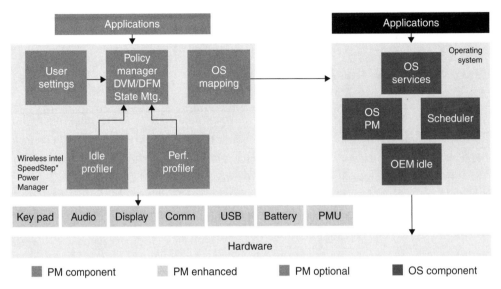

Figure 3.22: Power Manager software architecture[13]

3.10.4 Speedstep DVM

The processor implements DVM through its Voltage Manager. The Voltage Manager provides voltage management through use of an I2C unit that is dedicated to communication with an external PMIC regulator, and through use of a Voltage Change Sequencer.

When software initiates a voltage change mode, the Voltage Change Sequencer can automatically send commands via the I2C unit to an external PMIC regulator. The sequencer can send up to 32 commands, which can be categorized as dynamic commands and static commands:

- Dynamic commands are executed when the core is running.

- Static commands are executed after clocks to the processor are disabled.

Intel DVM augments DFM by enabling code to dynamically change the system voltage as well as frequency. Like many other leading companies, Intel's approach to energy conservation is holistic and includes process technology, SoC, and systems development.

3.11 Transmeta LongRun and LongRun2

For its Crusoe chip[14], Transmeta has introduced LongRun, a table-based set of multiple frequency–voltage points, similar to open loop DVS, which helps Crusoe track workloads more efficiently than by using a few steps.

Transmeta's Crusoe processor saves power not only by means of its LongRun Power Manager but also through its overall design which shifts many of the complexities in instruction execution from hardware into software. The Crusoe relies on a code morphing engine, a layer of software that interprets or compiles x86 programs into the processor's native VLIW instructions, saving the generated code in a Translation Cache so that they can be reused. This layer of software is also responsible for monitoring executing programs for hotspots and re-optimizing code on the fly.

As a result, features that are typically implemented in hardware are instead implemented in software. This reduces the total on-chip real estate, capacitance and its accompanying power dissipation. To further reduce the power dissipation, Crusoe's LongRun modulates the clock frequency and voltage according to workload demands. It identifies idle periods of the operating system and scales the clock frequency and voltage accordingly.

Transmeta's later generation, Efficeon processor, contained enhanced versions of the code morphing engine and the LongRun2 Power Manager[15]. LongRun2 includes techniques to reduce static leakage by dynamically adjusting a processor's threshold voltage.

LongRun2 is a broad solution to leakage problems that are emerging in chips made at the 90- and 65-nm nodes. LongRun2 adjusts threshold voltage on a CPU to virtually eliminate leakage current, taking power consumption in idle mode down from 144 to just 2 mW. The technique depends on both hardware circuitry already in Efficeon and the software.

LongRun2 has all the original LongRun techniques with multiple dynamic frequency and voltages. In addition it added dynamic threshold control and other power reducing technologies.

LongRun2 Vt control provides an approach to body bias that can be applied to standard bulk CMOS process technology. It adds capacitance that reduces body contact and power/ground noise. In addition it avoids body bias distribution in metal layers that may increase the SoC die area.

Body bias voltages can be employed to shift the threshold voltage of the SoC transistors toward the worst-case processing corner, resulting in lower leakage (Figure 3.23).

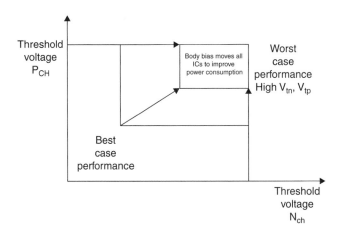

Figure 3.23: Body bias employed to shift transistor threshold voltage of the SoC to lowest leakage corner (top right) of the design target

A traditional body bias solution can provide some of the advantages of LongRun2 technologies; however, LongRun2 adds valuable tools and technologies to make body bias more practical to use. In a traditional approach, metal wires are required to distribute the PMOS and NMOS bias voltages across the chip, adding area and routing complexity. LongRun2 adds techniques to simplify the bias distribution reduce or eliminate the need for metal bias routing and allow retrofit of existing designs without a complete re-layout. Besides bias distribution, LongRun2 also offers innovations in dynamic bias control, feedback, and optimization, as well as supporting circuit structures and control techniques that may help to reduce minimum operating voltage.

3.11.1 LongRun2 IP

Transmeta has hardware blocks that semiconductor designers use to quickly implement LongRun2 Technologies into either existing SoC designs or new designs. There are four major categories of supporting blocks for LongRun2 Technologies.

3.11.2 Body Bias Controllers

The design of the threshold voltage controller is critical to achieve the greatest power, performance, and yield benefits. Transmeta is developing a range of controller designs that can be quickly integrated into existing or new designs to provide both yield distribution advantages and low power consumption. These controllers determine the combination of supply and body bias voltages that minimize power dissipation at a target frequency and activity. Advanced controller designs may monitor other factors such as the temperature of the device and degradation due to aging.

3.11.3 Body Bias Voltage Distribution

Traditional body bias voltage distribution requires semiconductor designers to utilize metal layers to route PMOS and NMOS bias voltages. Routing bias complicates global routing by introducing yet another global voltage supply pair, and also results in increased area. With Transmeta's proprietary body bias distribution semiconductor structures, it is possible to retrofit an existing design for body bias quickly with minimal changes to an existing design, reducing design cost and risk.

3.11.4 Body Bias Voltage Generators

NMOS and PMOS bias voltages may be generated off-chip using conventional regulator techniques or on-chip using dedicated bias voltage generators. Due to the low current drive required, on-chip generators consume very modest area and avoid the need to complicate board design and route bias pins externally. Transmeta is developing a range of bias voltage generator designs to enable designers to incorporate these blocks on-chip quickly and easily.

3.11.5 Monitor Circuits

Chip performance and leakage vary with changes in voltage, temperature, load, age, and other operating factors. Transmeta has developed a comprehensive set of monitor circuits that can provide continuous feedback on critical parameters which the Bias Controller can use to improve bias voltage control decisions and increase the benefits of LongRun2 Technologies.

3.12 Mobile Industry Processor Interface: System Power Management

The Mobile Industry Processor Interface (MIPI) Alliance is an open membership organization that includes leading companies in the mobile industry that share the objective of defining and promoting open specifications for interfaces in mobile terminals. MIPI Specifications establish standards for hardware and software interfaces between the processors and peripherals typically found in mobile terminal systems. By defining such standards and encouraging their adoption throughout the industry value chain, the MIPI Alliance intends to reduce fragmentation and improve interoperability among system components, benefiting the entire mobile industry.

The MIPI Alliance is intended to complement existing standards bodies such as the Open Mobile Alliance and 3GPP, with a focus on microprocessors, peripherals, and software interfaces.

3.12.1 System Power Management

MIPI has introduced System Power Management (SPM) Architectural Framework[16]. The document reflects the basis for future standardization activities of the MIPI Alliance System Power Management Working Group. In addition the document proposes a widely applicable abstract model for system-wide power management, defining a set of functional interfaces over which power management control can be performed. Furthermore, a power management request protocol and format are proposed.

MIPI's intention is to allow enough flexibility to enable creativity in power management, while ensuring backwards compatibility with hardware and software modules; and the framework should be extensible and scaleable to allow future power control methods. This allows the SPM Architectural Framework to still add significant value in systems where not all components are MIPI SPM aware, and be the foundation for future power management systems.

3.12.2 Power Management System Structure

A representative system model that provides structure to the SPM challenges has been developed. The key elements that represent such a system include:

- Devices and subsystem services

- Power management domains

- Policy managers

- Power management infrastructure

- Power management interfaces

- Protocols

- Clock generation and Vdd supply

- Application-supported and application-transparent power management

A Representative System Model, used to explain the SPM Architectural Framework in the following sections, is shown in Figure 3.24.

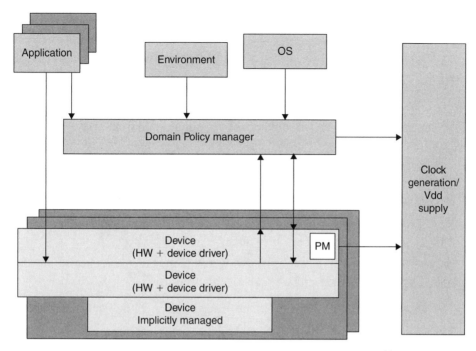

Figure 3.24: Representative system model of the SPM[16]

The SPM framework identifies categories of device management that have an effect on the system structure and vice versa. These management categories are where policy decisions are made. Categories include:

- Self-managed devices

- Other-managed devices

- Hybrid-managed devices

MIPI SPM architectural framework can be applied to a simple mobile device via a six step process:

1. Power/clock supply partitioning

2. Power management partitioning

3. Policy manager functionality

4. Power management control hierarchy and communication

5. Hardware/software partitioning

6. Device capabilities and CPW/device state mapping

The benefits to system integrators, operating system vendors, software developers, and device IP vendors of the SPM Architectural Framework include:

- Captures power management of whole system in a single framework.

- Provides a structure that enables the integration of devices from multiple sources into a power-managed system.

- Enables device IP vendors to deploy advanced power management capabilities in a portable fashion, and additionally enabling functionality that is decoupled from, and goes beyond that typically provided by different operating SPM models.

- Allows product differentiation by allowing power management innovations to be deployed without redesigning all parts of the system.

- Provides flexibility to fit many system organizations including centralized and de-centralized, hierarchical, and peer-to-peer.

- Provides flexibility to fit many different power management schemes and systems employing a mixture of schemes.

3.13 Summary

Combining high performance with low power consumption is one of the primary goals of portable devices. Historically developers have relied on Doze, Sleep, and Hibernate modes for conserving power in processors. However, an increasing number of system vendors, like Freescale, ARM, Intel, AMD, Transmeta, NSC, Texas Instruments, and others, have developed systems approaches that take advantage of the fact that reducing the frequency and voltage can yield approximately a quadratic decrease in energy used.

A low power system framework, EMS, employs techniques to minimize the drain from the battery or other limited power source. These include:

- The Platform Hardware with various PMICs.

- Performance Estimator employing DVFS, DPTC, and other energy conserving algorithms.

- Performance Setter that maps the estimations to specific platform "operating points" (OP), i.e., combinations of PMC settings.

The EMS dynamically adjusts the performance and power settings of the mobile device to levels that achieve enough performance to handle the task at hand with the minimum power consumed while meeting the real-time deadline.

Freescale has developed an EMS called eXtreme Energy Conservation or XEC. XEC is a framework-based system software targeted at various Freescale chipsets. Software components that comprise XEC include:

- *XEC-DVFS*, which is an OS-aware predictive DVFS that runs on applications processors. Power savings vary widely by use of cases but 40% for video player and audio player software is typical.

- *XEC-LCD*, which automatically regulates the LCD display backlighting uses either or both of two techniques and can save up to 50% of the total LCD power consumption.

The XEC solution has been ported to a number of mobile device platforms to meet the user expectations of high performance wireless and multimedia features with low power consumption.

Using intelligent software algorithms to predict, characterize and set the performance and power consumption of wireless and multimedia portable devices, XEC dynamically changes the power states of the system hardware, ensuring the necessary performance required to run applications while minimizing power wastage.

The ARM IEM provides continuous predictive monitoring of the processor demand. Its objective is to run the clock frequency at the lowest available value while still completing the task prior to its deadline. The performance level is set by predictive algorithms that are embedded in the operating system kernel to monitor all the processes.

The ARM IEM and NSC PowerWise technology take a systems view to control additional system components beyond the processor. Additional components such as memory controllers and graphics accelerators can also be monitored in order to control their performance level and power consumption. System-wide energy conservation is therefore possible with the combination of ARM's IEM and NSC's PowerWise technologies.

From a standards perspective, the MIPI SPM Working Group has presented the SPM architectural framework for interfaces that can be used to develop and build flexible and efficiently power-managed portable devices.

The SPM Architectural Framework provides a number of benefits to all parties of the portable device development ecosystem. These include capturing power management in a single framework and enabling device IP vendors to deploy advanced power management capabilities in a portable manner.

References

1. Morris P, Watson P. Automated low-power implementation methodology. *Information Quarterly* 2005; **4**(3).

2. Watts C. Intelligent energy manager. *ARM Forum*, October 2003.

3. Uttley P. ARM1176JZF-S intelligent energy management (IEM). *ECoFac*. Nice, France; April 6, 2006.

4. *Intelligent Energy Controller. Technical Overview, revision r0p1*; July 2005.

5. Lee HM. *National Semiconductors PowerWise brief*; June 2007.

6. Mudge T. Power: A first class architectural design constraint. *IEEE Computer* 2001; **34**(4).

7. Govil K, Chan E, Wasserman H. Comparing algorithms for dynamic speed-setting of a low-power CPU. *Proceedings of the First International Conference on Mobile Computing and Networking.* San Francisco, CA; November 1995.

8. Liu RH. How to create designs with dynamic/adaptive voltage scaling. *ARM Developers Conference.* Santa Clara, CA; 2004.

9. Dhar S, Maksimovic D, Kranzen B. Closed-loop adaptive voltage scaling controller for standard-cell ASICs. *Proceedings of the 2002 International Symposium on Low-Power Electronics and Design (ISLPED 2002).* Monterey, CA; August 2002.

10. *PowerWise Technology, Powering Next-Generation Portable Devices, National Semiconductors.* http://www.nsc.com; May 2004.

11. Watts C, Ambatipudi R. Dynamic management of energy consumption in embedded systems. *Information Quarterly* 2003; **2**(3).

12. Carlson B. *SmartReflex™ Power and Performance Management Technologies, White Paper*; 2005.

13. *Wireless Intel SpeedStep® Power Manager, White Paper.* http://www.intel.com; 2004.

14. Fleischmann M. LongRun™ Power Management; January 2001.

15. http://www.transmeta.com/tech/longrun2.html.

16. *MIPI Alliance SPM Architectural Framework, White Paper version 1.0.* http://www.mipi.org/docs/mipi-spm-framework-wp-2005.pdf

CHAPTER 4

Radio Communication Basics

Steve Rackley

Few if any technologies have impacted our lives more in the 21st century than wireless communications. In 2008 for the first time there were more cell phone subscribers worldwide than registered landlines. People now take being connected for granted. A wide range of RF interfaces enable you to communicate—whether by voice or data—almost anywhere on earth.

As in any period of rapid change, keeping up with the changes requires constant attention. In his book Wireless Networking Technology Steve Rackley does a particularly good job of sorting out a multitude of wireless networking technologies, explaining the protocols, implementations and applications. This chapter gives a good overview of the problems involved in implementing wireless networks, including multiplexing, multiple access and digital modulation techniques; and RF signal propagation and reception. He pays particular attention to newer techniques, including ultra-wideband (UWB) radio, multi-input multi-output (MIMO) antennas and near-field communications (NFC).

Rackley starts off with a detailed discussion of spread-spectrum techniques—direct sequence (DSSS), frequency hopping (FHSS) and time hopping (THSS)—since these are the basis for just about all of the low-power wireless interfaces used in portable designs. He shows how they are implemented in Wi-Fi, Bluetooth, Zigbee, UWB and the various cellular protocols.

The section on multiple access techniques addresses the interference issues that are starting to crop up in the crowded industrial, scientific and medical (ISM) bands where all coexist. TDMA separates signals by time, OFDMA by frequency, SDMA by space and CDMA by assigning users unique codes.

Rackley next turns to the various modulation techniques used to get the data stream onto an RF carrier. He examines each from the standpoint of spectral efficiency, bit error rate (BER) performance, power efficiency and implementation complexity. After a detailed discussion of signal propagation and reception issues, Rackley then shows how to calculate a link budget, *demonstrating how to juggle the numerous variables in order to maintain communications within an acceptable error rate.*

His explanations of UWB, MIMO and NFC are the best overviews I've seen, and well worth reading on their own. They're probably the only part of this chapter that actually is basic, but that section treats these complex topics with a clarity that is rarely found elsewhere.

—John Donovan

4.1 The RF Spectrum

The radio frequency, or RF, communication at the heart of most wireless networking operates on the same basic principles as everyday radio and TV signals. The RF section of the electromagnetic spectrum lies between the frequencies of 9 kHz and 300 GHz (Table 4.1), and different bands in the spectrum are used to deliver different services.

Recalling that the wavelength and frequency of electromagnetic radiation are related via the speed of light, so that wavelength (λ) = speed of light (c)/frequency (f), or wavelength in meters = 300/frequency in MHz.

Beyond the extremely high frequency (EHF) limit of the RF spectrum lies the infrared region, with wavelengths in the tens of micrometer range and frequencies in the region of 30 THz (30,000 GHz).

Table 4.1: Subdivision of the radio frequency spectrum

Transmission type	Frequency	Wavelength
Very low frequency (VLF)	9–30 kHz	33–10 km
Low frequency (LF)	30–300 kHz	10–1 km
Medium frequency (MF)	300–3000 kHz	1000–100 m
High frequency (HF)	3–30 MHz	100–10 m
Very high frequency (VHF)	30–300 MHz	10–1 m
Ultra high frequency (UHF)	300–3000 MHz	1000–100 mm
Super high frequency (SHF)	3–30 GHz	100–10 mm
Extremely high frequency (EHF)	30–300 GHz	10–1 mm

Virtually every hertz of the RF spectrum is allocated for one use or another (Figure 4.1), ranging from radio astronomy to forestry conservation, and some RF bands have been designated for unlicensed transmissions.

The RF bands which are used for most wireless networking are the unlicensed ISM or Instrument, Scientific and Medical bands, of which the three most important lie at 915 MHz (868 MHz in Europe), 2.4 GHz and 5.8 GHz (Table 4.2). As well as these narrow band applications, new networking standards such as ZigBee will make use of the FCC spectrum allocation for ultra wideband radio (UWB—see section 4.6) that permits very low power transmission across a broad spectrum from 3.1 to 10.6 GHz.

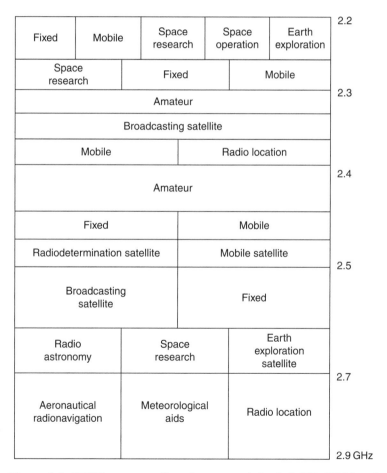

Figure 4.1: FCC Spectrum allocation around the 2.4-GHz ISM band

Table 4.2: Radio frequency bands in use for wireless networking

RF band	Wireless networking specification
915/868 MHz ISM	ZigBee
2.4 GHz ISM	IEEE 802.11b, g, Bluetooth, ZigBee
5.8 GHz	IEEE 802.11a

Table 4.3: Radio frequency spectrum regulatory bodies

Country/Region		Regulatory body
USA	FCC	Federal Communications Commission
Canada	IC	Industry Canada
Europe	ETSI	European Telecommunications Standards Institute
Japan	ARIB	Association of Radio Industries and Businesses

4.1.1 Radio Frequency Spectrum Regulation

The use of the radio frequency spectrum, in terms of the frequency bands that can be used for different licensed and unlicensed services, and the allowable transmission power levels for different signal formats, are controlled by regulatory authorities in individual countries or regions (Table 4.3).

Although there is an increasing trend towards harmonization of spectrum regulation across countries and regions, driven by the International Telecommunications Union's World Radio Communication Conference, there are significant differences in spectrum allocation and other conditions such as allowable transmitter power levels which have an impact on wireless networking hardware design and interoperability.

As an example, in the 5.8-GHz ISM band used for IEEE 802.11a networks, the FCC in the USA allows a maximum transmitted power of 1 W, while in Europe the ETSI permits a maximum EIRP (equivalent isotropic radiated power) of just 100 mW EIRP or 10 mW/MHz of bandwidth, with variations in other countries. Table 4.4 shows a range of other regulatory differences that apply to the 2.4 GHz ISM band used for IEEE 802.b/g networks.

The pace of regulatory change also differs from region to region. For example, the FCC developed regulations governing ultra wideband radio in 2002, while in Europe ETSI Task Group 31 was still working on similar regulations in 2006.

Table 4.4: 2.4-GHz ISM band regulatory differences by region

Regulator	2.4 GHz ISM specifications
FCC (USA)	1 W maximum transmitted power 2.402–2.472 GHz, 11 × 22 MHz channels
ETSI (Europe)	100 mW maximum EIRP 2.402–2.483 GHz, 13 × 22 MHz channels
ARIB (Japan)	100 mW maximum EIRP 2.402–2.497 GHz, 14 × 22 MHz channels

Table 4.5: Web sites of spectrum regulators

Regulator	Country/Region	URL
FCC	USA	www.fcc.gov
Industry Canada	Canada	www.ic.gc.ca
ETSI	Europe	www.etsi.org
ARIB	Japan	www.arib.or.jp/english

Although the regulatory bodies impose conditions on the unlicensed use of parts of the RF spectrum, unlike their role in the licensed spectrum, these bodies take no responsibility for or interest in any interference between services that might result from that unlicensed use. In licensed parts of the RF spectrum, the FCC and similar bodies have a role to play in resolving interference problems, but this is not the case in unlicensed bands. Unlicensed means in effect that the band is free for all, and it is up to users to resolve any interference problems. This situation leads some observers to predict that the 2.4 GHz ISM band will eventually become an unusable junk band, overcrowded with cordless phone, Bluetooth, 802.11 and a cacophony of other transmissions. This impending "tragedy of the commons" may be prevented by the development of spectrum agile radios.

Further information on current spectrum regulation and future developments, including the further development of regulations on ultra wideband radio outside the USA, can be found from the regulators web sites at the URLs shown in Table 4.5.

4.1.2 Radio Transmission as a Network Medium

Compared to traditional twisted-pair cabling, using RF transmission as a physical network medium poses a number of challenges, as outlined in Table 4.6. Security has been a

Table 4.6: The radio frequency networking challenge

Challenges	Considerations and solutions
Link reliability	Signal propagation, interference, equipment siting, link budget.
Media access	Sensing other users (hidden station and exposed station problems), Quality of service requirements.
Security	Wired equivalent privacy (WEP), Wi-Fi Protected Access (WPA), 802.11i, directional antennas.

significant concern since RF transmissions are far more open to interception than those confined to a cable.

Data link reliability, bit transmission errors resulting from interference and other signal propagation problems, are probably the second most significant challenge in wireless networks, and one technology that resulted in a quantum leap in addressing this problem (spread spectrum transmission) is the subject of the next section.

Controlling access to the data transmission medium by multiple client devices or stations is also a different type of challenge for a wireless medium, where, unlike a wired network, it is not possible to both transmit and receive at the same time. Two key situations that have the potential to degrade network performance are the so-called hidden station and exposed station problems.

The hidden station problem occurs when two stations A and C are both trying to transmit to an intermediate station B, where A and C are out of range and therefore one cannot sense that the other is also transmitting (see Figure 4.2). The exposed station problem occurs when a transmitting station C, prevents a nearby station B from transmitting although B's intended receiving station A is out of range of station C's transmission.

The later sections of this chapter look at digital modulation techniques and the factors affecting RF propagation and reception, as well as the practical implications of these factors in actual wireless network installations.

4.2 Spread Spectrum Transmission

Spread spectrum is a radio frequency transmission technique initially proposed for military applications in World War II with the intention of making wireless transmissions safe from interception and jamming. These techniques started to move into the commercial arena in the early 1980s. Compared to the more familiar amplitude or frequency modulated

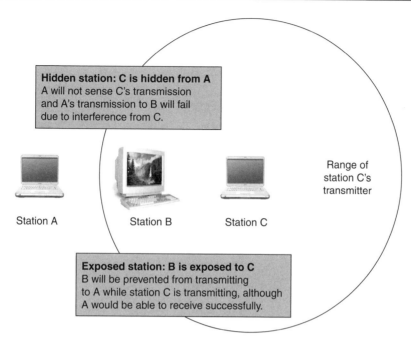

Hidden station: C is hidden from A
A will not sense C's transmission
and A's transmission to B will fail
due to interference from C.

Range of
station C's
transmitter

Station A   Station B   Station C

Exposed station: B is exposed to C
B will be prevented from transmitting
to A while station C is transmitting, although
A would be able to receive successfully.

Figure 4.2: Hidden and exposed station challenges for wireless media access control

radio transmissions, spread spectrum has the major advantage of reducing or eliminating interference with narrowband transmissions in the same frequency band, thereby significantly improving the reliability of RF data links.

Unlike simple amplitude or frequency modulated radio, a spread spectrum signal is transmitted using a much greater bandwidth than the simple bandwidth of the information being transmitted. Narrow band interference (the signal I in Figure 4.3) is rejected when the received signal is "de-spread". The transmitted signal also has noise-like properties and this characteristic makes the signal harder to eavesdrop on.

4.2.1 Types of Spread Spectrum Transmission

The key to spread spectrum techniques is some function, independent of the data being transmitted, that is used to spread the information signal over a wide transmitted bandwidth. This process results in a transmitted signal bandwidth which is typically 20 to several 100 times the information bandwidth in commercial applications, or 1000 to 1 million times in military systems.

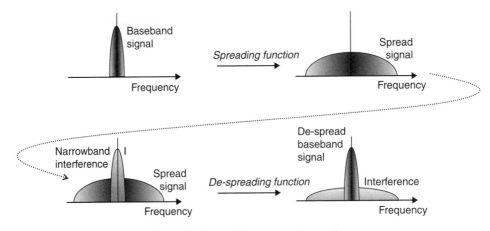

Figure 4.3: A simple explanation of spread spectrum

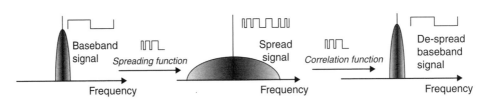

Figure 4.4: A simple explanation of DSSS

Several different methods of spread spectrum transmission have been developed, which differ in the way the spreading function is applied to the information signal. Two methods, direct sequence spread spectrum and frequency hopping spread spectrum, are most widely applied in wireless networking.

In Direct Sequence Spread Spectrum (DSSS) (Figure 4.4), the spreading function is a code word, called a chipping code, that is XOR'd with the input bit stream to generate a higher rate "chip stream" that is then used to modulate the RF carrier.

In Frequency Hopping Spread Spectrum (FHSS) (Figure 4.5), the input data stream is used directly to modulate the RF carrier while the spreading function controls the specific frequency slot of the carrier within a range of available slots spread across the width of the transmission band.

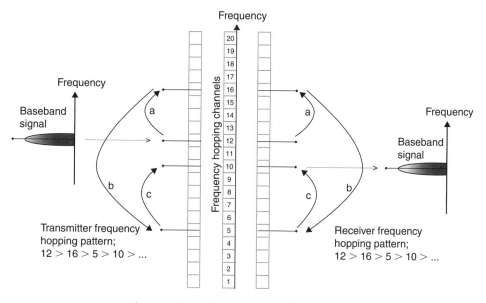

Figure 4.5: A simple explanation of FHSS

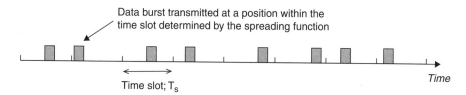

Figure 4.6: A simple explanation of THSS

Time Hopping Spread Spectrum (THSS) (Figure 4.6), is a third technique in which the input data stream is used directly to modulate the RF carrier which is transmitted in pulses with the spreading function controlling the timing of each data pulse. For example, impulse radio uses pulses that are so short, typically in the region of 1 nanosecond (ns), that the spectrum of the signal is very wide and meets the definition of an ultra wideband (UWB) system. The spectrum is effectively spread as a result of the narrowness of transmitted pulses, but time-hopping, with each user or node being assigned a unique hopping pattern, is a simple technique for impulse radio to allow multiple user access (see section 4.6.2).

Two other less common techniques are pulsed FM systems and hybrid systems (Figure 4.7). In pulsed FM systems, the input data stream is used directly to modulate the RF carrier,

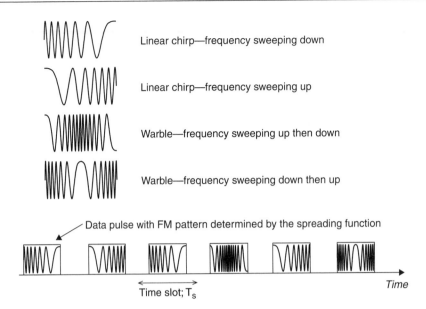

Linear chirp—frequency sweeping down

Linear chirp—frequency sweeping up

Warble—frequency sweeping up then down

Warble—frequency sweeping down then up

Data pulse with FM pattern determined by the spreading function

Time slot; T_s

Time

Figure 4.7: A simple explanation of pulsed FM systems

which is transmitted in frequency modulated pulses. The spreading function controls the pattern of frequency modulation, which could for example be a linear "chirp" with frequency sweeping up or down.

Hybrid systems also use combinations of spread spectrum techniques and are designed to take advantage of specific characteristics of the individual systems. For example, FHSS and THSS methods are combined to give the hybrid frequency division—time division multiple access (FDMA/TDMA) technique (see section 4.3).

Of these alternative spread spectrum techniques, DSSS and FHSS are specified in the IEEE 802.11 wireless LAN standards, although DSSS is most commonly used in commercial 802.11 equipment. FHSS is used by Bluetooth, and FHSS and chirp spread spectrum are optional techniques for the IEEE 802.15.4a (ZigBee) specification.

4.2.2 Chipping, Spreading and Correlating

The spreading function used in DSSS is a digital code, known as a chipping code or pseudo-noise (PN) code, which is chosen to have specific mathematical properties. One such property is that, to a casual listener on the broadcast band, the signal is similar to random noise, hence the "pseudo-noise" label.

Table 4.7: Barker codes of length 2 to 13

Length	Code
2	10 and 11
3	110
4	1011 and 1000
5	11101
7	1110010
11	11100010010
13	111100111001

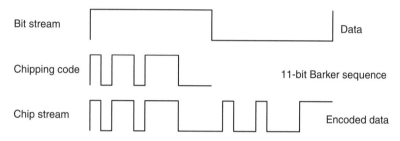

Figure 4.8: DSSS pseudo-noise encoding

Under the IEEE 802.11b standard, the specified PN code for 1 Mbps and 2 Mbps data rates is the 11-bit Barker code. Barker codes are binary sequences that have low auto-correlation, which means that the sequence does not correlate with a time-shifted version of itself. Barker codes of length 2 to 13 are shown in Table 4.7.

Figure 4.8 illustrates the direct sequence encoding of a data stream using this code. To distinguish a data bit from a code bit, each symbol (each 1 or 0) in the coded sequence is known as a chip rather than a bit.

This process results in a chip stream with a wider bandwidth than the original input data stream. For example, a 2 Mbps input data rate is encoded into a 22 Mcps (Mega chips per second) sequence, the factor of 11 coming about since the encoded sequence has 11 chips for each data bit. The resulting bandwidth of the transmitted RF signal will depend on the technique used to modulate the encoded data stream onto the RF carrier. As described in section 4.4, a simple modulation technique like binary phase-shift keying (BPSK) results in a modulated carrier with a bandwidth equal to twice the input bit rate (or in this case, the chip rate).

When the encoded signal is received, a code generator in the receiver recreates the same PN code and a correlator uses this to decode the original information signal in a process known as correlating or de-spreading. Since the correlator only extracts signals encoded with the same PN code, the receiver is unaffected by interference from narrow band signals in the same RF band, even if these signals have a higher power density (in watts/Hz) than the desired signal.

4.2.3 Chipping Codes

One of the desirable mathematical properties of PN codes is that it enables the receiver's PN code generator to very rapidly synchronize with the PN code in the received signal. This synchronisation is the first step in the de-spreading process. Fast synchronisation requires that the position of the code word can be quickly identified in a received signal, and this is achieved as a result of the low auto-correlation property of the Barker codes. Another benefit of low auto-correlation is that the receiver will reject signals that are delayed by more than one chip period. This helps to make the data link robust against multipath interference, which will be discussed in section 4.5.4.

A second key property of chipping codes that is important in applications where interference between multiple transmitters must be avoided, for example in mobile telephony, is low cross-correlation. This property reduces the chance that a correlator using one PN code will experience interference from a signal using a different code (i.e. that it will incorrectly decode a noisy signal that was encoded using a different chipping code). Ideally codes in use in this type of multiple access application should have zero cross-correlation, a property of the orthogonal codes used in CDMA (section 4.3.6).

Code orthogonality for multiple access control is not required for wireless networking applications, such as IEEE 802.11 networks, as these standards use alternative methods to avoid conflict between overlapping transmitted signals from multiple users, which are described in section 4.3.

4.2.4 Complementary Code Keying

An alternative to using a single chipping code to spread every bit in the input data stream is to use a set of spreading codes and to select one code from the set depending on the values of a group of input data bits. This scheme is known as complementary code keying (CCK).

CCK was proposed to the IEEE by Lucent Technologies and Harris Semiconductor (now part of Intersil Corp.) in 1998, as a means to raise the IEEE 802.11b data rate to 11 Mbps. Instead of using the Barker code, they proposed to use a set of codes called Complementary Sequences, based on the Walsh/Hadamard transforms (see section 4.3.6).

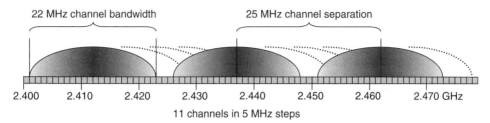

Figure 4.9: 802.11 DSSS channels

Using CCK, a chipping code word is chosen from a set of 64 unique codes depending on the value of each 6-bit segment of the input data stream. The encoded data sequence comprises a series of code words, and this chip sequence is modulated onto the RF carrier using one of a variety of modulation techniques that will be described in section 4.4.

The main advantage of CCK modulation is spectral efficiency, since each transmitted code word represents 6 input data bits instead of the single bit represented by the Barker code. CCK can achieve 11 Mbps using the same 22 MHz bandwidth used to transmit 1 Mbps with the Barker code. However, the price of this high data rate is complexity. A receiver using the Barker code requires just one correlator to pick out the chipping code, while a CCK system needs 64 correlators, one on the lookout for each of the complementary codes.

4.2.5 Direct Sequence Spread Spectrum in the 2.4 GHz ISM Band

As noted above, in DSSS the data signal is combined with a code word, the chipping code, and the combined signal is used to modulate the RF carrier, resulting in a transmitted signal spread over a wide bandwidth. For example, in the 2.4 GHz ISM band, a spread bandwidth of 22 MHz is specified for IEEE 802.11 networks, as shown in Figure 4.9.

The 2.4 GHz ISM band has a total allowed width of 83.5 MHz and is divided into a number of channels (11 in the USA, 13 in Europe, 14 in Japan), with 5 MHz steps between channels. To fit 11 or more 22 MHz wide channels into an 83 MHz wide band results in considerable overlap between the channels (as shown in Figure 4.10), resulting in the potential for interference between signals in adjacent channels. The 3 nonoverlapping channels allow 3 DSSS networks to operate in the same physical area without interference.

4.2.6 Frequency Hopping Spread Spectrum in the 2.4 GHz ISM Band

In frequency hopping spread spectrum transmission (FHSS) the data is modulated directly onto a single carrier frequency, but that carrier frequency hops across a number of channels

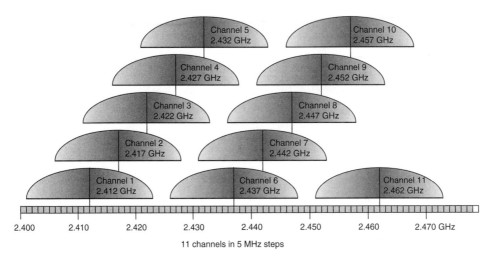

Figure 4.10: DSSS channels in the 2.4-GHz ISM band (US)

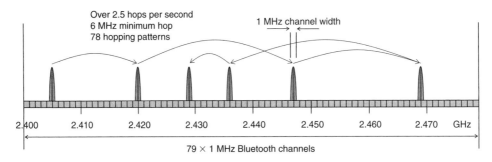

Figure 4.11: FHSS channels within the 2.4-Ghz ISM band

within the RF band using a pseudo-random hopping pattern. In the 2.4 GHz ISM band for example, a maximum channel width of 1 MHz is specified for FHSS systems, and 79 such channels are available. A transmitter switches between these channels many times a second, moving on to the next channel in its sequence after a predetermined time, known as the "dwell time".

The IEEE 802.11b standard specifies that the hop must be to a new channel a minimum of 6 MHz from the previous channel, and that hops must occur at least 2.5 time per second (Figure 4.11). The spectrum regulators specify the allowable limits for transmission parameters such as the maximum dwell time and individual standards, like IEEE 802.11, have to work within these boundary conditions.

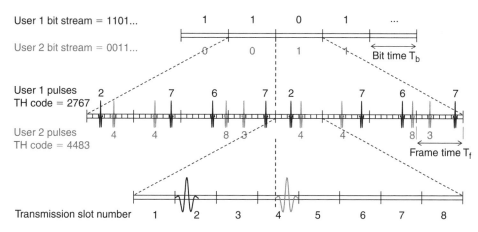

Figure 4.12: Time hopping spread spectrum

In the receiver, a PN generator recreates the same hopping pattern. This allows the receiver to make the same channel-to-channel hops as the transmitter, so that the data signal can be decoded.

Because the probability of two networks selecting the same channel at the same time is very low, many more FHSS networks can overlap physically without interference than is the case for DSSS networks.

Frequency hopping spread spectrum in the 2.4 GHz ISM band is specified alongside DSSS as an option in the IEEE 802.11b standard and is also used in Bluetooth networks.

All 79 available channels are normally used in Bluetooth, although an alternative hopping sequence that uses 23 channels (2.454 to 2.476 GHz), is available for use in France where special regulatory conditions apply. Frequency hops occur after each data packet, which will be a multiple of 1-, 3- or 5-times the time slot duration of 625 microseconds (320 to 1600 hops/second). The frequency hopping pattern is determined by the unique 48-bit device ID of the master device in each Bluetooth piconet, and synchronization to the hopping pattern is part of the process of device discovery when a new device joins the piconet.

4.2.7 Time Hopping Spread Spectrum

In a time hopping spread spectrum system, time is divided into frames, with each frame divided into a number of transmission slots. Within each frame, data is transmitted only during one time slot and the specific time slot used during each frame is determined using a PN code. Figure 4.12 shows a THSS system with two clients using different hopping codes.

Table 4.8: Benefits of common spread spectrum techniques

Frequency hopping	Direct sequence
Simple to design and manufacture	Higher data speeds
Cheaper to implement	Increased range
Higher density of overlapping networks	Throughput is interference-tolerant up to a threshold level
Gradual degradation of throughput with interference	

Impulse radio is an ultra-wide band transmission technique that is a candidate for the IEEE 802.15.4a (ZigBee) physical layer specification. This is a time hopping spread spectrum technique where a very short pulse is transmitted in each transmission time slot. Information is encoded via pulse position or pulse amplitude modulation (PPM, PAM). The spreading effect of time hopping, together with the short pulse duration, results in a transmitted signal spread across an ultra-wide bandwidth.

4.2.8 Spread Spectrum in Wireless Networks—Pros and Cons

The advantages of spread spectrum techniques, such as resistance to interference and eavesdropping and the ability to accommodate multiple users in the same frequency band, make this an ideal technology for wireless network applications (Table 4.8). Although the good interference performance is achieved at the cost of relatively inefficient bandwidth usage, the available radio spectrum, such as the 2.4 GHz ISM band, still permits data rates of up to 11 Mbps using these techniques.

Since speed and range are important factors in wireless networking applications, DSSS is the more widely used of the two techniques although, because of its simpler and cheaper implementation, FHSS is used for lower rate, shorter-range systems like Bluetooth and the now largely defunct HomeRF.

4.3 Wireless Multiplexing and Multiple Access Techniques

4.3.1 Introduction

Multiplexing techniques aim to increase transmission efficiency by transmitting multiple signals or data streams on a single medium. The resulting increased capacity can be used

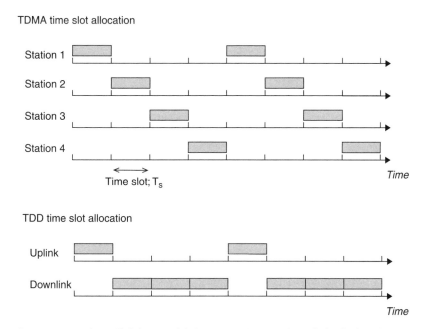

Figure 4.13: Time division multiple access (TDMA) and duplexing (TDD)

either to deliver a higher data rate to a single user, or to allow multiple users to access the medium simultaneously without interference.

User access to the bandwidth can be separated by a numbers of means: in time (TDMA), in frequency (FDMA or OFDMA), in space (SDMA) or by assigning users unique codes (CDMA). These methods will be described in turn in the following sections.

4.3.2 Time Division Multiple Access

Time division multiple access (TDMA) allows multiple users to access a single channel without interference by allocating specific time slots to each user. As shown in Figure 4.13, the time axis is divided into time slots that are assigned to users according to a slot allocation algorithm.

A simple form of TDMA is time division duplex (TDD), where alternate transmit periods are used for uplink and downlink in a duplex communication system. TDD is used in cordless phone systems to accommodate two-way communication in a single frequency band.

TDMA is used in Bluetooth piconets. The master device provides the system clock that determines the timing of slots and, within each time slot, the master first polls slave devices to

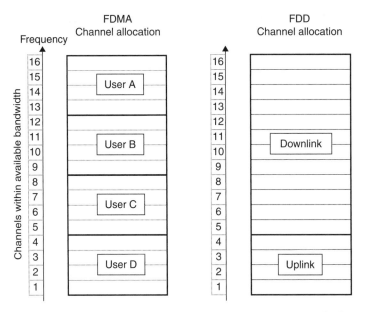

Figure 4.14: Frequency division multiple access (FDMA) and duplexing (FDD)

see which devices need to transmit and then allocates transmission time slots to devices that are ready to transmit.

4.3.3 Frequency Division Multiple Access

In contrast to TDMA, frequency division multiple access (FDMA) provides each user with a continuous channel that is restricted to a fraction of the total available bandwidth. This is done by dividing the available bandwidth into a number of channels that are then allocated to individual users as shown in Figure 4.14.

Frequency division duplex (FDD) is simple form of FDMA in which the available bandwidth is divided into two channels to provide continuous duplex communication. Cellular phone systems such as GSM (2G) and UMTS (3G) use FDD to provide separate uplink and downlink channels, while 1G cellular phone systems used FDMA to allocate bandwidth to multiple callers.

In practice, FDMA is often used in combination with TDMA or CDMA to increase capacity on a single channel in an FDMA system. As shown in Figure 4.15, FDMA/TDMA divides the available bandwidth into channels and then divides each channel into time slots that are allocated to individual users.

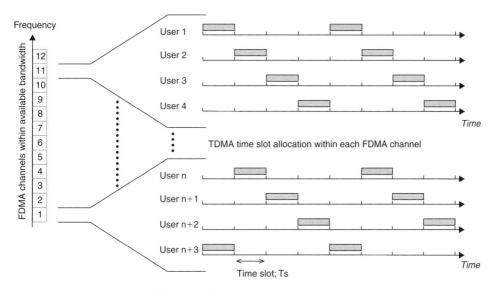

Figure 4.15: FDMA/TDMA multiple access system as used in GSM cellular phones

FDMA/TDMA is used by GSM cellular phones, with eight time slots available in each 200 kHz radio channel.

4.3.4 Orthogonal Frequency Division Multiplexing

Orthogonal frequency division multiplexing (OFDM) is a variant of frequency division multiplexing (FDM), in which a number of discrete subcarrier frequencies are transmitted within a band with frequencies chosen to ensure minimum interference between adjacent subcarriers.

This is achieved by controlling the spectral width of the individual subcarriers (also called tones) so that the frequencies of subcarriers coincide with minima in the spectra of adjacent subcarriers, as shown in Figure 4.16.

In the time domain, the orthogonality of OFDM tones means that the number of subcarrier cycles within the symbol transmission period is an integer, as illustrated in Figure 4.17. This condition can be expressed as:

$$T_s = n_i / v_i \quad or \quad v_i = n_i / T_s$$

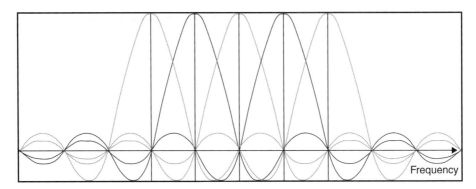

Figure 4.16: Orthogonality of OFDM subcarriers in the frequency domain

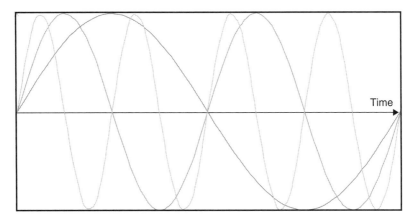

Figure 4.17: Orthogonality of OFDM subcarriers in the time domain

where T_s is the symbol transmission period and v_i is the frequency of the ith subcarrier. The subcarriers are therefore evenly spaced in frequency, with separation equal to the reciprocal of the symbol period.

There are a number of ways in which the multiple subcarriers of OFDM can be used:

- OFDM can be used as a multiple access technique (OFDMA), by assigning single subcarriers or groups of subcarriers to individual users according to their bandwidth needs.

- A serial bit stream can be turned into a number of parallel bit streams each one of which is encoded onto a separate subcarrier. All available subcarriers are used by a single user to achieve a high data throughput.

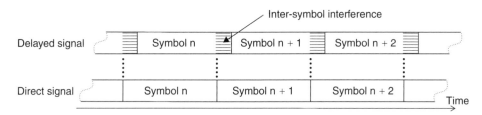

Figure 4.18: Inter Symbol Interference (ISI)

- A bit stream can be spread using a chipping code and then each chip can be transmitted in parallel on a separate subcarrier. Since the codes can allow multiple user access, this system is known as Multi-Carrier CDMA (MC-CDMA). MC-CMDA is under consideration by the WIGWAM project as one of the building blocks of the 1 Gbps wireless LAN.

A significant advantage of OFDM is that, since the symbol rate is much lower when spread across multiple carriers than it would be if the same total symbol rate were transmitted on a single carrier, the wireless link is much less susceptible to inter-symbol interference (ISI). ISI occurs when, as a result of multi-path propagation, two symbols transmitted at different times arrive together at the receiving antenna after traversing different propagation paths (Figure 4.18). Although OFDM is inherently less susceptible to ISI, most OFDM systems also introduce a guard interval between each symbol to further reduce ISI.

OFDM radios also use a number of subcarriers, called pilot tones, to gather information on channel quality to aid demodulation decisions. These subcarriers are modulated with known training data at the start of each transmitted data packet. Decoding this known data enables the receiver to determine and adaptively correct for the frequency offset and phase noise between the reference oscillators in the transmitter and the receiver and for fading during propagation.

Figure 4.19 shows a schematic block diagram of a simple OFDM transmitter and receiver. From the left, the input bit stream at a rate of R bps passes through a series to parallel converter and is split into N bit stream of rate R/N bps. Each of these bit streams drives one modulator, which maps each bit or symbol onto a point in the modulation constellation being used (section 4.4). The N resulting amplitude and phase points drive the inputs of an inverse fast Fourier transform (IFTT), the output of which is the sum of the subcarriers, each modulated according to the individual input bit streams.

At the receiver, after removing any guard interval, a fast Fourier transform (FFT) determines the amplitude and phase of each subcarrier in the received signal. The amplitude and phase

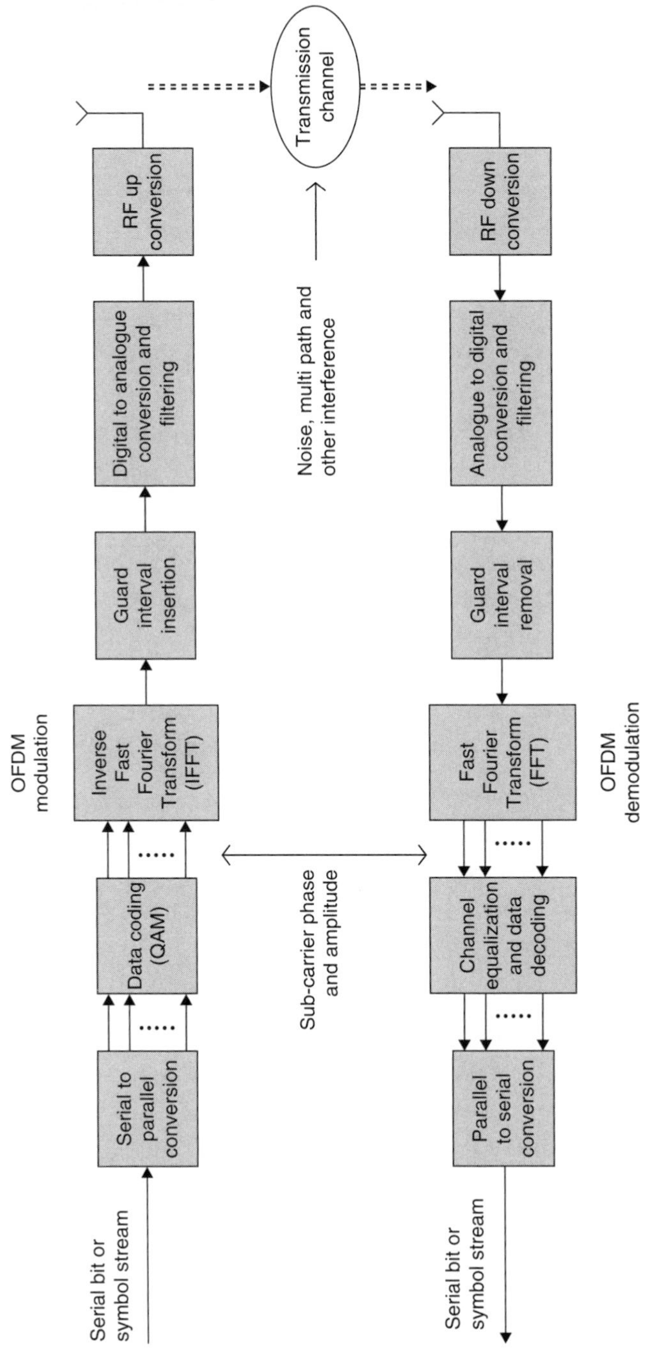

Figure 4.19: Schematic block diagram of an OFDM transmitter and receiver

are adjusted using information gathered from the pilot tones. A demodulation decision is made by mapping this amplitude and phase onto the modulation constellation and the corresponding input bit or bits are generated. The resulting N parallel R/N bps bit streams are then combined in a parallel to series converter to give the original R bps bit stream.

The IEEE 802.11a/g standards use OFDM in the unlicensed 2.4 and 5 GHz ISM bands respectively to provide data rates up to 54 Mbps.

The system uses 52 subcarriers of which 48 are used to carry data and are modulated using binary or quadrature phase shift keying (BPSK/QPSK), 16-quadrature amplitude modulation (QAM) or 64-QAM. The remaining four subcarriers are used as pilot tones.

4.3.5 Space Division Multiple Access

Space division multiple access (SDMA) is a technique which aims to multiply the data throughput of a wireless network by using spatial position as an additional parameter to control user access to the transmission medium. As a simple example, if a base station is equipped with sector antennas with a 30° horizontal beamwidth, it can separate users into twelve spatial segments or channels depending on their location around the base station. This arrangement would enable the network to achieve a potential twelve-fold increase in data capacity compared with a base station using a single isotropic antenna.

As well as simple sector antennas, smart antenna systems are being developed which combine an array of antennas with digital signal processing capabilities in order to achieve spatial control of transmission and reception. Smart antenna systems can adapt their directional characteristics in response to the signal environment and system demands, and can provide the basis for SDMA.

Generally a second multiple access technique, such as TDMA or CDMA, is also used in combination with SDMA in order to allow multiple user access within a single spatial segment.

Space division multiplexing (SDM), as opposed to SDMA, is based on the use of multiple propagation paths to simultaneously transmit multiple data channels using the same RF spectrum. This is the basis of MIMO radio which is specified in the IEEE 802.11n standard.

4.3.6 Code Division Multiple Access

CDMA is closely related to DSSS, where a pseudo-noise code is used to spread a data signal over a wide bandwidth in order to increase its immunity to interference. As noted above, if

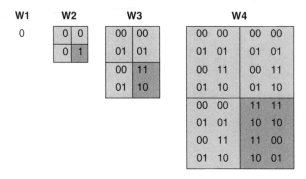

Figure 4.20: Construction of the Walsh codes

two or more transmitters use different, orthogonal pseudo-noise (PN) codes in DS spread spectrum, they can operate on the same frequency band and in the same physical area without interfering. This is because a correlator using one PN code will not detect a signal encoded using another orthogonal code, since orthogonal codes by definition do not correlate with each other.

Examples of an orthogonal code set are the Walsh codes, which can be easily generated from the procedure called the Hadamard transform. With each step to the right in Figure 4.20, the three light matrices are the same as the full matrix to the left, while the darker shaded matrix is the inverse of the one to the left. The Walsh codes can be read off as the lines in each matrix, so the Walsh codes of length 4 are: 0000, 0101, 0011 and 0110.

The property of orthogonality is the basis of CDMA and is used in 3G mobile telephony to ensure that many users, each assigned a unique orthogonal access code, can transmit and receive without interference within a single network cell.

4.4 Digital Modulation Technique

4.4.1 Introduction

Modulation is the step in the digital signal processing sequence that transforms and encodes the data stream onto the transmitted RF or infrared signal. The spectrum spreading and multiple access techniques will result in a bit-stream transformed into a chip-stream which must now be modulated onto either a single or multiple carrier frequencies, or used to modulate the position or shape of a transmitted RF or Ir pulse.

Table 4.9: Spectral efficiency of typical modulation techniques

Modulation technique	Spectral efficiency (Bits/Hz)
BPSK	0.5
QPSK	1.0
16-QAM	2.0
128-QAM	3.5
256-QAM	4.0

A wide variety of modulation techniques are used in wireless networking. These range from the simple return to zero inverted (RZI) used in IrDA at low data rates, through a variety of phase shift and code keying methods of increasing complexity, such as BPSK and CCK, used for example in IEEE 802.11b at intermediate data rates, to more complex methods, such as the HHH (1,13) code used in IrDA at high data rates.

The selection of the best digital modulation technique for a specific application is driven by a number of criteria, the most important being:

- Spectral efficiency—achieving the desired data rate within the available spectral bandwidth (see Table 4.9).

- Bit error rate (BER) performance—achieving the required error rate given the specific factors causing performance degradation in the particular application (interference, multipath fading, etc.).

- Power efficiency—particularly important in mobile applications where battery life is an important user acceptance factor.

- Modulation schemes with higher spectral efficiency (in terms of data bits per hertz of bandwidth) require higher signal strength for error-free detection.

- Implementation complexity—which translates directly into the cost of hardware to apply a particular technique. Some aspects of modulation complexity can be implemented in software, which has less impact on end-user costs.

4.4.2 Simple Modulation Techniques

On/Off keying (OOK) is perhaps the simplest modulation technique, where the carrier is turned off during a 0-bit and turned on during a 1-bit. OOK is a special case of amplitude shift keying (ASK) in which two amplitude levels represent 0- and 1-bits. The magnitude of the amplitude shift between these two levels is called the modulation index.

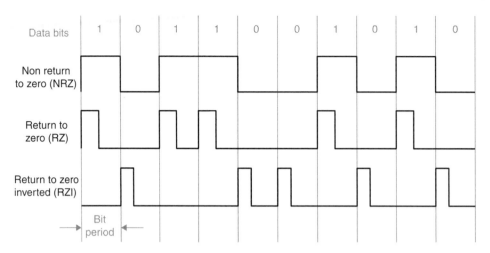

Figure 4.21: NRZ, RZ and RZI modulation techniques

Return to zero inverted (RZI) is the modulation technique used in IrDA for data rates up to 1.152 Mbps. It is a derivation of the non-return to zero (NRZ) modulation used in UART data transmission (Figure 4.21), in which a 1-bit is represented by a high state, a 0-bit by a low state, and the transition from high state to low state only occurs when a 1-bit is followed by a 0-bit.

In contrast, a return to zero (RZ) transmission has a low-high-low pulse during the bit time for each 1-bit, while the RZI scheme inverts this to give a pulse for each zero bit or symbol.

When the RZI modulated signal is received, the bit stream is recovered by triggering a high to low transition for each received pulse, as shown in Figure 4.22. The low state of the decoded signal returns to a high state and a high state remains high at the end of each bit period, unless another pulse is received.

The advantage of the RZI scheme for IrDA is that it allows the transmitting LED to be off for most of the bit time, in order to conserve battery power.

4.4.3 Phase Shift Keying

Phase shift keying is a modulation technique in which the phase of the carrier is determined by the input bit or chip stream. There are several types of phase shift key (PSK) modulation including binary (BPSK) and quadrature (QPSK).

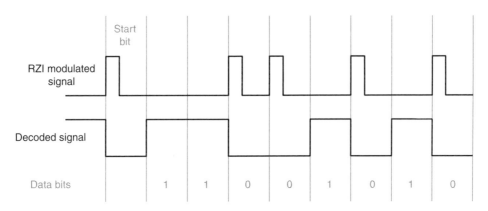

Figure 4.22: RZI bit stream decoding

Table 4.10: Binary phase shift keying

Symbol	Carrier phase
0	0 degrees
1	180 degrees

4.4.3.1 Binary Phase Shift Keying

BPSK is the simplest technique in this class, with the carrier phase taking one of two states, as shown in Table 4.10. A 0 input symbol (whether it is a bit or a chip) corresponds to a zero phase carrier while a 1-symbol corresponds to a $180°$ phase shifted carrier, resulting in the output waveform shown in Figure 4.23.

BPSK modulation is used by IEEE 802.11b at a data rate of 1 Mbps, and by IEEE 802.11a, in combination with OFDM, to achieve data rates of 6 and 9 Mbps.

4.4.3.2 Quadrature Phase Shift Keying

Instead of the two phase states used in BPSK, QPSK uses four distinct carrier phases, each of which is used to encode a symbol comprised of two input bits or chips.

These four carrier phases are illustrated in Figure 4.24, which represents the phase of the carrier signal in the IQ plane (I = In phase, Q = Quadrature or 90 degrees out of phase). The angle of a given point on the plane from the I axis represents the phase angle and the distance of a point from the origin represents the signal amplitude. The four points 00, 01, 11, and 10 shown in Table 4.11, are known as the modulation constellation, and represent the four carrier phases each with unit amplitude.

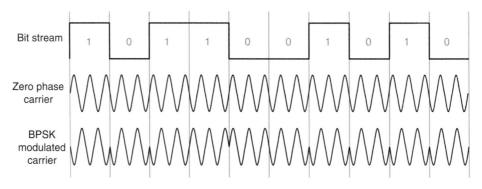

Figure 4.23: Binary Phase Shift Keying Modulation (BPSK)

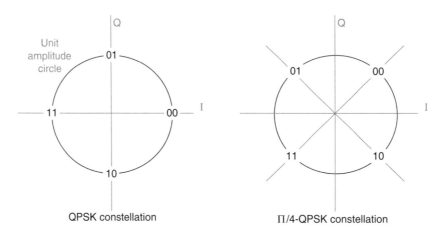

Figure 4.24: QPSK Phase Constellation

Table 4.11: Quadrature Phase Shift Keying

Symbol	Carrier phase
00	0 degrees
01	90 degrees
11	180 degrees
10	270 degrees

Table 4.12: Differential quadrature phase shift keying

Symbol	Phase change
00	0 degrees
01	90 degrees
11	180 degrees
10	270 degrees

QPSK modulation is used by IEEE 802.11b at a data rate of 2 Mbps, and by IEEE 802.11a, in combination with OFDM, to achieve data rates of 12 and 18 Mbps. π/4-QPSK is a variant of QPSK, as shown in Figure 4.24, which uses carrier phases offset by 45 degrees (i.e., 45, 135, 225 and 315 degrees).

Offset QPSK (O-QPSK) is another variation on QPSK in which transmission of the quadrature phase is delayed by half a symbol period. The consequence is that, unlike QPSK, carrier phase transitions can never be more than 90 degrees and, as a result, the carrier phase and amplitude never passes through zero. The advantage is a narrower spectral width, which is important in applications where interference between adjacent channels must be avoided.

O-QPSK is part of the IEEE 802.15.4 radio specification used by ZigBee, where 16 channels, each 5 MHz wide, are used in the 2.4 GHz ISM band to enable 16 co-located networks. Use of O-QPSK helps to reduce interference between these closely spaced channels.

4.4.3.3 Differential Phase Shift Keying

Differential phase shift keying is a variation on BPSK and QPSK in which the input symbol results in a differential change of phase instead of defining the absolute phase of the carrier. With BPSK, a 0-symbol corresponds to a period of zero phase carrier, while in DBPSK, a 0-symbol corresponds to no change in carrier phase from the previous bit-period.

Similarly in DQPSK, each symbol translates to a change of phase rather than an absolute carrier phase, as shown in Table 4.12.

Although BPSK or QPSK are conceptually simpler, differential phase shift keying, whether DBPSK or DQPSK, has the practical advantage that the receiver only needs to detect relative changes in carrier phase. The BPSK or QPSK receiver always needs to know the absolute phase reference of the carrier and this reference can be difficult to maintain, for example, if the phase of the received signal is varying due to multipath interference.

Table 4.13: Binary frequency shift keying

Symbol	Carrier frequency
0	$f_0 - f_1$
1	$f_0 + f_1$

Other variants on PSK and DPSK include 8-DPSK, which extends the DQPSK keying table to encode 8 data symbols using phase changes separated by 45 degrees rather than by 90 degrees, and $\pi/4$-DQPSK which, by analogy with $\pi/4$-QPSK, uses carrier phase changes similar to Table 4.12 but offset by 45 degrees (i.e. 45, 135, 225 and 315 degrees). $\pi/4$-DQPSK and 8-DPSK are used in the enhanced data rate (EDR) Bluetooth 2.0 radio for 2 Mbps and 3 Mbps data rates, respectively.

4.4.3.4 Frequency Shift Keying

Frequency shift keying (FSK) is a simple frequency modulation method in which data symbols correspond to different carrier frequencies, as shown in Table 4.13 for BFSK.

The sudden carrier waveform changes in simple FSK generate significant out-of-band frequencies and, as a result, FSK is inefficient in terms of spectrum usage. This situation can be improved by passing the input bit stream through a filter to make the frequency transitions more gradual. A Gaussian filter is one type of filter with a specific mathematical form, and use of this as a pre-modulation filter results in Gaussian frequency shift keying (GFSK).

GFSK is used in the Bluetooth radio for standard data rate transmission, with a carrier frequency f_0 of 2.40 to 2.48 GHz and frequency deviation f_1 of between 145 and 175 kHz. Spectral efficiency is particularly important as the FHSS frequency hopping channels are only separated by 1 MHz.

4.4.4 Quadrature Amplitude Modulation

Quadrature amplitude modulation (QAM) is a composite modulation technique that combines both phase modulation and amplitude modulation.

In BPSK or QPSK, a constant carrier amplitude with 2 or 4 different phases is used to represent the input data symbols, as described above. Instead of using 2 or 4 points, QAM defines a constellation of 16, 64 or more points, each with a particular phase and amplitude, and each representing a 4- or 6-bit (or chip) data symbol.

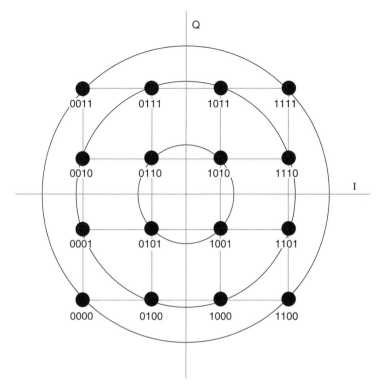

Figure 4.25: 16-QAM constellation

16-QAM and 64-QAM modulation techniques are used in the IEEE 802.11a and g specification, together with OFDM, to achieve data rates of 24 to 54 Mbps. Figure 4.25 shows the 16-QAM constellation—the 16 points on the IQ plane—used to achieve data rates of 24 Mbps and 36 Mbps.

The points in the 16-QAM constellation can be alternatively numbered according to a Gray coding in which adjacent points differ only in the switching of one bit, as shown in Figure 4.26. Using this numbering reduces the chance of two-bit errors in the receiver—if a point is erroneously detected as a neighboring point only one bit will be incorrect. This makes it easier to recover the bit error using error correction techniques.

The next step, a 256-QAM modulation scheme, would further improve achievable data rate with no increase in the occupied bandwidth, but generating and processing 256-QAM modulated signals is currently a significant challenge for hardware performance and cost.

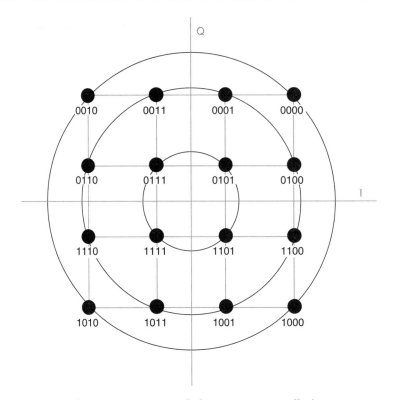

Figure 4.26: Gray coded 16-QAM constellation

4.4.5 Dual Carrier Modulation

Dual carrier modulation is a technique applied in multi-carrier systems such as OFDM to combat the loss of data due to the destructive interference, or fading, of individual carrier signals in a multi-path environment. By modulating data onto two carrier frequencies rather than one, the transmission can be made more robust, although at the cost of using additional bandwidth.

In multi-band OFDM (see section 4.6.4) a 4-bit symbol is mapped onto two different 16-QAM constellations and the resulting symbols are transmitted on two OFDM carrier tones separated by at least 200 MHz. If reception of one of the tones is affected by fading, the data can be recovered from the other tone, the wide separation assuring that the probability of both tones being affected is very small.

Table 4.14: Data symbols for 4-PPM modulation

Input data symbol	4-PPM data symbol
00	1000
01	0100
10	0010
11	0001

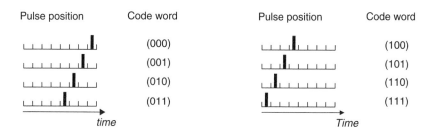

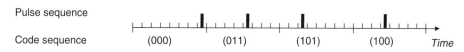

Figure 4.27: 8-PPM modulation

4.4.6 Pulse Modulation Methods

Several wireless network standards specify the use of pulsed rather than continuous transmission of a carrier wave, and a number of specific modulation techniques are used in these systems.

4.4.6.1 Pulse Position Modulation

In pulse position modulation (PPM), each pulse is transmitted within a reference time frame, and the information carried by the pulse is determined by the specific transmission time of the pulse within its frame. For example, a 4-PPM system will define four possible positions for a pulse within the reference frame, with each possible position coding one of four input data symbols (Table 4.14).

More generally, an m-PPM system will have m possible pulse transmission slots within a frame. An 8-PPM modulation system is shown in Figure 4.27. PPM is specified in the IrDA standard at a 4 Mbps data rate, and is also used in Impulse Radio (IR as opposed to Ir!).

Table 4.15: PAM encoding table

Input data symbol	Pulse amplitude
00	0
01	1
10	2
11	3

4.4.6.2 Pulse Shape Modulation

Pulse shape modulation (PSM) encodes the input data stream in the shape of the transmitted pulse. The simplest form of PSM is pulse amplitude modulation (PAM) in which, typically, two or four distinct pulse amplitudes are used to encode data symbols, as shown in Table 4.15.

Similarly, pulse width modulation (PWM) uses the width of transmitted pulses and more generally, PSM may use some other pulse shape characteristics—such as the derivative of the pulse waveform—to encode data onto the pulse train.

Pulse amplitude and pulse shape modulation are candidates for use in the ultra wideband radio physical layer of the ZigBee specification.

4.5 RF Signal Propagation and Reception

The first part of this chapter described the various techniques that are used to encode and modulate an input data stream onto a radio frequency carrier. The following four sections discuss the various factors that impact on the transmission, propagation and reception of radio waves, which will enable an estimate to be made of the power requirements for a given wireless networking application.

The key factors are transmitter power, antenna gain at the transmitter, propagation or link losses, antenna gain once again at the receiving station, and finally receiver sensitivity. Taken together these factors make up the link budget—the balance of power plus gain required to compensate for losses in the link so that sufficient signal strength is available at the receiver to allow data decoding at an acceptable error rate.

4.5.1 Transmitter Power

Every RF transmitter generates a certain amount of power (P_{TX}), which is the first major factor in determining the range of a radiated signal. Transmitter power is measured in one

Table 4.16: Power in mW and dBm

Power (mW)	Power (dBm)
0.01	−20
0.1	−10
0.5	−3
1	0
10	10
20	13
100	20
1000	30

of two ways, either in the familiar unit of watts (or milliwatts) or alternatively using a relative unit called "dBm". The power in dBm is calculated as dBm = $10 \times \log_{10}$ (power in milliwatts), so a transmitter of 100 mW (0.1 watts) is equivalent to 20.0 dBm (Table 4.16).

The dB (or dBm) unit is useful for two reasons. First, when considering the various factors affecting signal strength, these effects can be easily combined when using dB units by simply adding the relevant dB numbers together. Second, it is easy to translate dB into relative power by remembering that +3 dB represents a doubling of power and −3 dB similarly a halving of power. The additive rule applies here too, so −6 dB is $\frac{1}{4}$ the power, −9 dB is $\frac{1}{8}$ and so on.

Transmitter power levels for typical wireless networking products are in the region of 100 milliwatts to 1 watt (20 to 30 dBm). For example, in the US the FCC specifies a maximum transmitter power of 1 watt for FHSS and DSSS transmitters in the 2.4 and 5.8 GHz ISM bands. In the UK the Radio Communications Agency (RA) specifies that these devices must have a maximum effective isotropic radiated power (EIRP) in the 2.4 GHz band of 100 mW or 20 dBm. As described below, EIRP is a combination of the transmitter's power and the antenna gain.

4.5.2 Antenna Gain

An antenna converts the power from the transmitter into electro-magnetic waves that are radiated to the receiver, and the type of antenna affects the pattern and power density of this radiation, and therefore the strength of signal seen by a receiver. For example, a simple dipole antenna emits radiation relatively evenly in all directions apart from along its axis, while a

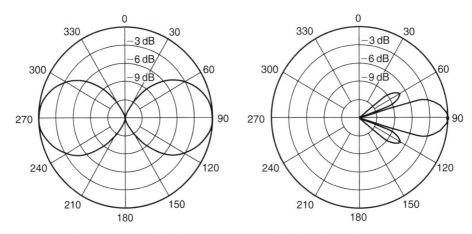

Figure 4.28: Radiation pattern from dipole and Yagi antennas

directional antenna emits radio waves in a narrow beam. Typical radiation patterns of a dipole and a directional antenna are shown in Figure 4.28.

The ratio of the maximum power density at the centre of the radiation pattern of any antenna to the power density of the radiation from a reference isotropic antenna is known as the antenna gain (G_{TX} or G_{RX}), and is measured in dBi units. The effective isotropic radiated power, or EIRP, of a radiating antenna is then, the sum of the dBm power arriving at the antenna from the transmitter plus the dBi antenna gain. The types of antenna that can be used in wireless networks are antennas with gain ranging from 0 dBi for an omnidirectional dipole antenna to +20 dBi or more for a narrow beam directional antenna.

The cables and connectors that link the transmitter or receiver to the antenna also introduce a loss into the system that can range from a few dB to tens of dB depending primarily on the length and quality of cabling. An equally important aspect of transmitter to antenna or receiver to antenna connections, is the matching of impedance between these components. For example, maximum power will only be transmitted if the impedances of the transmitter, connecting cable and antenna are equal, otherwise power will be lost as a result of reflections at the connections between components. In the case of equipment with integrated antennas, this will be part of the design dealt with by the equipment manufacturer. However, this aspect will have to be considered when attaching an external antenna to a wireless network adapter or access point.

4.5.3 Receiver Sensitivity

As the strength of the signal reaching the receiver input drops, the decoding of data will be increasingly affected by noise and, as a result, will become increasingly error prone. The sensitivity limit of the receiver is determined by the allowable bit error rate and the receiver noise floor.

As these factors are discussed in the following sections, an example will be worked through, based on the following parameters:

- 802.11b DSSS system (2.4 GHz, 22 MHz spread bandwidth)
- DQPSK modulation
- 2 Mbps data rate or 2 MHz de-spread bandwidth
- required bit error rate of 1 in 10^5
- receiver noise figure of 6 dB
- 20°C ambient temperature.

4.5.3.1 Bit Error Rate

The rate at which decoding errors occur is measured by the bit error rate (BER), with a BER of 1 in 10^5 being typical at the receiver sensitivity limit. Since data is transmitted in packets containing several hundreds or thousands of bits of data, even a 1 in 10^5 chance of an error in decoding any single bit will multiply up to a significant probability of an error in a large data packet, and the resulting packet error rate (PER) can be in the range of several percent. For example, with a BER of 1 in 10^5 the PER for a 100 bit data packet will be:

$$(1 - PER) = (1 - 10^{-5})^{100}$$

or PER $= 0.1\%$, rising to 1% for a 1 kb data packet.

The BER is a function of the signal-to-noise ratio in the receiver, and also depends on the specific type of modulation method being used. The signal-to-noise ratio of a communication channel is given by:

$$SNR = (E_b / N_o) * (f_b / W) \tag{4.1}$$

where E_b is the energy required per bit of information (joules), f_b is the bit rate (Hz), N_o is the noise power density (watts/Hz) and W is the bandwidth of the modulated carrier signal (Hz).

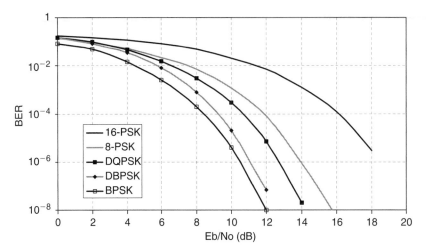

Figure 4.29: Bit error rate (BER) for some common modulation methods

Note that for our example, considering a DSSS system, it is the de-spread bandwidth that is considered in equation 4.1. Using a spread spectrum rather than a narrow band transmission results in an additional gain known as the processing gain.

$$\text{Processing Gain} = 10 \log_{10}(C) \text{dB} \qquad (4.2)$$

where C is the code length in chips (11 for the Barker code discussed above). This processing gain is effectively included in the calculation of channel SNR by using the de-spread bandwidth in equation 4.1.

The bit rate per Hz of bandwidth, f_b/W, is a function of the modulation method employed. BER is then given by:

$$\text{BER} = \tfrac{1}{2}\,erfc\,(\text{SNR})^{1/2} \qquad (4.3)$$

where *erfc* is the so-called complementary error function which can be looked up in mathematical tables. Figure 4.29 shows BER as a function of SNR for some of the common modulation methods.

The figure shows that for the example DQPSK modulated signal, with 1 bit per Hz of bandwidth, a signal-to-noise ratio of 10.4 dB is required to achieve a bit error rate of 1 in 10^5.

4.5.3.2 Receiver Noise Floor

The receiver noise floor has two components, the theoretical thermal noise floor (N) for an ideal receiver, and the receiver noise figure (NF) which is a measure of the additional noise and losses in a particular receiver. The thermal noise is given as:

$$N = kTW \qquad (4.4)$$

where k is the Boltzmann constant (1.38×10^{-23} joules/°K), T is the ambient temperature in °K and W is the bandwidth of the transmission (Hz)[1]. Receivers for wireless networking will typically have noise figures in the range from 6 to 15 dB.

The receiver noise floor (RNF) is then the sum of these two terms:

$$RNF = kTW + NF \qquad (4.5)$$

For the example 802.11b receiver with a 2 MHz despread bandwidth, operating at 20°C (290°K) and with a noise figure of 10 dB:

$$N = 1.38 \times 10^{-23}\,\text{J/K} \times 290°\text{K} \times 2 \times 10^6\,\text{Hz}$$
$$= 8.8 \times 10^{-12}\,\text{mW}$$
$$= -110.6\,\text{dBm}$$

$$RNF = -110.6\,\text{dBm} + 10\,\text{dB}$$
$$= -100.6\,\text{dBm}$$

4.5.3.3 Receiver Sensitivity

The receiver sensitivity, P_{RX}, is the sum of the receiver noise floor (RNF) and the signal-to-noise ratio (SNR) required to achieve the desired bit error rate:

$$P_{RX} = RNF + SNR \qquad (4.6)$$

For the example;

$$P_{RX} = -100.6 + 10.4\,\text{dBm}$$
$$= -90.2\,\text{dBm}$$

---

[1]Again, the despread bandwidth is used here.

Table 4.17: P_{RX} versus data rate for a typical 802.11b receiver

Data Rate (Mbps)	Modulation technique	P_{RX} (dBm)
11	256 CCK+DQPSK	−85
5.5	16 CCK+DQPSK	−88
2	Barker+DQBSK	−89
1	Barker+DBPSK	−92

From this discussion it can be seen that as the data rate in the example increases from 2 Mbps towards the 802.11b maximum of 11 Mbps, different modulation methods will be needed to achieve the higher bandwidth efficiency (more bits per Hz of bandwidth, or f_b/W in equation 4.1).

This will result in higher signal-to-noise ratio requirements for the same bit error rate, so that the receiver sensitivity will decrease at higher data rates. This is shown in Table 4.17 for a typical 802.11b receiver.

This dependence of P_{RX} on data rate underlies the gradual deterioration in wireless network throughput as signal strength decreases. There is no abrupt cut-off in performance, but rather a gradual reduction in throughput as the transmitter and receiver switch to a lower data rate at which a low BER can be maintained.

4.5.4 RF Signal Propagation and Losses

Between the transmitting and receiving antennas, the RF signal is subject to a number of factors that affect signal strength. These are considered in the following sections.

4.5.4.1 Free Space Loss

Once the signal is radiating outwards from the antenna, the signal power falls off with distance due to the spreading out of the radio waves. This is known as free space loss, and overall is the most significant factor affecting received signal strength.

Free space loss is measured in dB, and depends on the signal frequency and transmission distance according to the formula:

$$L_{FS} = 20 \log_{10} (4\pi D / \lambda) \tag{4.7}$$

where D is the transmitter to receiver distance in meters. λ is the wavelength of the radio signal in meters which can also be expressed as:

$$\lambda = c/f \tag{4.8}$$

where c is the speed of light (3×10^8 m/s) and f is the signal frequency in Hz.

Expressing these quantities in more convenient units, f in MHz and D in km, L_{FS} can be calculated from the formula:

$$L_{FS} = 32.45 + 20\log_{10}(f) + 20\log_{10}(D) \tag{4.9}$$

So, at 2.4 GHz (2400 MHz) the free space loss at 100 m (0.1 km) will be:

$$\begin{aligned} L_{FS} &= 32.4 + 20\log_{10}(2400) + 20\log_{10}(0.1) \\ &= 32.4 + 67.6 - 20 \\ &= 80\,\text{dB} \end{aligned}$$

The third term in equation 4.9 shows that L_{FS} increases by 20 dB for every factor of 10 increase in range. This gives the useful rule of thumb for the 802.11b 2.4 GHz band that L_{FS} is 60 dB at 10 meters, 80 dB at 100 meters and so on.

From the second term in equation 4.9 it can be seen that the frequency dependence results in an increase in L_{FS} for transmissions at 5.8 GHz of 20 $\log_{10}(5800/2400)$ or 7.7 dB (see Figure 4.30). While this may be important in open-air applications, in practical indoor situations, this difference is often small compared with other environmental effects.

This calculation of free space loss assumes that there is a clear line-of-sight between the transmitting and receiving antennas, which means that the receiver can effectively "see" the transmitter. However, to maximize RF propagation range in the open-air it is not enough just to be able to see in a straight line between the two antennas. The volume of space around this straight line affects signal propagation as well, and any obstructions that come close to the direct line-of-sight will also cause signal loss.

4.5.4.2 Fresnel Zone Theory

The theory used to calculate the effect of obstructions is called Fresnel zone theory. The Fresnel zone is a region between the two antennas with an oval shape similar to a rugby ball.

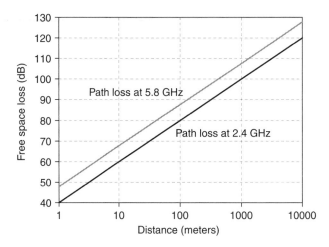

Figure 4.30: Free space loss at 2.4 GHz and 5.8 GHz

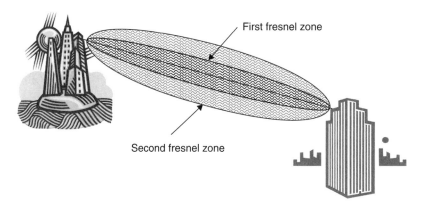

Figure 4.31: The Fresnel zones around a propagation path

There are actually a series of such regions, called the 1st, 2nd, 3rd, etc. Fresnel zones (Figure 4.31), and at the mid point between transmitter and receiver, the radius of the nth Fresnel zone in meters is calculated from the formula:

$$R = 0.5(n \times \lambda \times D) \tag{4.10}$$

where the wavelength and the transmitter to receiver distance are also in meters. For a 2.4 GHz signal, with a wavelength of 12.5 cm (0.125 m), the first Fresnel zone has a mid-point

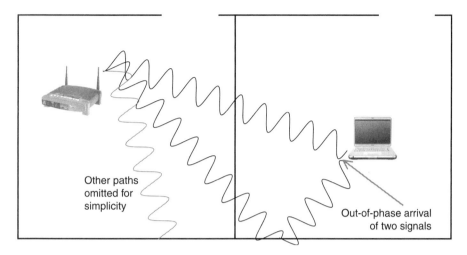

Figure 4.32: Multi-path fading in an indoor environment

radius of 1.8 m for a 100 m range, or 5.6 m for a 1 km range. Any obstructions within the first Fresnel zone will cause signal loss through reflection, refraction or diffraction.

4.5.4.3 Multipath Fading

Multipath fading occurs when reflected, refracted or diffracted signals travel to the receiver along different paths, resulting in a range of different arrival times known as the multipath delay spread. Signals arriving along different paths will be phase-shifted with respect to the direct path signal, as shown in Figure 4.32, and will therefore cause some degree of destructive interference at the receiving antenna. This is familiar in UHF TV reception as a ghost image caused by an interfering signal reflected from a nearby building or other large object.

Interference between these multiple delayed signals can substantially reduce the signal strength at the receiving antenna, introducing a loss that can be as much as 20 to 30 dB.

It is possible to compute multipath losses using complex ray tracing or other algorithms, but this is rarely done in practice.

4.5.4.4 Signal Attenuation Indoors

For a typical wireless network in a home or small office, multiple obstructions such as walls, floors, furniture and other objects will obstruct the propagation path from transmitter to receiver, and signal reception will tend to be very variable. Depending on its construction,

Table 4.18: Typical attenuation for building materials at 2.4 GHz

Attenuation range	Materials	Loss (dB)
Low	Nontinted glass, wooden door, cinder block wall, plaster.	2–4
Medium	Brick wall, marble, wire mesh or metal tinted glass.	5–8
High	Concrete wall, paper, ceramic bullet-proof glass.	10–15
Very high	Metal, silvering (mirrors).	>15

transmission through a wall can introduce a loss of 3 to 6 dB or more, as shown in Table 4.18, and an additional allowance will be required in the link budget to account for this loss.

In a multi-story building, losses between floors will also depend on the building materials used, and will be very high in buildings with sheet steel construction. More typically, a loss of approximately 6 dB is seen between adjacent floors, rising to around 10 dB per additional floor for separations of two to three floors. The typical losses shown in Table 4.18 are highly dependent on the specific materials used and methods of construction. For example, even a stud wall can introduce a significant loss if it contains a fire retarding foil membrane.

In common with multi-path fading, complex algorithms are required to calculate the various types of losses indoors, and it is therefore convenient to combine these loss components to give a single additional term in the link budget. This term, the fade margin (L_{FM}), will generally be estimated by a rule-of-thumb, or determined by an on-site survey.

4.5.5 Link Budget

The factors considered above, transmitter power (P_{TX}), antenna gain at the transmitter (G_{TX}) and receiver (G_{RX}), receiver sensitivity (P_{RX}), free-space loss (L_{FS}) and other losses combined in the fade margin (L_{FM}), together define the link budget that is available to bring the data signal successfully from transmission to detection (Table 4.19).

It is convenient to express the link budget in terms of the transmitter power (P_{TX}) required to deliver a signal to the receiver at its sensitivity limit (P_{RX}). Expressed in dBm, this is:

$$P_{TX} = P_{RX} - G_{TX} - G_{RX} - L_{FX} - L_{FM} \, \text{dBm} \tag{4.11}$$

Table 4.19: Balancing factors in the link budget

Reducing required P_{TX}	Increasing required P_{TX}
Lower receiver sensitivity (bigger negative) P_{RX}	Higher free space loss L_{FS}
Higher transmitter antenna gain G_{TX}	Higher fade margin L_{FM}
Higher transmitter antenna gain G_{RX}	

For example, a system comprising a directional transmitting antenna with a gain of 14 dBi, a patch receiving antenna (6 dBi), and a receiver with a sensitivity of -90 dBm, operating over 100 meters at 2.4 GHz ($L_{FS} = 80$ dB) with a 36 dB fade margin (L_{FM}) results in a required transmitter power of:

$$P_{TX} = -90\,\text{dBm} - 14\,\text{dBi} - 6\,\text{dBi} + 80\,\text{dB} + 36\,\text{dBm}$$
$$= +6\,\text{dBm}$$

To ensure that the signal at the receiving antenna is above the receiver sensitivity, the required transmitter power is therefore $+6$ dBm (4 mW), as shown graphically in Figure 4.33. This configuration would be comfortably achieved with a 100 mW (20 dBm) transmitter, with an extra 14 dB link margin for unaccounted losses or noise.

4.5.6 Ambient Noise Environment

As well as the receiver noise floor, which defines the limit of receiver sensitivity, other sources of external RF noise will also have an impact on the reliability of RF signal detection and data decoding.

The total RF noise entering a radio antenna at any particular location is termed the ambient noise environment and is made up of two components:

- Ambient noise floor—the aggregate background noise from distant sources such as car ignition, power distribution and transmission systems, industrial equipment, consumer products, distant electrical storms and cosmic sources.

- Incidental noise—the aggregate background noise from localized man-made sources.

The ambient noise floor is generally "white noise", with constant power per unit bandwidth, while incidental noise may be either broadband or narrow band. In implementing local or

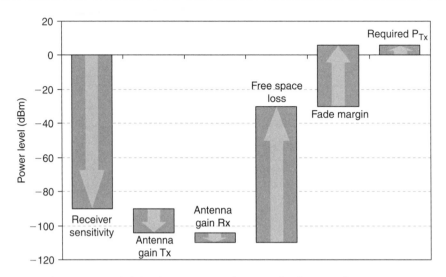

Figure 4.33: Link budget expressed as required transmitter power

metropolitan area wireless networks, the ambient noise floor will be established during an RF site survey, and should be explicitly included in the link budget if it is above the receiver sensitivity of planned equipment.

For example, in the link budget calculation above, if an ambient noise floor of −85 dBm was measured, using this in place of the receiver sensitivity of −90 dBm would result in a required transmitter power of 11 dBm. This type of environmental noise will limit the range that can be achieved for a given equipment configuration (transmitter power, antenna gains, etc.), but will not degrade the performance of wireless networks when operating within that limit.

Narrow band incidental noise, from nearby man-made sources such as a microwave oven or a narrow band transmitter, is more likely to result in unpredictable and unreliable network performance.

4.5.6.1 Interference Mitigation Techniques

Wireless networking specifications are increasingly including a range of measures to mitigate the effect of interference on network performance. Wireless USB is a good example of the approach which starts with establishing information about the quality of the RF link and then provides measures to control various link characteristics.

In wireless USB, the host and other devices can maintain statistical information on packet error rate and on link indicators such as received signal strength (RSSI) and link quality

Table 4.20: Wireless USB interference mitigation controls

Control	Description
Transmit power (TPC)	Host can control its own transmit power level as well as querying and controlling transmit power of devices in the cluster.
Transmitted bit rate	Host can adjust the transmitted bit rate for both outward (host to device) and inward (device to host) transfers.
Data payload size	When interference causes PER to rise, reducing packet size can improve throughput by reducing uncorrectable errors.
RF channel selection	Wireless USB's MB–OFDM radio provides multiple alternative channels which can be used by a host if supported by all devices in the cluster.
Host schedule control	Allowing isochronous data transfers to temporarily use channel time allocated for asynchronous transfers, in order to retransmit failed isochronous data packets.
Dynamic bandwidth control	Host control of the spectral shaping capabilities of the MB–OFDM UWB radio, described in the following section.

(LQI). The latter indicator measures the error in the received modulation of successfully decoded symbols.

The main RF link controls available in wireless USB are described in Table 4.20.

Transmit power control and RF channel selection are also included in the network optimization measures introduced in the 802.11k extension to the Wi-Fi networking standard.

4.6 Ultra Wideband Radio

4.6.1 Introduction

Ultra Wideband wireless communication systems are based on impulse radar technology that was developed for military applications by the USA and USSR in the 1960s. Impulse radar or radio transmits extremely short electromagnetic pulses, typically less than 1 ns (nanosecond) in length, with no underlying carrier signal. Such short pulses result in an effective bandwidth of the transmission that may be from 500 MHz up to several GHz.

In 2002 the FCC in the USA opened 7.5 GHz of radio spectrum for UWB applications, from 3.1 to 10.6 GHz, and adopted a definition of UWB as any intentional transmission in which the bandwidth to -3 dB points was at least 20% of the mean frequency of the transmission, with a minimum bandwidth of 500 MHz.

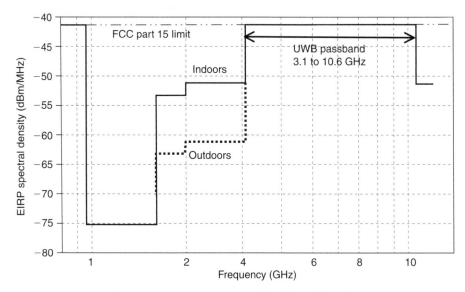

Figure 4.34: FCC UWB passband specification

Since UWB transmissions cover a wide swath of the radio spectrum, an important requirement is that they do not result in harmful interference with other RF transmitted services, whether current or planned. To ensure this coexistence, the FCC has defined strict EIRP limits on UWB transmission, as shown in Figure 4.34. The maximum permitted power density (EIRP) of −41.3 dBm/MHz is below the FCC Part 15 noise power limit for unintentional emitters such as computers and other electronic devices. As a result of this very low EIRP specification, UWB wireless is suited for applications where very long battery life is required.

A second characteristic of some UWB implementations is spectral shaping—the capability to control the radiated power spectrum in order to avoid transmission at particular narrow band frequencies.

UWB comes in three varieties for data communication applications:

- time-hopping pulse position modulation (or Impulse radio)

- direct sequence spread spectrum–UWB (DS–UWB)

- multiband–UWB (such as multiband OFDM).

Of these, MB–OFDM offers the greatest flexibility in spectral shaping, with a wide range of course and fine control options easily implemented in software.

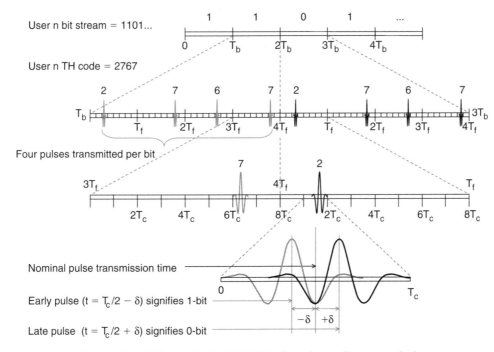

Figure 4.35: Pulse train in a TH-PPM impulse radio transmission

4.6.2 Time Hopping PPM UWB (Impulse Radio)

Impulse radio (IR) is the name given to UWB radio based on time-hopping, pulse position modulation. Data is transmitted as a discontinuous series of very short pulses, with one pulse per user in each time hopping frame of length T_f. The nominal transmission time of a pulse in a given frame is determined by a pseudo-noise (PN) code that is specific for each user of the communication channel.

Finally, whether a pulse represents a 1-bit or a 0-bit depends on the actual transmission time relative to the nominal transmission time (the pulse position modulation). For example, in an early/late PPM system, if the pulse is transmitted a time offset δ ahead of the nominal time it represents a 1-bit, or if an offset δ after the nominal time then it represents a 0-bit.

In the example shown in Figure 4.35, a TH code of length 4 is used, so that four pulses are transmitted for each bit, each pulse in one of the eight code slots (T_c) in each of four successive frames.

Pulse amplitude modulation (PAM) or pulse shape modulation (PSM) can be used as alternatives to PPM, with 1-bit and 0-bit then being determined by the amplitude or shape of each individual pulse.

Impulse radio is one of two optional physical layer specifications selected in March 2005 by the IEEE 802.15 Task Group 4a as part of the enhancement of the 802.15.4 standard. (The other optional PHY is a chirp spread spectrum operating in the 2.4 GHz ISM band.)

4.6.3 Direct Sequence UWB (DS-UWB)

Direct sequence, the spread spectrum technique underlying the IEEE 802.11b and 802.11g physical layer, can also be applied in UWB radios. Instead of the chipping code being used to spread the carrier spectrum by increasing the symbol transmission rate, the spectrum is spread to UWB proportions as a result of the very narrow pulse that is used to transmit each symbol.

The chipping code then plays a multiple access role (CDMA) with individual user codes determining the exact times at which individual users of the channel will transmit or receive a pulse. A variety of modulation methods (PAM, PSM) can be used to code the data stream onto the pulse stream. So far, DS-UWB has not been identified as a target technology for any wireless networking applications.

4.6.4 Multiband UWB

In multiband UWB, ultra-wide bandwidth is achieved by dividing the frequency band of interest into multiple overlapping or adjacent bands, and operating simultaneously on all available bands. Currently, the most important example of this technique is multiband (MB) OFDM, which is being promoted by the MB–OFDM Alliance (MBOA) and has been adopted as the basis for Wireless USB and Wireless FireWire (Wireless 1394).

MB–OFDM, as proposed by the MBOA, uses a bandwidth from 3.168 GHz to 10.560 GHz, which is divided into 14 bands of 528 MHz full width—thus meeting the FCC's 500 MHz minimum bandwidth specification. The 14 bands are grouped into 5 band groups or channels, as shown in Figure 4.36.

Frequency hopping between bands within a band group can be used to enable overlapping piconets to be formed but unlike Bluetooth, which makes 1600 hops per second across 79 frequencies, the MBOA radio as specified for wireless USB makes 3 million hops per second, one hop after every transmitted symbol, across just 3 frequencies.

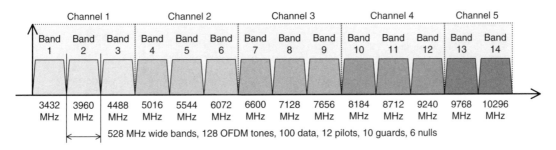

Figure 4.36: MB–OFDM frequency bands and channels

Table 4.21: MBOA time-frequency codes

Code number	Code type	Band number (Band group 1)					
1	TFI	1	2	3	1	2	3
2	TFI	1	3	2	1	3	2
3	TFI	1	1	2	2	3	3
4	TFI	1	1	3	3	2	2
5	FFI	1	1	1	1	1	1
6	FFI	2	2	2	2	2	2
7	FFI	3	3	3	3	3	3

MBOA specifies two types of time-frequency codes (TFC) as shown in Table 4.21. Time-frequency interleaving (TFI) codes define frequency hopping patterns, while fixed frequency interleaving (FFI) codes define continuous transmission on a single OFDM band. The FFI option can be used to improve the performance of two or more simultaneously operating piconets, by assigning a single OFDM band to each piconet.

Within each 528 MHz band, 128 OFDM subcarriers are transmitted, with data modulated onto 100 of these and the remainder used as pilot, guard and null tones. For data rates of up to 200 Mbps, MBOA specifies data modulation using QPSK, while rates of 320 to 480 Mbps use dual carrier modulation (DCM).

Spectral shaping is used to avoid interference with other RF services and can be changed under software control to respond to specific local regulations or time-varying conditions. Coarse control can be achieved by dropping whole bands (or in extreme circumstances

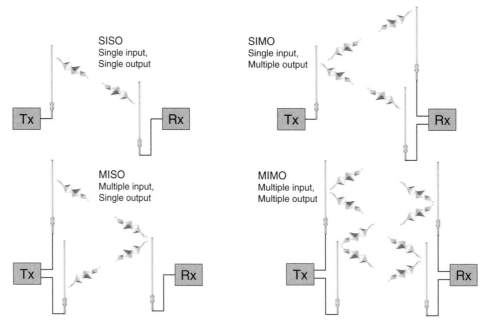

Figure 4.37: MIMO radio definition

whole band groups), but extremely precise shaping is also possible by "nulling out" a certain number of tones within a single band.

4.7 MIMO Radio

The multiple paths that a radio signal takes between transmitter and receiver often lead to a degradation of signal strength through multi-path fading. Multi-input multi-output (MIMO) radio takes advantage of this characteristic of RF propagation by sending multiple data streams across multiple transmitters to receiver paths in order to achieve a higher data capacity (Figure 4.37). Mathematical modelling of the propagation paths, using a channel calibration period during each transmitted data packet, allows the different signal paths and data streams to be identified and correctly recombined in the receiver.

This technique, space division multiplexing (SDM), is analogous to FDM in the frequency domain but instead of different frequencies carrying data in parallel, here different spatial paths carry data in parallel.

Effectively the same bandwidth is being used simultaneously to create multiple communication paths. If these paths are equally strong and can be perfectly separated, the overall capacity of the communication channel increases linearly with the number of independent paths used. In a system with M transmitters and N receivers, the number of independent paths is the minimum of M and N.

In practice, all paths will not be equally strong or perfectly separated, and performance will be determined by coefficients, known as singular values, which characterize each path between a transmitting and receiving antenna. These singular values are determined by including a short "training period" in the preamble of each transmitted data packet, during which known and different signals are transmitted from each antenna. These signals provide information about the transmission channel (so-called Channel State Information or CSI), and with this information the receiver can compute the singular values that are used to decode the remainder of the data packet.

The increased capacity of MIMO radio can be used to achieve a higher data rate or to increase link robustness or range for a given data rate. The IEEE 802.11n specification will use MIMO to increase the PHY layer data capacity of the 802.11a/g radio from 54 Mbps to in excess of 200 Mbps.

Space time block coding (STBC) is a related technique which combines space and time diversity to increase the robustness or range of an RF link. STBC breaks the transmitted data into blocks and transmits multiple time-shifted copies of each block of data from each transmitting antenna to the receiving antenna. STBC is thus a Multi-Input Single-Output (MISO) technique (see Figure 4.37), although multiple receiving antennas can further improve performance.

4.8 Near Field Communications

4.8.1 Introduction

Near field communications (NFC) is a very short range radio frequency communications technology that has been extensively developed for use in RF identification (RFID) tags and other smart labelling applications. These applications have typically employed a RF carrier frequency of 13.56 MHz, which is internationally allocated as an unlicensed ISM band.

NFC is distinct from so-called far field RF communication used in personal area and longer range wireless networks, since it relies on direct magnetic field coupling between transmitting and receiving devices.

There are two types of NFC devices, active and passive, which operate quite differently. Passive devices do not have an internal power source, but derive their power from an active initiating device by inductive coupling. A passive device also does not transmit data by generating a magnetic field as an active device does. Instead a passive device transfers data back to an active device through a process called load modulation. These concepts are described in the following sections.

4.8.2 Near Field and Far Field Communication

The space around an antenna can be divided into two regions based on the differing nature of the electromagnetic fields generated by the antenna. The boundary between the two regions is known as the radian sphere and has a radius of $\lambda/2\pi$, where λ is the wavelength of the propagated electromagnetic wave.

The primary magnetic field begins at the antenna and oscillations in this field induce an electric field in the surrounding space. This region, inside the radian sphere, is within the influence of the primary magnetic field and is called the near field of the antenna. The electromagnetic field equations in this region reflect energy storage in the magnetic field and are described by near field coupling volume theory.

The region outside the radian sphere is called the far field of the antenna and here the fields separate from the antenna and propagate into space as an electromagnetic wave. The electromagnetic field equations here represent energy propagation rather than storage.

For NFC operating at 13.56 MHz, $\lambda = 22$ meters, so that the radius of the radian sphere is $\lambda/2\pi = 3.5$ meters. In the near field region, the magnetic field strength is inversely proportional to the cube of the distance between the antennas while the power in the magnetic field, which is used to energize passive NFC, decreases as the inverse sixth power of the separation. This is equivalent to an attenuation of 60 dB for a ten-fold increase in distance.

4.8.3 Inductive Coupling

Near-field inductive coupling uses an oscillating magnetic field to transfer RF energy between devices. Each device includes a resonant circuit tuned to the RF carrier frequency, and a loosely coupled "space transformer" is established when the coil windings or "antenna loops" of the two devices are brought into range (Figure 4.38). The effective range is comparable to the actual physical dimensions of the transmitting antenna loop.

When the resonant circuit in the transmitting device is energized by a RF power source, the resulting magnetic flux linkage results in energy transfer between the two resonant coil windings.

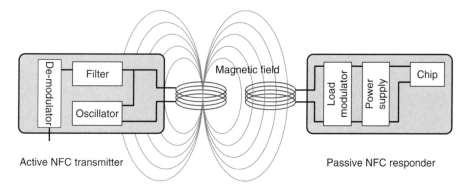

Figure 4.38: Inductive coupling between NFC antenna loops

Inductive coupling is only effective in the near-field region of the transmitting antenna loop. In the far-field region, where the electromagnetic field separates from the antenna and propagates as an electromagnetic wave, it can no longer have a direct effect through inductive coupling.

4.8.4 Load Modulation

When an NFC target device in passive mode is within range of an active NFC transmitter (or initiator) its resonant circuit draws energy from the magnetic field created by the initiating device. This additional consumption results in a voltage perturbation that can be measured in the resonant circuit of the initiator. If an additional load resistance in the target is periodically switched on and off, this has the effect of an amplitude modulation of the carrier wave voltage in the initiator.

By using the data stream to be transmitted from the target device to control this load switching, the data stream is transferred from the target to the initiator. This technique is called load modulation.

Load modulation creates amplitude modulated sidebands on the 13.56 MHz carrier frequency, and the data stream is recovered by demodulating these sidebands in the initiating device's RF signal processing circuits.

CHAPTER 5

Applications and Technologies

Alan Bensky

In Chapter 4, Rackley provided an introduction to radio communication basics, showing at a high level how a wide range of WLAN and cellular protocols are implemented. In this chapter Alan Bensky takes it the next step, doing a detailed drill-down on Wi-Fi, Bluetooth, Zigbee and UWB.

Keeping up with these protocols hasn't been easy; they've all taken a long time to come to market, and they've evolved considerably in the process. Wi-Fi is the granddaddy of all wireless networking protocols, dating from the original 802.11 spec in 1997. Since then it's developed variants a/b/g/n/y, with different modulation and access protocols.

Bluetooth took a number of twists and turns over the decade it took to come to market; now, having settled into its niche in cellular headphones, Bluetooth is adding a high-speed channel using UWB, with the idea of moving high-speed data (read: videos) between consumer devices. Here they'll be going head-to-head with Wireless USB, which currently has a lot more buzz than traction.

High-speed Bluetooth has a problem: as of this writing UWB has failed to provide its promised throughput and 6 GHz operation, forcing the Bluetooth camp to look to Wi-Fi for high-speed operation. Unfortunately, Wi-Fi seems to interfere with WiMAX (802.16), which may or may not appear in 3G/LTE handsets—and Wi-Fi is a power hog, the last thing you want in a handset. All designs involve trade-offs, but some are uglier than others.

Meanwhile, while Bluetooth was taking its time to get to market, part of the pack broke off and decided to boil it down to a low-power stack for mesh networks. Several years later, ZigBee is starting to get traction in wireless thermostats and motor controllers that enable both homeowners and the utility company to regulate power consumption during peak hours. Still trying to break out of the headset market, the Bluetooth camp is taking aim at their former comrades with a low-power version of Bluetooth (low-speed channel?) targeting the same markets.

As promised, Bensky covers each of these technologies in detail, from the architectural level down to the MAC and PHY layers. This level of understanding—combined with his explanation of conflict and compatibility between the various protocols—will help the reader sort out the issues as these RF technologies continue to morph into each others' markets.

—John Donovan

An important factor in the widespread penetration of short-range devices into the office and the home is the basing of the most popular applications on industry standards. In this chapter, we take a look at some of these standards and the applications that have emerged from them. Those covered pertain to HomeRF, Wi-Fi, HIPERLAN/2, Bluetooth, and Zigbee. In order to be successful, a standard has to be built so that it can keep abreast of rapid technological advancements by accommodating modifications that don't obsolete earlier devices that were developed to the original version. A case in point is the competition between the WLAN (wireless local area network) standard that was developed by the HomeRF Working Group based on the SWAP (shared wireless access protocol) specification, and IEEE specification 802.11, commonly known as Wi-Fi. The former used frequency-hopping spread-spectrum exclusively, and although some increase of data rate was provided for beyond the original 1 and 2 Mbps, it couldn't keep up with Wi-Fi, which incorporated new bandwidth efficient modulation methods to increase data rates 50-fold while maintaining compatibility with first generation DSSS terminals. Other reasons why HomeRF lost out to Wi-Fi are given below.

Many of the new wireless short-range systems are designed for operation on the 2.4 GHz ISM band, available for license-free operation in North America and Europe, as well as virtually all other regions in the world. Most systems have provisions for handling errors due to interference, but when the density of deployment of one or more systems is high, throughput, voice intelligibility, or quality of service in general is bound to suffer. We will look at some aspects of this problem and methods for solving it in relation to Bluetooth and Wi-Fi.

A relatively new approach to short-range communications with unique technological characteristics is ultra-wideband (UWB) signal generation and detection. UWB promises to add applications and users to short-range communication without impinging on present spectrum use. Additionally, it has other attributes including range finding and high power efficiency that are derived from its basic principles of operation. We present the main features of UWB communication and an introduction to how it works.

5.1 Wireless Local Area Networks (WLAN)

One of the hottest applications of short-range radio communication is wireless local area networks. While the advantage of a wireless versus wired LAN is obvious, the early versions of WLAN had considerably inferior data rates so conversion to wireless was often not worthwhile, particularly when portability is not an issue. However, advanced modulation techniques have allowed wireless throughputs to approach and even exceed those of wired networks, and the popularity of highly portable laptop and handheld computers, along with the decrease in device prices, have made computer networking a common occurrence in multi-computer offices and homes.

There are still three prime disadvantages to wireless networks as compared to wired: range limitation, susceptibility to electromagnetic interference, and security. Direct links may be expected to perform at a top range of 50 to 100 meters depending on frequency band and surroundings. Longer distances and obstacles will reduce data throughput. Greater distances between network participants are achieved by installing additional access points to bridge remote network nodes. Reception of radio signals may be interfered with by other services operating on the same frequency band and in the same vicinity. Wireless transmissions are subject to eavesdropping, and a standardized security implementation in Wi-Fi called WEP (wired equivalent privacy), has been found to be breachable with relative ease by persistent and knowledgeable hackers. More sophisticated encryption techniques can be incorporated, although they may be accompanied by reduction of convenience in setting up connections and possibly in performance.

Various systems of implementation are used in wireless networks. They may be based on an industrial standard, which allows compatibility between devices by different manufacturers, or a proprietary design. The latter would primarily be used in a special purpose network, such as in an industrial application where all devices are made by the same manufacturer and where performance may be improved without the limitations and compromises inherent in a widespread standard.

5.1.1 The HomeRF Working Group

The HomeRF Working Group was established by prominent computer and wireless companies that joined together to establish an open industry specification for wireless digital communication between personal computers and consumer electronic devices anywhere in and around the home. It developed the SWAP specification—Shared Wireless Access Protocol, whose major application was setting up a wireless home network that connects one or more computers with peripherals for the purposes of sharing files, modems, printers, and other electronic devices, including telephones. In addition to acting as a transparent wire replacement medium, it also permitted integration of portable peripherals into a computer network. The originators expected their system to be accepted in the growing number of homes that have two or more personal computers.

Following are the main system technical parameters:

- Frequency-hopping network: 50 hops per second
- Frequency range: 2.4 GHz ISM band
- Transmitter power: 100 milliwatt
- Data rate: 1 Mbps using 2FSK modulation
  2 Mbps using 4FSK modulation
- Range: Covers typical home and yard
- Supported stations: Up to 127 devices per network
- Voice connections: Up to 6 full-duplex conversations
- Data security: Blowfish encryption algorithm (over 1 trillion codes)
- Data compression: LZRW3-A (Lempel-Ziv) algorithm
- 48-bit network ID: Enables concurrent operation of multiple co-located networks

The HomeRF Working Group ceased activity early in 2003. Several reasons may be cited for its demise. Reduction in prices of its biggest competitor, Wi-Fi, all but eliminated the advantage HomeRF had for home networks—low cost. Incompatibility with Wi-Fi was a liability, since people who used their Wi-Fi equipped laptop computer in the office also needed to use it at home, and a changeover to another terminal accessory after work hours was not an option. If there were some technical advantages to HomeRF, support of voice and connections between peripherals for example, they are becoming insignificant with the development of voice interfaces for Wi-Fi and the introduction of Bluetooth.

5.1.2 Wi-Fi

Wi-Fi is the generic name for all devices based on the IEEE specification 802.11 and its derivatives. It is promoted by the Wi-Fi Alliance that also certifies devices to ensure their interoperability. The original specification is being continually updated by IEEE working groups to incorporate technical improvements and feature enhancements that are agreed upon by a wide representation of potential users and industry representatives. 802.11 is the predominant industrial standard for WLAN and products adhering to it are acceptable for marketing all over the world.

802.11 covers the data link layer of lower-level software, the physical layer hardware definitions, and the interfaces between them. The connection between application software and the wireless hardware is the MAC (medium access control). The basic specification defines three types of wireless communication techniques: DSSS (direct sequence spread spectrum), FSSS (frequency-hopping spread spectrum) and IR (infra-red). The specification is built so that the upper application software doesn't have to know what wireless technique is being used—the MAC interface firmware takes care of that. In fact, application software doesn't have to know that a wireless connection is being used at all and mixed wired and wireless links can coexist in the same network.

Wireless communication according to 802.11 is conducted on the 2.400 to 2.4835 GHz frequency band that is authorized for unlicensed equipment operation in the United States and Canada and most European and other countries. A few countries allow unlicensed use in only a portion of this band. A supplement to the original document, 802.11b, adds increased data rates and other features while retaining compatibility with equipment using the DSSS physical layer of the basic specification. Supplement 802.11a specifies considerably higher rate operation in bands of frequencies between 5.2 and 5.8 GHz. These data rates were made available on the 2.4 GHz band by 802.11g that has downward compatibility with 802.11b.

5.1.3 Network Architecture

Wi-Fi architecture is very flexible, allowing considerable mobility of stations and transparent integration with wired IEEE networks. The transparency comes about because upper application software layers (see below) are not dependent on the actual physical nature of the communication links between stations. Also, all IEEE LAN stations, wired or wireless, use the same 48-bit addressing scheme so an application only has to reference source and destination addresses and the underlying lower-level protocols will do the rest.

Three Wi-Fi network configurations are shown in Figures 5.1 through 5.3. Figure 5.1 shows two unattached basic service sets (BSS), each with two stations (STA). The BSS is the basic building block of an 802.11 WLAN. A station can make ad hoc connections with other stations within its wireless communication range but not with those in another BSS that is outside of this range. In order to interconnect terminals that are not in direct range one with the other, the distributed system shown in Figure 5.2 is needed. Here, terminals that are in range of a station designated as an access point (AP) can communicate with other terminals not in direct range but who are associated with the same or another AP. Two or more such access points communicate between themselves either by a wireless or wired medium, and therefore data exchange between all terminals in the network is supported. The important thing here is that the media connecting the STAs with the APs, and connecting the APs among themselves are totally independent.

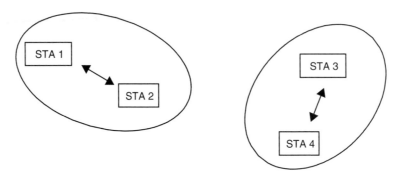

Figure 5.1: Basic service set

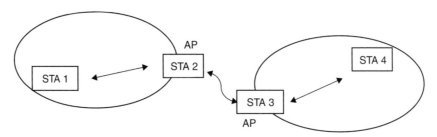

Figure 5.2: Distribution system and access points

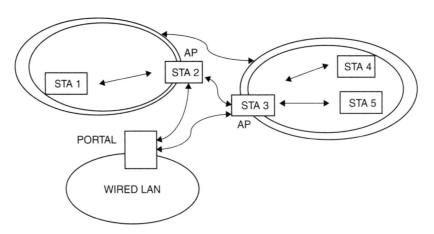

Figure 5.3: Extended service set

A network of arbitrary size and complexity can be maintained through the architecture of the extended service set (ESS), shown in Figure 5.3. Here, STAs have full mobility and may move from one BSS to another while remaining in the network. Figure 5.3 shows another element type—a portal. The portal is a gateway between the WLAN and a wired LAN. It connects the medium over which the APs communicate to the medium of the wired LAN— coaxial cable or twisted pair lines, for example.

In addition to the functions Wi-Fi provides for distributing data throughout the network, two other important services, although optionally used, are provided. They are authentication and encryption. Authentication is the procedure used to establish the identity of a station as a member of the set of stations authorized to associate with another station. Encryption applies coding to data to prevent an eavesdropper from intercepting it. 802.11 details the implementation of these services in the MAC. Further protection of confidentiality may be provided by higher software layers in the network that are not part of 802.11.

The operational specifics of WLAN are described in IEEE 802.11 in terms of defined protocols between lower-level software layers. In general, networks may be described by the communication of data and control between adjacent layers of the Open System Interconnection Reference Model (OSI/RM), shown in Figure 5.4, or the peer-to-peer communication between like layers of two or more terminals in the network. The bottom layer, physical, represents the hardware connection with the transmission medium that connects the terminals of the network—cable modem, radio transceiver and antenna, infrared transceiver, or power line transceiver, for example. The software of the upper layers is wholly independent of the transmission medium and in principle may be used unchanged no matter what the nature of the medium and the physical connection to it. IEEE 802.11 is concerned only with the two lowest layers, physical and data link.

IEEE 802.11 prescribes the protocols between the MAC sublayer of the data layer and the physical layer, as well as the electrical specifications of the physical layer. Figure 5.5 illustrates the relationship between the physical and MAC layers of several types of networks with upper-layer application software interfaced through a commonly defined logical link control (LLC) layer. The LLC is common to all IEEE local area networks and is independent of the transmission medium or medium access method. Thus, its protocol is the same for wired local area networks and the various types of wireless networks. It is described in specification ANSI/IEEE standard 802.2.

The Medium Access Control function is the brain of the WLAN. Its implementation may be as high-level digital logic circuits or a combination of logic and a microcontroller or a digital

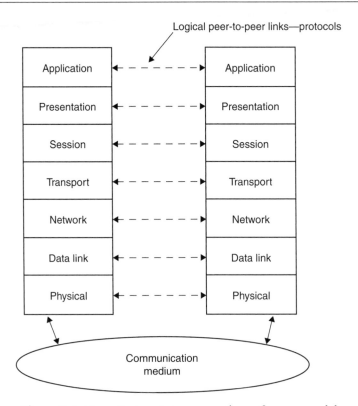

Figure 5.4: Open System Interconnection reference model

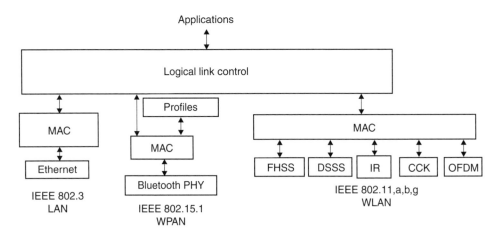

Figure 5.5: Data Link and Physical Layers (PHY)

signal processor. IEEE 802.11 and its supplements, (which may be generally designated 802.11x), prescribe various data rates, media (radio waves or infrared), and modulation techniques (FHSS, DSSS, CCK, ODFM). These are the principle functions of the MAC:

- Frame delimiting and recognition,

- Addressing of destination stations,

- Transparent transfer of data, including fragmentation and defragmentation of packets originating in upper layers,

- Protection against transmission error,

- Control of access to the physical medium,

- Security services—authentication and encryption.

An important attribute of any communications network is the method of access to the medium. 802.11 prescribes two possibilities: DCF (distributed coordination function) and PCF (point coordination function).

The fundamental access method in IEEE 802.11 is the DCF, more widely known as CSMA/CA (carrier sense multiple access with collision avoidance). It is based on a procedure during which a station wanting to transmit may do so only after listening to the channel and determining that it is not busy. If the channel is busy, the station must wait until the channel is idle. In order to minimize the possibility of collisions when more than one station wants to transmit at the same time, each station waits a random time-period, called a back off interval, before transmitting, after the channel goes idle. Figure 5.6 shows how this method works.

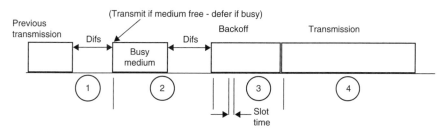

Figure 5.6: CSMA/CA access method

The figure shows activity on a channel as it appears to a station that is attempting to transmit. The station may start to transmit if the channel is idle for a period of at least a duration of DIFS (distributed coordination function interframe space) since the end of any other transmission (Section 1 of the figure). However, if the channel is busy, as shown in Section 2 of the figure, it must defer access and enter a back off procedure. The station waits until the channel is idle, and then waits an additional period of DIFS. Now it computes a time-period called a back off window that equals a pseudo-random number multiplied by constant called the "slot time." As long as the channel is idle, as it is in Section 3 of the figure, the station may transmit its frame at the end of the back off window, Section 4. During every slot time of the back off window the station senses the channel, and if it is busy, the counter that holds the remaining time of the back off window is frozen until the channel becomes idle and the back off counter resumes counting down.

Actually, the back off procedure is not used for every access of the channel. For example, acknowledgement transmissions and RTS and CTS transmissions, (see below), do not use it. Instead, they access the channel after an interval called SIFS (short interframe space) following the transmission to which they are responding. SIFS is shorter than DIFS, so other stations waiting to transmit cannot interfere since they have to wait a longer time, after the previous transmission, and by then the channel is already occupied.

In waiting for a channel to become idle, a transmission contender doesn't have to listen continuously. When one hears another station access the channel, it can interpret the frame length field that is transmitted on every frame. After taking into account the time of the acknowledgement transmission that replies to a data transmission, the time that the channel will become idle is known even without physically sensing it. This is called a virtual carrier sense mechanism.

The procedure shown in Figure 5.6 may not work well under some circumstances. For example, if several stations are trying to transmit to a single access point, two or more of them may be positioned such that they all are in range of the access point but not of each other. In this case, a station sensing the activity of the channel may not hear another station that is transmitting on the same network. A refinement of the described CSMA/SA procedure is for a station thinking the channel is clear to send a short RTS (request to send) control frame to the AP. It will then wait to receive a CTS (clear to send) reply from the AP, which is in range of all contenders for transmission, before sending its data transmission. If the originating station doesn't hear the CTS it assumes the channel was busy and so it must try to access the channel again. This RTS/CTS procedure is also effective when not all stations on the network have compatible modulation facilities for high rate communication and one station may not be able to detect the transmission length field of another. RTS and CTS transmissions are always sent at a basic rate that is common to all participants in the network.

The PCS is an optional access method that uses a master-slave procedure for polling network members. An AP station assumes the role of master and distributes timing and priority information through beacon management transmissions, thus creating a contention free access method. One use of the PCS is for voice communications, which must use regular time slots and will not work in a random access environment.

5.1.4 Physical Layer

The discussion so far on the services and the organization of the WLAN did not depend on the actual type of wireless connection between the members of the network. 802.11 and its additions specify various bit rates, modulation methods, and operating frequency channels, on two frequency bands, which we discuss in this section.

5.1.4.1 IEEE 802.11 Basic

The original version of the 802.11 specification prescribes three different air interfaces, each having two data rates. One is infrared and the others are based on frequency-hopping spread spectrum (FHSS) and direct-sequence spread-spectrum, each supporting raw data rates of 1 and 2 Mbps. Below is a short description of the IR and FHSS links, and a more detailed review of DSSS.

5.1.4.2 Infrared PHY

Infrared communication links have some advantages over radio wave transmissions. They are completely confined within walled enclosures and therefore eavesdropping concerns are greatly relieved, as are problems from external interference. Also, they are not subject to intentional radiation regulations. The IEEE 802.11 IR physical layer is based on diffused infrared links, and the receiving sensor detects radiation reflected off ceilings and walls, making the system independent of line-of-site. The range limit is on the order of 10 meters. Baseband pulse position modulation is used, with a nominal pulse width of 250 nsec. The IR wavelength is between 850 and 950 nm. The 1 Mbps bit rate is achieved by sending symbols representing 4 bits, each consisting of a pulse in one of 16 consecutive 250 nsec slots. This modulation method is called 16-PPM. Optional 4-PPM modulation, with four slots per two-bit symbol, gives a bit rate of 2 Mbps.

Although part of the original IEEE 802.11 specification and having what seems to be useful characteristics for some applications, products based on the infrared physical layer for WLAN have generally not been commercially available. However, point-to-point, very short-range infrared links using the IrDA (Infrared Data Association) standard are very widespread (reputed to be in more than 300 million devices). These links work reliably line-of-site at one meter and are found, for example, in desktop and notebook computers, handheld PC's, printers, cameras and toys. Data rates range from 2400 Bps to 16 Mbps. Bluetooth devices will take over some of the applications but for many cases IrDA imbedding will still have an advantage because of its much higher data rate capability.

5.1.4.3 FHSS PHY

While overshadowed by the DSSS PHY, acquaintance with the FHSS option in 802.11 is still useful since products based on it may be available. In FHSS WLAN, transmissions occur on carrier frequencies that hop periodically in pseudo-random order over almost the complete span of the 2.4 GHz ISM band. This span in North America and most European countries is 2.400 to 2.4835 GHz, and in these regions there are 79 hopping carrier frequencies from 2.402 to 2.480 GHz. The dwell on each frequency is a system-determined parameter, but the recommended dwell time is 20 msec, giving a hop rate of 50 hops per second. In order for FHSS network stations to be synchronized, they must all use the same pseudo-random sequence of frequencies, and their synthesizers must be in step, that is, they must all be tuned to the same frequency channel at the same time. Synchronization is achieved in 802.11 by sending the essential parameters—dwell time, frequency sequence number, and present channel number—in a frequency parameter set field that is part of a beacon transmission (and other management frames) sent periodically on the channel. A station wishing to join the network can listen to the beacon and synchronize its hop pattern as part of the network association procedure.

The FHSS physical layer uses GFSK (Gaussian frequency shift keying) modulation, and must restrict transmitted bandwidth to 1 MHz at 20 dB down (from peak carrier). This bandwidth holds for both 1 Mbps and 2 Mbps data rates. For 1 Mbps data rate, nominal frequency deviation is ±160 kHz. The data entering the modulator is filtered by a Gaussian (constant phase delay) filter with 3 dB bandwidth of 500 kHz. Receiver sensitivity must be better than −80 dBm for a 3% frame error rate.

In order to keep the same transmitted bandwidth with a data rate of 2 Mbps, four-level frequency shift-keying is employed. Data bits are grouped into symbols of two bits, so each symbol can have one of four levels. Nominal deviations of the four levels are ±72 kHz and ±216 kHz. A 500 kHz Gaussian filter smoothes the four-level 1 Megasymbols per second at the input to the FSK modulator. Minimum required receiver sensitivity is −75 dBm.

Although development of Wi-Fi for significantly increased data rates has been along the lines of DSSS, FHSS does have some advantageous features. Many more independent networks can be collocated with virtually no mutual interference using FHSS than with DSSS. As we will see later, only three independent DSSS networks can be collocated. However, 26 different hopping sequences (North America and Europe) in any of three defined sets can be used in the same area with low probability of collision. Also, the degree of throughput reduction by other 2.4 GHz band users, as well as interference caused to the other users is lower with FHSS. FHSS implementation may at one time also have been less expensive. However, the updated versions of 802.11—specifically 802.11a, 802.11b, and 802.11g—have all based their methods of increasing data rates on the broadband channel characteristics of DSSS in 802.11, while being downward compatible with the 1 and 2 Mbps DSSS modes (except for 802.11a which operates on a different frequency band).

5.1.4.4 DSSS PHY

The channel characteristics of the direct sequence spread spectrum physical layer in 802.11 are retained in the high data rate updates of the specification. This is natural, since systems based on the newer versions of the specification must retain compatibility with the basic 1 and 2 Mbps physical layer. The channel spectral mask is shown in Figure 5.7, superimposed on the simulated spectrum of a filtered 1 Mbps transmission. It is 22 MHz wide at the -30 dB points. Fourteen channels are allocated in the 2.4 GHz ISM band, whose center frequencies

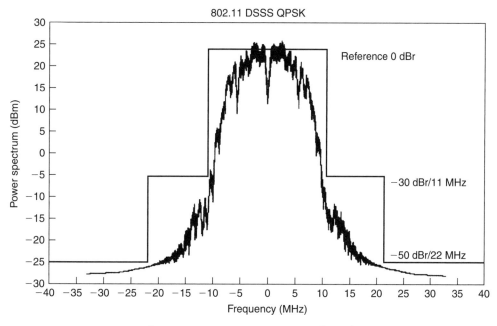

Figure 5.7: 802.11 DSSS spectral mask

are 5 MHz apart, from 2.412 GHz to 2.484 GHz. The highest channel, number fourteen, is designated for Japan where the allowed band edges are 2.471 GHz and 2.497 GHz. In the US and Canada, the first eleven channels are used. Figure 5.8 shows how channels one, six and eleven may be used by three adjacent independent networks without co-interference. When there are no more than two networks in the same area, they may choose their operating channels to avoid a narrow-band transmission or other interference on the band.

In 802.11 DSSS, a pseudo-random bit sequence phase modulates the carrier frequency. In this spreading sequence, bits are called chips. The chip rate is 11 megachips per second (Mcps). Data is applied by phase modulating the spread carrier. There are eleven chips per data symbol. The chosen pseudo-random sequence is a Barker sequence, represented as $1, -1,$ $1,1, -1,1,1,1, -1, -1, -1$. Its redeeming property is that it is optimally detected in a receiver by a matched filter or correlation detector. Figure 5.9 is one possible implementation of the modulator. The DSSS PHY specifies two possible data rates—1 and 2 Mbps. The differential encoder takes the data stream and produces two output streams at 1 Mbps that represent changes in data polarity from one symbol to the next. For a data rate of 1 Mbps, differential binary phase shift keying is used. The input data rate of 1 Mbps results in two identical output data streams that represent the changes between consecutive input bits. Differential quadrature phase shift keying handles 2 Mbps of data. Each sequence of two input bits

Figure 5.8: DSSS noninterfering channels

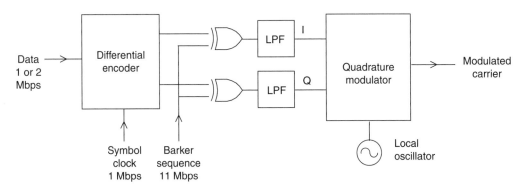

Figure 5.9: DSSS modulation

creates four permutations on two outputs. The differential encoder outputs the differences from symbol to symbol on the lines that go to the inputs of the exclusive OR gates shown in Figure 5.9. The outputs on the I and Q lines are the Barker sequence of 11 Mcps inverted or sent straight through, at a rate of 1 Msps, according to the differentially encoded data at the exclusive OR gate inputs. These outputs are spectrum shifted to the RF carrier frequency (or an intermediate frequency for subsequent up-conversion) in the quadrature modulator.

Reception of DSSS signals is represented in Figure 5.10. The downconverted I and Q signals are applied to matched filters or correlation detectors. These circuits correlate the Barker sequence with the input signal and output an analog signal that represents the degree of correlation. The following differential decoder performs the opposite operation of the differential encoder described above and outputs the 1 or 2 Mbps data.

The process of despreading the input signal by correlating it with the stored spreading sequence requires synchronization of the receiver with transmitter timing and frequency. To facilitate this, the transmitted frame starts with a synchronization field (SYNC), shown at the beginning of the physical layer protocol data unit in Figure 5.11. Then a start frame delimiter (SFD) marks out the commencement of the following information bearing fields. All bits in the indicated preamble are transmitted at a rate of 1 Mbps, no matter what the subsequent data rate will be. The signal field specifies the data rate of the following fields in the frame so that the receiver can adjust itself accordingly. The next field, SERVICE, contains all zeros for

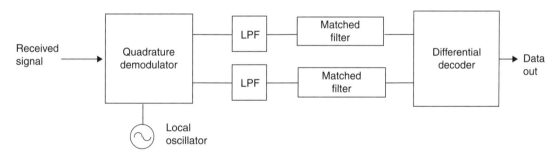

Figure 5.10: DSSS reception

Figure 5.11: DSSS frame format

devices that are only compliant with the basic version of 802.11, but some of its bits are used in devices conforming with updated versions. The value of the length field is the length, in microseconds, required to transmit the data-carrying field labeled MPDU (MAC protocol data unit). An error check field, labeled CRC, protects the integrity of the SIGNAL, SERVICE, and LENGTH fields. The last field MPDU (MAC protocol data unit) is the data passed down from the MAC to be sent by the physical layer, or to be passed up to the MAC after reception. All bits in the transmitted frame are pseudo-randomly scrambled to ensure even power distribution over the spectrum. Data is returned to its original form by descrambling in the receiver.

5.1.4.5 802.11b

The "b" supplement to the original 802.11 specification supports a higher rate physical layer for the 2.4 GHz band. It is this 802.11b version that provided the impetus for Wi-Fi proliferation. With it, data rates of 5.5 Mbps and 11 Mbps are enabled, while retaining downward compatibility with the original 1 and 2 Mbps rates. The slower rates may be used not only for compatibility with devices that aren't capable of the extended rates, but also for fall back when interference or range conditions don't provide the required signal-to-noise ratio for communication using the higher rates.

As previously stated, the increased data rates provided for in 802.11b do not entail a larger channel bandwidth. Also, the narrow-band interference rejection, or jammer resisting qualities of direct sequence spread-spectrum are retained. The classical definition of processing gain for DSSS as being the chip rate divided by the data bandwidth doesn't apply here. In fact, the processing gain requirement that for years was part of the FCC Rules paragraph 15.247 definition of direct sequence spread-spectrum was deleted in an update from August 2002, and at the same time reference to DSSS was replaced by "digital modulation."

The mandatory high-rate modulation method of 802.11b is called complementary code keying (CCK). An optional mode called packet binary convolutional coding (PBCC) is also described in the specification. Although there are similarities in concept, the two modes differ in implementation and performance. First the general principle of high-rate DSSS is presented below, applying to both CCK and PBCC, then the details of CCK are given.

As in the original 802.11, a pseudo-random noise sequence at the rate of 11 Mcps is the basis of high-rate transmission in 802.11b. It is this 11 Mcps modulation that gives the 22 MHz null-to-null bandwidth. However, in contrast to the original specification, the symbol rate when sending data at 5.5 or 11 Mbps is 1.375 Msps. Eight chips per symbol are transmitted instead of eleven chips per symbol as when sending at 1 or 2 Mbps. In "standard" DSSS as used in 802.11, the modulation, BPSK or QPSK, is applied to the group of eleven chips constituting a symbol. The series of eleven chips in the symbol is always the same (the Barker sequence previously defined). In contrast, high-rate DSSS uses a different 8-chip sequence in each symbol, depending on the sequence of data bits that is applied to each symbol. Quadrature modulation is used, and each chip has an I value and a Q value which represent a complex number having a normalized amplitude of one and some angle, α, where α=arctangent (Q/I). α can assume one of four values divided equally around 360 degrees. Since each complex bit has four possible values, there are a total of $4^8 = 65536$ possible 8-bit complex words. For the 11 Mbps data rate, 256 out of these possibilities are actually used—which one being determined by the sequence of 8 data bits applied to a particular symbol. Only 16-chip sequences are needed for the 5.5 Mbps rate, determined by four data bits per symbol. The high-rate algorithm describes the manner in which the 256 code words, or 16 code words, are chosen from the 65536 possibilities. The chosen 256 or 16 complex words have the very desirable property that when correlation detectors are used on the I and Q lines of the received signal, downconverted to baseband, the original 8-bit (11 Mbps rate) or 4-bit (5.5 Mbps rate) sequence can be decoded correctly with high probability even when reception is accompanied by noise and other types of channel distortion.

The concept of CCK modulation and demodulation is shown in Figures 5.12 and 5.13. It's explained below in reference to a data rate of 11 Mbps. The multiplexer of Figure 5.12 takes a block of eight serial data bits, entering at 11 Mbps, and outputs them in parallel, with updates at the symbol rate of 1.375 MHz. The six latest data bits determine 1 out of 64 (2^6) complex code words. Each code word is a sequence of eight complex chips, having phase angles α_1

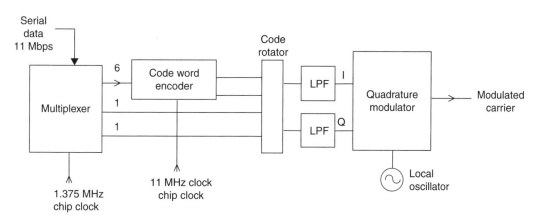

Figure 5.12: High-rate modulator—11 Mbps

through α_8 and a magnitude of unity. The first two data bits, d_0 and d_1, determine an angle, α_8' which, in the code rotator (see Figure 5.12), rotates the whole code word relative to α_8 of the previous code word. This angle of rotation becomes the absolute angle α_8 of the present code word. The normalized I and Q outputs of the code rotator, which after filtering are input to a quadrature modulator for up-conversion to the carrier (or intermediate) frequency, are:

$$I_i = \cos(\alpha_i), Q_i = \sin(\alpha_i) \quad i = 1\ldots8.$$

Figure 5.13 is a summary of the development of code words a for 11 Mbps rate CCK modulation. High rate modulation is applied only to the payload —MPDU in Figure 5.11. The code word described in Figure 5.13 is used as shown for the first symbol and then every other symbol of the payload. However, it is modified by adding $180°$ to each element of the code word of the second symbol, fourth symbol, and so on.

The development of the symbol code word or chip sequence may be clarified by an example worked out per Figure 5.13. Let's say the 8-bit data sequence for a symbol is $\boldsymbol{d}=d_0 \ldots d_7=1$ 0 1 0 1 1 0 1. From the phase table of Figure 5.13 we find the angles ϕ: $\phi_1=180°$, $\phi_2=180°$, $\phi_3=-90°$, $\phi_4=90°$. Now summing up these values to get the angle α_i of each complex chip, then taking the cosine and sine to get I_i and Q_i, we summarize the result in the following table:

i	1	2	3	4	5	6	7	8
α	0	180	90	90	-90	90	180	180
I	1	-1	0	0	0	0	-1	-1
Q	0	0	1	1	-1	1	0	0

The code words for 5.5 Mbps rate CCK modulation are a subset of those for 11 Mbps CCK. In this case, there are four data bits per symbol which determine a total of 16 complex chip sequences. Four 8-element code words (complex chip sequences) are determined using the last two data bits of the symbol, d_2 and d_3. The arguments (angles) of these code words are shown in Table 5.1. Bits d_0 and d_1 are used to rotate the code words relative to the preceding code word as in 11 Mbps modulation and shown in the phase table of Figure 5.13. Code words are modified by $180°$ every other symbol, as in 11 Mbps modulation.

The concept of CCK decoding for receiving high rate data is shown in Figure 5.14. For the 11 Mbps data rate, a correlation bank decides which of the 64 possible codes best fits each received 8-bit symbol. It also finds the rotation angle of the whole code relative to the previous symbol (one of four values). There are a total of 256 (64 $\times$ 4) possibilities and the chosen one is output as serial data. At the 5.5 Mbps rate there are four code words to choose from and after code rotation a total of 16 choices from which to decide on the output data.

Data symbol: $d_0\ d_1\ d_2\ d_3\ d_4\ d_5\ d_6\ d_7$

Phase Table		
d_i	d_{i+1}	φ
0	0	0°
1	0	180°
0	1	90°
1	1	−90°

Phase $(d_0, d_1) = \varphi_1$
Phase $(d_2, d_3) = \varphi_2$
Phase $(d_4, d_5) = \varphi_3$
Phase $(d_6, d_7) = \varphi_4$

$\alpha_1 = \varphi_1 + \varphi_2 + \varphi_3 + \varphi_4$
$\alpha_2 = \varphi_1 + \varphi_3 + \varphi_4$
$\alpha_3 = \varphi_1 + \varphi_2 + \varphi_4$
$\alpha_4 = \varphi_1 + \varphi_4 + 180°$
$\alpha_5 = \varphi_1 + \varphi_2 + \varphi_3$
$\alpha_6 = \varphi_1 + \varphi_3$
$\alpha_7 = \varphi_1 + \varphi_2 + 180°$
$\alpha_8 = \varphi_1$

$I_i = \cos(\alpha_i)$
$Q_i = \sin(\alpha_i)$
$i = 1...8$

Figure 5.13: Derivation of code word

Table 5.1: 5.5 Mbps CCK decoding

d_3, d_2	α_1	α_2	α_3	α_4	α_5	α_6	α_7	α_8
00	90°	0°	90°	180°	90°	0°	−90°	0°
10	−90°	180°	−90°	0°	90°	0°	−90°	0°
01	−90°	0°	−90°	180°	−90°	0°	90°	0°
11	90°	180°	90°	0°	−90°	0°	90°	0°

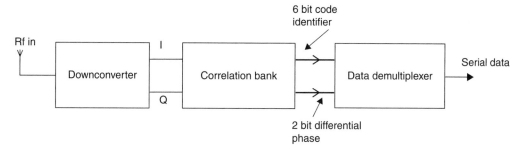

Figure 5.14: CCK decoding

To maintain compatibility with earlier non-high-rate systems, the DSSS frame format shown in Figure 5.11 is retained in 802.11b. The 128-bit preamble and the header are transmitted at 1 Mbps while the payload MPDU can be sent at a high rate of 5.5 or 11 Mbps. The long and slow preamble reduces the throughput and cancels some of the advantage of the high data rates. 802.11b defines an optional short preamble and header which differ from the standard frame by sending a preamble with only 72 bits and transmitting the header at 2 Mbps, for a total overhead of 96 μsec instead of 192 μsec for the long preamble and header. Devices using this option can only communicate with other stations having the same capability.

Use of higher data rates entails some loss of sensitivity and hence range. The minimum specified sensitivity at the 11 Mbps rate is −76 dBm for a frame-error rate of 8% when sending a payload of 1024 bytes, as compared to a sensitivity of −80 dBm for the same frame-error rate and payload length at a data rate of 2 Mbps.

5.1.4.6 802.11a and OFDM

In the search for ways to communicate at even higher data rates than those applied in 802.11b, a completely different modulation scheme, OFDM (orthogonal frequency division multiplexing) was adopted for 802.11a. It is not DSSS yet it has a channel bandwidth similar to the DSSS systems already discussed. The 802.11a supplement is defined for channel frequencies between 5.2 and 5.85 GHz, obviously not compatible with 802.11b signals in the 2.4 GHz band. However, since the channel occupancy characteristics of its modulation are similar to that of DSSS Wi-Fi, the same system was adopted in IEEE 802.11g for enabling the high data rates of 802.11a on the 2.4 GHz band, while allowing downward compatibility with transmissions conforming to 802.11b.

802.11a specifies data rates of 6, 9, 12, 18, 24, 36, 48, and 54 Mbit/s. As transmitted data rates go higher and higher, the problem of multipath interference becomes more severe. Reflections in an indoor environment can result in multipath delays on the order of 100 nsec but may be as long as 250 nsec, and a signal with a bit rate of 10 Mbps (period of 100 nsec) can be completely overlapped by its reflection. When there are several reflections, arriving at the receiver at different times, the signal may be mutilated beyond recognition. The OFDM transmission system goes a long way to solving the problem. It does this by sending the data partitioned into symbols whose length in time is several times the expected reflected path length time differences. The individual data bits in a symbol are all sent in parallel on separate subcarrier frequencies within the transmission channel. Thus, by sending many bits during the same time, each on a different frequency, the individual transmitted bit can be lengthened so that it won't be affected by the multipath phenomenon. Actually, the higher bit rates are accommodated by representing a group of data bits by the phase and amplitude of a particular transmitted carrier. A carrier modulated using quadrature phase shift keying (QPSK) can represent two data bits and 64-QAM (quadrature amplitude modulation) can present six data bits as a single data unit on a subcarrier.

Naturally, transmitting many subcarriers on a channel of given width brings up the problem of interference between those subcarriers. There will be no interference between them if all the subcarriers are orthogonal—that is, if the integral of any two different subcarriers over the symbol period is zero. It is easy to show that this condition exists if the frequency difference between adjacent subcarriers is the inverse of the symbol period.

In OFDM, the orthogonal subcarriers are generated mathematically using the inverse Fourier transform (IFT), or rather its discrete equivalent, the inverse discrete Fourier transform (IDFT). The IDFT may be expressed as:

$$x(n) = \frac{1}{N} \sum_{m=0}^{N-1} X(m)[\cos(2\pi mn/N) + j \cdot \sin(2\pi mn/N)]$$

$x(n)$ are complex sample values in the time domain, $n = 0...N - 1$, and $X(m)$ are the given complex values, representing magnitude and phase, for each frequency in the frequency domain. The IDFT expression indicates that the time domain signal is the sum of N harmonically related sine and cosine waves each of whose magnitude and phase is given by $X(m)$. We can relate the right side of the expression to absolute frequency by multiplying the arguments $2\pi mn/N$ by f_s/f_s to get

$$x(n) = \frac{1}{N} \sum_{n=0}^{N-1} X(m)[\cos(2\pi mf_1 nt_s) + j \cdot \sin(2\pi mf_1 nt_s)] \qquad (5.1)$$

where f_1 is the fundamental subcarrier and the difference between adjacent subcarriers, and t_s is the sample time $1/f_s$. In 802.11a OFDM, the sampling frequency is 20 MHz and $N = 64$, so $f_1 = 312.5$ kHz. Symbol time is $Nt_s = 64/f_s = 3.2$ μsec.

In order to prevent intersymbol interference, 802.11a inserts a guard time of 0.8 μsec in front of each symbol, after the IDFT conversion. During this time, the last 0.8 μsec of the symbol is copied, so the guard time is also called a circular prefix. Thus, the extended symbol time that is transmitted is 3.2 + .8 = 4 μsec. The guard time is deleted after reception and before reconstruction of the transmitted data.

Although the previous equation, where $N = 64$, indicates 64 possible subcarriers, only 48 are used to carry data, and four more for pilot signals to help the receiver phase lock to the transmitted carriers. The remaining carriers that are those at the outside of the occupied bandwidth, and the DC term ($m = 0$ in Eq. (5.1)), are null. It follows that there are 26 ((48 + 4)/2) carriers on each side of the nulled center frequency. Each channel width is 312.5 kHz, so the occupied channels have a total width of 16.5625 (53 × 312.5 kHz) MHz.

For accommodating a wide range of data rates, four modulation schemes are used—BPSK, QPSK, 16-QAM and 64-QAM, requiring 1, 2, 4, and 6 data bits per symbol, respectively. Forward error correction (FEC) coding is employed with OFDM, which entails adding code bits in each symbol. Three coding rates: 1/2, 2/3, and 3/4, indicate the ratio of data bits to the total number of bits per symbol for different degrees of coding performance. FEC permits reconstruction of the correct message in the receiver, even when one or more of the 48 data channels have selective interference that would otherwise result in a lost symbol. Symbol bits are interleaved so that even if adjacent subcarrier bits are demodulated with errors, the error correction procedure will still reproduce the correct symbol. A block diagram of the OFDM transmitter and receiver is shown in Figure 5.15. Blocks FFT and IFFT indicate the fast Fourier transform and its inverse instead of the mathematically equivalent (in terms of results) discrete Fourier transform and inverse discrete Fourier transform (IFDT) that we used above because it is much faster to implement. Table 5.2 lists the modulation type and coding rate used for each data rate, and the total number of bits per OFDM symbol, which includes data bits and code bits.

The available frequency channels in the 5 GHz band in accordance with FCC paragraphs 15.401–15.407 for unlicensed national information infrastructure (U-NII) devices are shown in Table 5.3. Channel allocations are 5 MHz apart and 20 MHz spacing is needed to prevent co-channel interference. Twelve simultaneous networks can coexist without mutual interference. Power limits are also shown in Table 5.4.

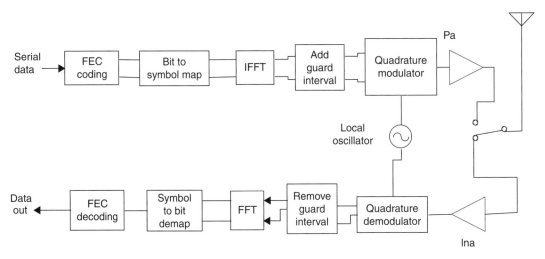

Figure 5.15: OFDM system block diagram

Extension of the data rates of 802.11b to those of 802.11a, but on the 2.4 GHz band is covered in supplement 802.11g. The OFDM physical layer defined for the 5 GHz band is applied essentially unchanged to 2.4 GHz. Equipment complying with 802.11g must also have the lower-rate features and the CCK modulation technique of 802.11b so that it will be downward compatible with existing Wi-Fi systems.

Table 5.2: OFDM characteristics according to data rate

Data Rate Mbps	Modulation	Coding Rate	Coded Bits per Subcarrier	Coded Bits per OFDM Symbol	Data Bits per OFDM Symbol
6	BPSK	1/2	1	48	24
9	BPSK	3/4	1	48	36
12	QPSK	1/2	2	96	48
18	QPSK	3/4	2	96	72
24	16-QAM	1/2	4	192	96
36	16-QAM	3/4	4	192	144
48	64-QAM	2/3	6	288	192
54	64-QAM	3/4	6	288	216

Table 5.3: Channel allocations and maximum power for 802.11a in United States

Band	Operation Channel Numbers	Channel Center Frequencies (MHz)	Maximum Power with up to 6 dBi antenna gain (mW)
U-NII lower band (5.15–5.25 GHz)	36	5180	40
	40	5200	
	44	5220	
	48	5240	
U-NII middle band (5.25–5.35 GHz)	52	5260	200
	56	5280	
	60	5300	
	64	5320	
U-NII upper band (5.725–5.825 GHz)	149	5745	800
	153	5765	
	157	5785	
	161	5805	

Table 5.4: HIPERLAN/2 frequency channels and power levels (Reference 22, ETSI TS 101 475 V1.3.1 (2001–12))

Center Frequency (MHz)	Radiated Power (mean EIRP) (dBm)
Every 20 MHz from 5180 to 5320	23
Every 20 MHz from 5500 to 5680	30
5700	23

5.1.5 HIPERLAN/2

While 802.11b was designed for compliance with regulations in the European Union and most other regions of the world, 802.11a specifically refers to the regulations of the FCC and the Japanese MPT. ETSI (European Telecommunications Standards Institute) developed a high-speed wireless LAN specification, called HIPERLAN/2 (high performance local area network), which meets the European regulations and in many ways goes beyond the capabilities of 802.11a. HIPERLAN/2 defines a physical layer essentially identical to that of 802.11a, using coded OFDM and the same data rates up to 54 Mbps. However, its second layer software level is very different from the 802.11 MAC and the two systems are not compatible. Built-in features of HIPERLAN/2 that distinguish it from IEEE 802.11a are the following:

- Quality of service (QOS). Time division multiple access/time division duplex (TDMA/TDD) protocol permits multimedia communication.

- Dynamic frequency selection (DFS). Network channels are selected and changed automatically to maintain communication reliability in the presence of interference and path disturbances.

- Transmit power control (TPC). Transmission power is automatically regulated to reduce interference to other frequency band users and reduce average power supply consumption.

- High data security. Strong authentication and encryption procedures.

All of the above features of HIPERLAN/2 are being dealt with by IEEE task groups for implementation in 802.11. Specifically, the features of DFS and TPC are necessary for conformance of 802.11a to European Union regulations.

Frequency channels and power levels of HIPERLAN/2 are shown in Table 5.4.

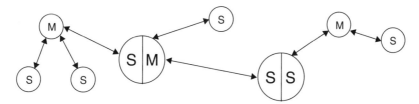

Figure 5.16: Bluetooth Scatternet

5.2 Bluetooth

There are two sources of the Bluetooth specification. One is the Bluetooth Special Interest Group (SIG). The current version at this writing is Version 1.1. It is arranged in two volumes—Core and Profiles. Volume 1, the core, describes the physical, or hardware radio characteristics of Bluetooth, as well as low-level software or firmware which serves as an interface between the radio and higher level specific user software. The profiles in Volume 2 detail protocols and procedures for several widely used applications. The other Bluetooth source specification is IEEE 802.15.1. It is basically a rewriting of the SIG core specification, made to fit the format of IEEE communications specifications in general.

Bluetooth is an example of a wireless personal area network (WPAN), as opposed to a wireless local area network (WLAN). It's based on the creation of ad hoc, or temporary, on-the-fly connections between digital devices associated with an individual person and located in the vicinity of around ten meters from him. Bluetooth devices in a network have the function of a master or a slave, and all communication is between a master and one or more slaves, never directly between slaves. The basic Bluetooth network is called a piconet. It has one master and from one to seven slaves. A scatternet is an interrelated network of piconets where any member of a piconet may also belong to an adjacent piconet. Thus, conceptually, a Bluetooth network is infinitely expandable. Figure 5.16 shows a scatternet made up of three piconets. In it, a slave in one piconet is a master in another. A device may be a master in one piconet only.

The basic RF communication characteristics of Bluetooth are shown in Table 5.5.

A block diagram of a Bluetooth transceiver is shown in Figure 5.17. It's divided into three basic parts: RF, baseband, and application software. A Bluetooth chip set will usually include the RF and baseband parts, with the application software being contained in the system's computer or controller. The user data stream originates and terminates in the application software. The baseband section manipulates the data and forms frames or data bursts for transmission. It also controls the frequency synthesizer according to the Bluetooth frequency-hopping protocol.

Table 5.5: Bluetooth technical parameters

Characteristic	Value	Comment
Frequency Band	2.4 to 2.483 GHz	May differ in some countries
Frequency Hopping Spread Spectrum (FHSS)	79 1-MHz channels from 2402 to 2480 MHz	May differ in some countries
Hop Rate	1600 hops per second	
Channel Bandwidth	1 MHz	20 dB down at edges
Modulation	Gaussian Frequency Shift Keying (GFSK)	
	Filter BT = 0.5	Gaussian Filter bandwidth = 500 kHz
	Nominal modulation index = 0.32	Nominal deviation = 160 kHz
Symbol Rate	1 Mbps	
Transmitter Maximum Power		
Class 1	100 mW	Power control required
Class 2	2.5 mW	Must be at least 0.25 mW
Class 3	1 mW	No minimum specified
Receiver Sensitivity	−70 dBm for BER = 0.1%	

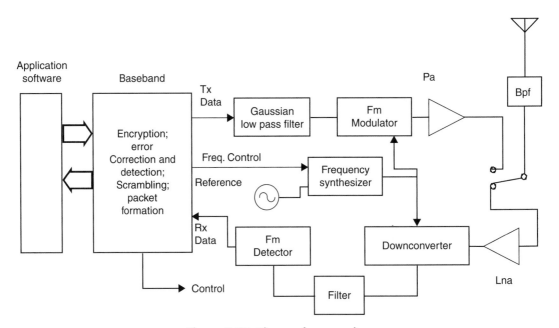

Figure 5.17: Bluetooth transceiver

The blocks in Figure 5.17 are general and various transmitter and receiver configurations are adopted by different manufacturers. The Gaussian low-pass filter block before the modulator, for example, may be implemented digitally as part of a complex signal I/Q modulation unit or it may be a discrete element filter whose output is applied to the frequency control line of a VCO. Similarly, the receiver may be one of several types. If a superheterodyne configuration is chosen, the filter at the output of the downconverter will be a bandpass type. A direct conversion receiver will use low pass filters in complex I and Q outputs of the downconverter. While different manufacturers employ a variety of methods to implement the Bluetooth radio, all must comply with the same strictly defined Bluetooth specification, and therefore the actual configuration used in a particular chipset should be of little concern to the end user.

The Bluetooth protocol has a fixed-time slot of 625 microseconds, which is the inverse of the hop rate given in Table 5.5. A transmission burst may occur within a duration of one, three, or five consecutive slots on one hop channel. As mentioned, transmissions are always between the piconet master and a slave, or several slaves in the case of a broadcast, or point-to-multipoint transmission. All slaves in the piconet have an internal timer synchronized to the master device timer, and the state of this timer determines the transmission hop frequency of the master and that of the response of a designated slave. Figure 5.18 shows a sequence of transmissions between a master and two slaves. Slots are numbered according to the state, or phase, of the master clock, which is copied to each slave when it joins the piconet. Note that master transmissions take place during even numbered clock phases and slave transmissions during odd numbered phases. Transmission frequency depends on the clock phase, and if a device makes a three or five slot transmission (slave two in the diagram), the intermediate frequencies that would have been used if only single slots were transmitted are omitted (f_4 and f_5 in this case). Note that transmissions do not take up a whole slot. Typically, a single-slot transmission burst lasts 366 microseconds, leaving 259 microseconds for changing the frequency of the synthesizer, phase locked loop settling time, and for switching the transceiver between transmit and receive modes.

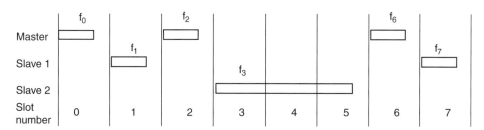

Figure 5.18: Bluetooth timing

There are two different types of wireless links associated with a Bluetooth connection. An asynchronous connectionless link (ACL) is used for packet data transfer while a synchronous connection oriented link (SCO) is primarily for voice. There are two major differences between the two link types. When an SCO link is established between a master and a slave, transmissions take place on dedicated slots with a constant interval between them. Also, unlike an ACL link, transmitted frames are not repeated in the case of an error in reception. Both of these conditions are necessary because voice is a continuous real-time process whose data rate cannot be randomly varied without affecting intelligibility. On the other hand, packet data transmission can use a handshaking protocol to regulate data accumulation and the instantaneous rate is not usually critical. Thus, for ACL links the master has considerable leeway in proportioning data transfer with the slaves in its network. An ARQ (automatic repeat request) protocol is always used, in addition to optional error correction, to ensure the highest reliability of the data transfer.

Bluetooth was conceived for employment in mobile and portable devices, which are more likely than not to be powered by batteries, so power consumption is an important issue. In addition to achieving low-power consumption due to relatively low transmitting power levels, Bluetooth incorporates power saving features in its communication protocol. Low average power is achieved by reducing the transmission duty cycle, and putting the device in a low-power standby mode for as long a period as possible relative to transmit and receive times while still maintaining the minimum data flow requirements.

5.2.1 Power Consumption Modes

Bluetooth has three modes for achieving different degrees of power consumption during operation: sniff, hold, and park. Even in the normal active mode, some power saving can be achieved, as described below.

5.2.1.1 Active Mode

During normal operation, a slave can transmit in a particular time slot only if it is specifically addressed by the master in the proceeding slot. As soon as it sees that its address is not contained in the header of the master's message, it can "go to sleep," or enter a low-power state until it's time for the next master transmission. The master also indicates the length of its transmission (one, three, or five slots) in its message header, so the slave can extend its sleep time during a multiple slot interval.

5.2.1.2 Sniff Mode

In this mode, sleep time is increased because the slave knows in advance the time interval between slots during which the master may address the slave. If it's not addressed during the agreed slot, it returns to its low-power state for the same period and then wakes up and listens again. When it is addressed, the slave continues listening during subsequent master transmission slots as long as it is addressed, or for an agreed time-out period.

5.2.1.3 Hold Mode

The master can put a slave in the hold mode when data transfer between them is being suspended for a given period of time. The slave is then free to enter a low-power state, or do something else, like participate in another piconet. It still maintains its membership in the original piconet, however. At the end of the agreed time interval, the slave resynchronizes with the traffic on the piconet and waits for instructions from the master.

5.2.1.4 Park Mode

Park has the greatest potential for power conservation, but as opposed to hold and sniff, it is not a directly addressable member of the piconet. While it is outside of direct calling, a slave in park mode can continue to be synchronized with the piconet and can rejoin it later, either on its own initiative or that of the master, in a manner that is faster than if it had to join the piconet from scratch. In addition to saving power, park mode can also be considered a way to virtually increase the network's capacity from eight devices to 255, or even more. When entering park mode, a slave gives up its active piconet address and receives an 8-bit parked member address. It goes into low-power mode but wakes up from time to time to listen to the traffic and maintain synchronization. The master sends beacon transmissions periodically to keep the network active. Broadcast transmissions to all parked devices can be used to invite any of them to rejoin the network. Parked units themselves can request re-association with the active network by way of messages sent during an access window that occurs a set time after what is called a "beacon instant." A polling technique is used to prevent collisions.

5.2.2 Packet Format

In addition to the data that originates in the high-level application software, Bluetooth packets contain fields of bits that are created in the baseband hardware or firmware for the purpose of acquisition, addressing, and flow control. Packet bits are also subjected to data whitening (randomization), error-correction coding, and encryption as defined for each particular data type. Figure 5.19 shows the standard packet format.

Access code	Header	Payload
72 bits	54 bits	0 to 2745 bits

Figure 5.19: Bluetooth packet

The access code is used for synchronization, d-c level compensation, and identification. Each Bluetooth device has a unique address, and it is the address of the device acting as master that is used to identify transmitted packets as belonging to a specific piconet. A 64-bit synchronization word sandwiched between a four-bit header and four-bit trailer, which provide d-c compensation, is based on the master's address. This word has excellent correlation properties so when it is received by any of the piconet members it provides synchronization and positive identification that the packet of which it is a part belongs to their network. All message packets sent by members of the piconet use the same access code.

The header contains six fields with link control information. First, it has a three-bit active member address which identifies to which of the up to seven slaves a master's message is destined. An all zero address signifies a broadcast message to all slaves in the piconet. The next field has four bits that define the type of packet being sent. It specifies, for example, whether one, three, or five slots are occupied, and the level of error correction applied. The remaining fields involve flow control (handshaking), error detection and sequencing. Since the header has prime importance in the packet, it is endowed with forward-error correction having a redundancy of times three.

Following the header in the packet is the payload, which contains the actual application or control data being transferred between Bluetooth devices. The contents of the payload field depend on whether the link is an ACL or SCO. The payload of ACL links has a payload header field that specifies the number of data bytes and also has a handshaking bit for data-buffering control. A CRC (cyclic redundancy check) field is included for data integrity. As stated above, SCO links don't retransmit packets so they don't include a CRC. They don't need a header either because the SCO payload has a constant length.

The previous packet description covers packets used to transfer user data, but other types of packets exist. For example, the minimum length packet contains only the access code, without the four-bit trailer, for a total of 68 bits. It's used in the inquiry and paging procedures for initial frequency-hopping synchronization. There are also NULL and POLL packets that have an access code and header, but no payload. They're sent when slaves are being polled to maintain synchronization or confirm packet reception (in the case of NULL) in the piconet but there is no data to be transferred.

5.2.3 Error Correction and Encryption

The use of forward error correction (FEC) improves throughput on noisy channels because it reduces the number of bad packets that have to be retransmitted. In the case of SCO links that don't use retransmission, FEC can improve voice quality. However, error correction involves bit redundancy so using it on relatively noiseless links will decrease throughput. Therefore, the application decides whether to use FEC or not.

As already mentioned, there are various types of packets, and the packet type defines whether or not FEC is used. The most redundant FEC method is always used in the packet header, and for the payload in one type of SCO packet. It simply repeats each bit three times, allowing the receiver to decide on the basis of majority rule what data bit to assign to each group of incoming bits.

The other FEC method, applied in certain type ACL and SCO packets, uses what's called a (15,10) shortened Hamming code. For every ten data bits, five parity bits are generated. Since out of every 15 transmitted bits only ten are retrieved, the data rate is only two-thirds what it would be without coding. This code can correct all single errors and detect all double errors in each 15-bit code word.

Wireless communication is susceptible to eavesdropping so Bluetooth incorporates optional security measures for authentication and encryption. Authentication is a procedure for verifying that received messages are actually from the party we expect them to be and not from an outsider who is inserting false messages. Encryption prevents an eavesdropper from understanding intercepted communications, since only the intended recipient can decipher them. Both authentication and implementation routines are implemented in the same way. They involve the creation of secret keys that are generated from the unique Bluetooth device address, a PIN (personal identification number) code, and a random number derived from a random or pseudo-random process in the Bluetooth unit. Random numbers and keys are changed frequently. The length of a key is a measure of the difficulty of cracking a code. Authentication in Bluetooth uses a 128-bit key, but the key size for encryption is variable and may range from 8 to 128 bits.

5.2.4 Inquiry and Paging

A distinguishing feature of Bluetooth is its ad hoc protocol and connections are often required between devices that have no previous knowledge of their nature or address. Also, Bluetooth networks are highly volatile, in comparison to WLAN for example, and connections are made and dissolved with relative frequency. To make a new connection, the initiator—the master— must know the address of the new slave, and the slave has to synchronize its clock to the master's in order to align transmit and receive channel hop-timing and frequencies. The inquiry and paging procedures are used to create the connections between devices in the piconet.

By use of the inquiry procedure, a connection initiator creates a list of Bluetooth devices within range. Later, desired units can be summoned into the piconet of which the initiator is master by means of the paging routine.

As mentioned previously, the access code contains a synchronization word based on the address of the master. During inquiry, the access code is a general inquiry access code (GIAC) formed from a reserved address for this purpose. Dedicated inquiry access codes (DIAC) can also be used when the initiator is looking only for certain types of devices. Now a potential slave can lock on to the master, provided it is receiving during the master's transmission time and on the transmission frequency. To facilitate this match-up, the inquiry procedure uses a special frequency hop routine and timing. Only 32 frequency channels are used and the initiator transmits two burst hops per standard time slot instead of one. On the slot following the transmission inquiry bursts, the initiator listens for a response from a potential slave on two consecutive receive channels whose frequencies are dependent on the previously transmitted frequencies.

When a device is making itself available for an inquiring master, it remains tuned to a single frequency for a period of 1.28 seconds and at a defined interval and duration scans the channel for a transmission. At the end of the 1.28-second period, it changes to another channel frequency. Since the master is sending bursts over the whole inquiry frequency range at a fast rate—two bursts per 1250 microsecond interval—there's a high probability the scanning device will catch at least one of the transmissions while it remains on a single frequency. If that channel happens to be blocked by interference, then the slave will receive a transmission after one of its subsequent frequency changes. When the slave does hear a signal, it responds during the next slot with a special packet called FHS (frequency hop synchronization) in which is contained the slave's Bluetooth address and state of its internal clock register. The master does not respond but notes the slave's particulars and continues inquiries until it has listed the available devices in its range. The protocol has provisions for avoiding collisions from more than one scanning device that may have detected a master on the same frequency and at the same time.

The master makes the actual connection with a new device appearing in its inquiry list using the page routine. The paging procedure is quite similar to that of the inquiry. However, now the master knows the paged device's address and can use it to form the synchronization word in its access code. The designated slave does its page scan while expecting the access code derived from its own address. The hopping sequence is different during paging than during inquiry, but the master's transmission bursts and the slave's scanning routine are very similar.

A diagram of the page state transmissions is given in Figure 5.20. When the slave detects a transmission from the master (Step 1), it responds with a burst of access code based on its own Bluetooth address. The master then transmits the FHS, giving the slave the access code information (based on the master's address), timing and piconet active member address (between one and seven) needed to participate in the network. The slave acknowledges FHS receipt in Step 4. Steps 5 and 6 show the beginning of the network transmissions which use the normal 79 channel hopping-sequence based on the master's address and timing.

5.3 Zigbee

Zigbee is the name of a standards-based wireless network technology that addresses remote monitoring and control applications. Its promotion and development is being handled on two levels. A technical specification for the physical and data link layers, IEEE 802.15.4, was drawn up by a working group of the IEEE as a low data rate WPAN (wireless personal area network). An association of committed companies, the Zigbee Alliance, is defining the network, security, and application layers above the 802.15.4 physical and medium access control layers, and will deal with interoperability certification and testing.

The distinguishing features of Zigbee to which the IEEE standard addresses itself are

- Low data rates—throughput between 10 and 115.2 Kbps

- Low power consumption—several months up to two years on standard primary batteries

- Network topology appropriate for multisensor monitoring and control applications

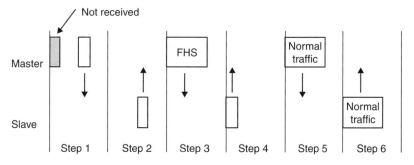

Figure 5.20: Paging transmissions

- Low complexity for low cost and ease of use

- Very high reliability and security

These will lend themselves to wide-scale use embedded in consumer electronics, home and building automation and security systems, industrial controls, PC peripherals, medical and industrial sensor applications, toys and games and similar applications. It's natural to compare Zigbee with the other WPAN standard, Bluetooth, and there will be some overlap in implementations. However, the two systems are quite different, as is evident from the comparison in Table 5.6.

5.3.1 Architecture

The basic architecture of Zigbee is similar to that of other IEEE standards, Wi-Fi and Bluetooth for example, a simplified representation of which is shown in Figure 5.21. On the bottom are the physical layers, showing two alternative options for the RF transceiver functions of the specification. Both of these options are never expected to exist in a single device, and indeed their transmission characteristics—frequencies, data rates, modulation system—are quite different. However, the embedded firmware and software layers above them will be essentially the same no matter what physical layer is applied. Just above the

Table 5.6: Comparison of Zigbee and Bluetooth

	Bluetooth	Zigbee
Transmission Scheme	FHSS (Frequency Hopping Spread Spectrum)	DSSS (Direct Sequence Spread Spectrum)
Modulation	GFSK (Gaussian Frequency Shift Keying)	QPSK (Quadrature Phase Shift Keying) or BPSK (Binary Phase Shift Keying)
Frequency Band	2.4 GHz	2.4 GHz, 915 MHz, 868 MHz
Raw Data Bit Rate	1 MBPS	250 KBPS, 40 KBPS or 20 KBPS (depends on frequency band)
Power Output	Maximum 100 mW, 2.5 mW, or 1 mW, depending on class	Minimum capability 0.5 mW; maximum as allowed by local regulations
Minimum Sensitivity	−70 dBm for 0.1% BER	−85 dBm (2.4 GHz) or −92 dBm (915/868 MHz) for packet error rate <1%
Network topology	Master-Slave 8 active nodes	Star or Peer-Peer 255 active nodes

physical layers is the data link layer, consisting of two sublayers: medium access control, or MAC, and the logical link control, LLC. The MAC is responsible for management of the physical layer and among its functions are channel access, keeping track of slot times, and message delivery acknowledgement. The LLC is the interface between the MAC and physical layer and the upper-application software.

Application software is not a part of the IEEE 802.15.4 specification and it is expected that the Zigbee Alliance will prepare profiles, or programming guidelines and requirements for various functional classes in order to assure product interoperability and vendor independence. These profiles will define network formation, security, and application requirements while keeping in mind the basic Zigbee features of low power and high reliability.

5.3.2 Communication Characteristics

In order to achieve high flexibility of adaptation to the range of applications envisioned for Zigbee, operation is being specified for three unlicensed bands—2.4 GHz, 915 MHz and 868 MHz; the latter two being included in the same physical layer. Those two bands are

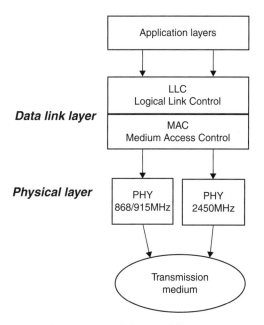

Figure 5.21: Zigbee architecture

generally mutually exclusive, their use being determined by geographic location and regional regulations. The following 27 transmitting channels are defined:

Channel Number	Center Frequency Range	Channel Width
0	868.3 MHz	600 kHz
1 to 10	906 to 924 MHz	2 MHz
11 to 27	2405 to 2480	5 MHz

Data rates and modulation types for each of the bands are shown in Table 5.7.

In both physical layers, the modulation is DSSS (direct sequence spread spectrum). The spreading parameters are defined to meet communication authority regulations in the various regions as well as desired data rates. For example, the chip rate of 600 Kbps on the 902–928 band allows the transmission to meet the FCC paragraph 15.247 requirement of minimum 500 kHz bandwidth at 6 dB down for digital modulation. However the chip rate, and with it the data rate, has to be reduced on Channel 0 in order to meet the confines of the 868 to 868.6 MHz channel allowed under ERC recommendation 70-03 and ETSI specification EN 300-220. On the 2400 to 2483.5 MHz band, the bit rate of 250 Kbps allows a throughput, after considering the overheads involved in packet transmissions, to attain 115.2 Kbps, a rate used for some PC peripherals for example.

The spreading modulation used on the 2450 MHz physical layer has similarity in principle to that used on IEEE 802.11b (high-rate Wi-Fi) to increase the data bit rate without raising the chip rate, thereby achieving a desired carrier bandwidth. Sixteen different, almost orthogonal 32-bit long spreading sequences are available for transmission at 2 Mchips/second. Each consecutive sequence of four data bits determines which of the sixteen spreading sequences is sent. On reception, the receiver can identify the spreading sequence and thus decode the data bits. The modulation used, O-QPSK (offset quadrature phase shift keying) with half-sine wave

Table 5.7: Data rates and modulation

PHY (MHz)	Frequency Band (MHz)	Spreading Parameters		Data Parameters		
		Chip Rate (kcps)	Modulation	Bit Rate (kbps)	Symbol Rate (ksps)	Symbols
868/915	868–868.6	300	BPSK	20	20	Binary
	902–928	600	BPSK	40	40	Binary
2450	2400–2483.5	2000	Offset-QPSK	250	62.5	16-ary Orthogonal

pulse shaping is essentially equivalent to a form of frequency shift keying, MSK (minimum shift keying). It is fairly easy to generate and has a relatively narrow bandwidth for the given chip rate. This latter feature allows a large number of nonoverlapping channels that can be used, with proper upper layer software, on the crowded 2.4 GHz band to avoid interference.

Other physical layer characteristics of Zigbee are output power and receiver sensitivity. The devices must be capable of radiating at least -3 dBm although output may be reduced to the minimum necessary in order to limit interference to other users. Maximum power is determined by the regulatory authorities. While much higher powers are allowed, it may not be practical to transmit over, say 10 dBm, because of absolute limits on spurious radiation and the general objective of low-cost and low-power consumption. Minimum receiver sensitivity for the 868/915 MHz physical layer is specified as -92 dBm and -85 dBm on 2.4 GHz. These limits are for a packet error rate of one percent.

5.3.3 Device Types and Topologies

Two device types, of different complexities, are defined. A full function device (FFD) will be able to implement the full protocol set and can act as a network coordinator. Devices capable of minimal protocol implementation are reduced function devices (RFD). Due to the distinction between device types, networks in which most members require only minimum functionality, such as switches and sensors, can be made significantly less costly and have lower power consumption than if all devices were constrained to have maximum capability.

Flexibility in network configuration is achieved through two topologies—star and peer-to-peer that are depicted in Figure 5.22. A network may have as many as 255 members, one of which is a PAN (personal area network) coordinator. The function of the PAN coordinator, in addition to any specific application it may have, is to initiate, terminate, or route communication around the network. It also provides synchronization services. In a star network, each device communicates directly with the coordinator. The coordinator must be a FFD, and the others can be FFDs or RFDs. Relatively simple applications, like PC peripherals and toys, would typically use the star topology.

In the peer-to-peer topology, any device can communicate with any other device as long as it is in range. RFDs cannot participate, since an RFD can only communicate with a FFD. More complicated structures can be set up as a combination of peer-to-peer groups and star configurations. There is still just one PAN coordinator in the whole network. One example of such a structure is a cluster-tree network shown in Figure 5.23. In this arrangement devices on the network extremities may well be out of radio range of each other, but they can still communicate by relaying messages through the individual clusters.

5.3.4 Frame Structure, Collision Avoidance, and Reliability

Zigbee frame construction and channel access are similar to those of WLAN 802.11 (Wi-Fi) but are less complex. The transmitted packet has the basic construction shown in Figure 5.24. The purpose of the preamble is to permit acquisition of chip and symbol timing. The PHY header, which is signaled by a delimiter byte, notifies the baseband software in the receiver

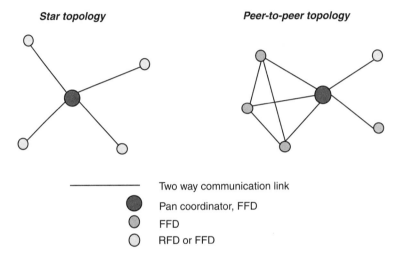

Figure 5.22: Network topologies

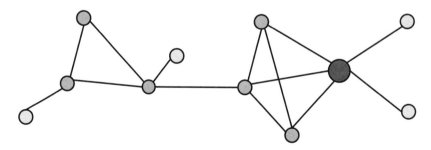

Figure 5.23: Cluster-tree topology

32 bits	8 bits	8 bits	127 bytes maximum
Preamble	Start-of-frame deliminator	Frame length	PSDU — PHY Service Data Unit

Figure 5.24: Transmission packet

of the length of the subsequent data. The PSDU (PHY service data unit) is the message that has been passed down through the higher protocol layers. As shown, it can have a maximum of 127 bytes although monitoring and control applications will typically be much shorter. Included in the PSDU are information on the format of the message frame, a sequence number, address information, the data payload itself, and at the end, two bytes that serve as a frame check sequence. Reliability is assured since the receiver performs an independent calculation of this frame check sequence and compares it with the number received. If any bits have been changed by interference or noise, the numbers will not match. Only if a match occurs, the receiving side returns an acknowledgement to the originator of the message. Lacking an acknowledgement, the transmission will be repeated until it is successfully received.

In order to avoid two or more stations trying to transmit at the same time, a carrier sense multiple access with collision avoidance (CSMA-CA) routine is employed, similar to that used in Wi-Fi, IEEE 802.11. The Zigbee receiver monitors the channel and only if it is idle it may initiate a transmission. If the channel is occupied, the terminal must wait a random back off period before it can again attempt access.

Acknowledgement messages are sent without using the collision avoidance mechanism.

5.3.5 Zigbee Applications

While the promoters of Zigbee aim to cover a very large market for those applications that require relatively low data rates, there will remain applications for which the compromises inherent in a general specification are not acceptable, and producers will continue to develop devices with proprietary specifications and characteristics. However, the open specification and a recognized certification of conformity are an advantage in many situations. For example, a home burglar alarm system would accept wireless sensors produced by different manufacturers, which will facilitate future expansion or allow installers to add sensors of types not available from the original system manufacturer. Use of devices approved according to a recognized standard gives the consumer some security against obsolescence.

Although Zigbee claims to be appropriate for most control applications, it will not fit all of them, and will not necessarily take advantage of all the possibilities of the unlicensed device regulations. Its declared maximum range of some 50 to 75 meters will fall short of the requirements of many systems. Given the meager maximum power allowed, greater range means reduced bandwidth and reduced data rate. In fact, a great many of the applications envisaged by Zigbee can get by very well with data rates of hundreds or a few thousand bits per second, and by matching receiver sensitivity to these rates, ranges of hundreds of meters can be achieved.

One partial answer to the range question is the deployment of the Zigbee network in a cluster-tree configuration, as previously described. Adjacent nodes serve as repeaters so that large areas can be covered, as long as the greatest distance between any two directly communicating nodes does not exceed Zigbee's basic range capability. For example, in a multi-floor building, sensors on the top floor can send alarms to the control box in the basement by passing messages through sensors located on every floor and operating as relay stations.

No doubt that there will be competition between Bluetooth and Zigbee for use in certain applications, but the overall deployment and the reliability of wireless control systems will increase. The proportion of wireless security and automation systems will increase because the new standard will provide a significant boost in reliability, security, and convenience, as compared to most present solutions.

5.4 Conflict and Compatibility

With the steep rise of Bluetooth product sales and the already large and growing use of wireless local area networks, there is considerable concern about mutual interference between Bluetooth-enabled and Wi-Fi devices. Both occupy the 2.4 to 2.4835 GHz unlicensed band and use wideband spread-spectrum modulating techniques. They will most likely be operating concurrently in the same environments, particularly office/commercial but also in the home.

Interference can occur when a terminal of one network transmits on or near the receiving frequency of a terminal in another collocated network with enough power to cause an error in the data of the desired received signal. Although they operate on the same frequency band, the nature of Bluetooth and Wi-Fi signals are very different. Bluetooth has a narrowband transmission of 1 MHz bandwidth which hops around pseudo-randomly over an 80 MHz band while Wi-Fi (using DSSS) has a broad, approximately 20 MHz, bandwidth that is constant in some region of the band. The interference phenomenon is apparent in Figure 5.25. Whenever there is a frequency and time coincidence of the transmission of one system and reception of the other, it's possible for an error to occur. Whether it does or not depends on the relative signal strengths of the desired and undesired signals. These in turn depend on the radiated power outputs of the transmitters and the distance between them and the receiver. When two

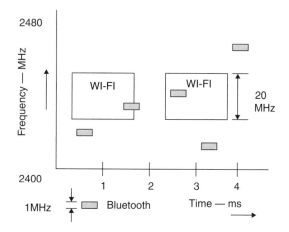

Figure 5.25: Wi-Fi and Bluetooth spectrum occupation

terminals are very close (on the order of centimeters), interference may occur even when the transmitting frequency is outside the bandwidth of the affected receiver.

Bluetooth and Wi-Fi systems are not synchronous and interference between them has to be quantified statistically. We talk about the probability of a packet error of one system caused by the other system. The consequence of a packet error is that the packet will have to be retransmitted once or more until it is correctly received, which causes a delay in message throughput. Voice transmissions generally don't allow packet retransmission because throughput cannot be delayed, so interference results in a decrease in message quality.

Following are parameters that affect interference between Bluetooth and Wi-Fi:

- *Frequency and time overlap.* A collision occurs when the interferer transmits at the same time as the desired transmitter and is strong enough to cause a bit or symbol error in the received packet.

- *Packet length.* The longer the packet length of the Wi-Fi system, relative to a constant packet length and hop rate of Bluetooth, the longer the victim may be exposed to interference from one or more collisions and the greater the probability of a packet error.

- *Bit rate.* Generally, the higher the bit rate, the lower the receiver sensitivity and therefore the more susceptible the victim will be to packet error for given desired and interfering signal strengths. On the other hand, higher bit rates usually result in reduced packet length, with the opposite effect.

- *Use factor.* Obviously, the more often the interferer transmits, the higher the probability of packet error. When both communicating terminals of the interferer are in the interfering vicinity of the victim the use factor is higher than if the terminals are further apart and one of them does not have adequate strength to interfere with the victim.

- *Relative distances and powers.* The received power depends on the power of the transmitter and its distance. Generally, Wi-Fi systems use more power than Bluetooth, typically 20 mW compared to 1 mW. Bluetooth Class 1 systems may transmit up to 100 mW, but their output is controlled to have only enough power to give a required signal level at the receiving terminal.

- Signal-to-interference ratio of the victim receiver, SIR, for a specified symbol or frame error ratio.

- Type of modulation, and whether error-correction coding is used.

A general configuration for the location of Wi-Fi and interfering Bluetooth terminals is given in Figure 5.26. In this discussion, only transmissions from the access point to the mobile terminal are considered. We can get an idea of the vicinity around the Wi-Fi mobile terminal in which operating Bluetooth terminals will affect transmissions from the access point to the mobile terminal by examining the following parameters:

CI_{cc}, CI_{ac}—Ratio of signal carrier power to co-channel or adjacent channel interfering power for a given bit or packet error rate (probability).

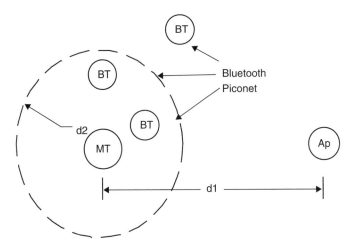

Figure 5.26: Importance of relative terminal location

P_{WF}, P_{BT}—Wi-Fi and Bluetooth radiated power outputs.

$PL = Kd^r$—Path loss which is a function of distance d between transmitting and receiving terminals, and the propagation exponent r. K is a constant.

$d1$—Distance between Wi-Fi mobile terminal and access point.

$d2$—Radius of area around mobile terminal within which an interfering Bluetooth transmitter signal will increase the Wi-Fi bit error rate above a certain threshold.

PR_{WF}, PR_{BT}—Received powers from the access point and from the Bluetooth interfering transmitters.

$d2$ as a function of $d1$ is found as follows, using power in dBm:

1. $PR_{WF} = P_{WF} - 10\log(Kd1^r); PR_{BT} = P_{BT} - 10\log(Kd2^r)$

2. $CI_{cc} = PR_{WF} - PR_{BT} = P_{WF} - 10\log(Kd1^r) - P_{BT} + 10\log(Kd2^r)$

3. $(CI_{cc} - P_{WF} + P_{BT})/10r = \log(d2/d1)$

4. $d2 = d1 \cdot 10^{(CI_{cc} - P_{WF} + P_{BT})/(10 \cdot r)}$ $\hfill(5.2)$

As an example, the interfering area radius d2 is now calculated from equation (5.2) using the following system parameters:

$$CI_{cc} = 10\,\text{dB}, P_{WF} = 13\,\text{dBm}, P_{BT} = 0\,\text{dBm}, r = 2(\text{free space})$$
$$\text{solving equation } (11 - 1): d2 = d1 \times .71$$

In this case, if a Wi-Fi terminal is located 15 meters from an access point, for example, all active Bluetooth devices within a distance of 10.6 meters from it have the potential of interfering. Only co-channel interference is considered. Adjacent channel interference, if significant, would increase packet error probability because many more Bluetooth hop channels would cause symbol errors. However, the adjacent channel CI_{ac} is on the order of 45 dB lower than CI_{cc} and would be noticed only when Bluetooth is several centimeters away from the Wi-Fi terminal.

The effect of an environment where path loss is greater than in free space can be seen by using an exponent r=3. For the same Wi-Fi range of 15 meters, the radius of Bluetooth interference becomes 11.9 meters.

While equation (5.2) does give a useful insight into the range where Bluetooth devices are liable to deteriorate Wi-Fi performance, its development did involve simplifications. It considered that the signal-to-interference ratio that causes the error probability to exceed a threshold is constant for all wanted signal levels, which isn't necessarily so. It also implies a step relationship between signal-to-interference ratio and performance degradation, whereas the effect of changing interference level is continuous. The propagation law used in the development is also an approximation.

5.4.1 Methods for Improving Bluetooth and Wi-Fi Coexistence

By dynamically modifying one or more system operating parameters according to detected interference levels, coexistence between Bluetooth and Wi-Fi can be improved. Some of these methods are discussed below.

5.4.1.1 Power Control

Limiting transmitter power to the maximum required for a satisfactory level of performance will reduce interference to collocated networks. Power control is mandatory for Class 1 Bluetooth systems, where maximum power is 100 mW. The effect of the power on the interference radius is evident in equation (5.2). For example, in a Bluetooth piconet established between devices located over a spread of distances from the master, the master will use only the power level needed to communicate with each of the slaves in the network. Lack of power control would mean that all devices would communicate at maximum power and the collocated Wi-Fi system would be exposed to a high rate of interfering Bluetooth packets.

5.4.1.2 Adaptive Frequency Hopping

Wi-Fi and Bluetooth share approximately 25 percent of the total Bluetooth hop-span of 80 MHz. Probably the most effective way to avoid interference between the two systems is to restrict Bluetooth hopping to the frequency range not used by Wi-Fi. When there is no coordination or cooperation between collocated networks, the Bluetooth piconet master would have to sense the presence of Wi-Fi transmissions and modify the frequency-hopping scheme of the network accordingly. A serious obstacle to this method was lifted by a change to the FCC regulations governing spread-spectrum transmissions in the 2.4 GHz band. Previously, frequency hopping devices were committed to hopping over at least 75 pseudo-randomly selected hop channels. In August 2002, paragraph 15.247, according to which Bluetooth and Wi-Fi devices are regulated, was changed to allow a minimum of 15 nonoverlapping channels in the 2400 to 2483 MHz band. In addition, the regulation allows employing intelligent hopping techniques, when less than 75 hopping frequencies are used, to avoid interference with other transmissions, and also suppression of transmission on an occupied channel provided that there are a minimum of 15 hops. The Bluetooth specification is due to be modified to take advantage of the adaptive frequency hopping method of avoiding interference.

There are situations where adaptive frequency hopping may not be effective or may have a negative effect. When two or more adjacent Wi-Fi networks are operating concurrently, they will utilize different 22 MHz sections of the 2.4 GHz band—three nonoverlapping Wi-Fi channels are possible. In this case, Bluetooth may not be able to avoid collisions while using a minimum of 15 hop frequencies. In addition, if there are several Bluetooth piconets in the same area, collisions among themselves will be much more frequent than when the full 79 channel hopping sequences are used.

5.4.1.3 Packet Fragmentation

The two interference-avoiding methods described above are applicable primarily for action by the Bluetooth network. One method that the Wi-Fi network can employ to improve throughput is packet fragmentation. By fragmenting data packets and sending more, but shorter transmission frames, each transmission will have a lower probability of collision with a Bluetooth packet. Although reducing frame size increases the percentage of overhead bits in the transmission, when interference is heavy the overall effect may be higher throughput than if fragmentation was not used. Increasing bit rate for a constant packet length will also result in a shorter transmitted frame and less exposure to interference.

The methods mentioned above for reducing interference presume no coordination between the two different types of collocated wireless networks. However, devices are now being produced, in laptop and notebook computers for example, that include both Wi-Fi and Bluetooth, sometimes even in the same chipset. In this case collaboration is possible in the device software to prevent inter-network collisions.

5.5 Ultra-wideband Technology

Ultra-wideband (UWB) technology is based on transmission of very narrow electromagnetic pulses at a low repetition rate. The result is a radio spectrum that is spread over a very wide bandwidth—much wider than the bandwidth used in the spread-spectrum systems previously discussed. Ultra-wideband transmissions are virtually undetectable by ordinary radio receivers and therefore can exist concurrently with existing wireless communications without demanding additional spectrum or exclusive frequency bands.

These are some of the advantages cited for ultra-wideband technology:

- Very low spectral density—Very low probability of interference with other radio signals over its wide bandwidth,

- High immunity to interference from other radio systems,

- Low probability of interception/detection by other than the desired communication link terminals,

- High multipath immunity,

- Many high data rate ultra-wideband channels can operate concurrently,

- Fine range-resolution capability,

- Relatively simple, low-cost construction, based on nearly all digital architectures.

Transmission and reception methods are unique, and are described briefly below.

Differing from conventional radio communication systems, which use up conversion and down conversion to pass information signals between baseband and bandpass frequency channels where wireless propagation occurs, UWB signal generation and detection use baseband techniques. An example of a UWB "carrier" is a Gaussian monopulse, shown in Figure 5.27 (see reference). Its power spectrum is shown in Figure 5.28. If the time scale in Figure 5.27 is in nanoseconds, then the width of the pulse is 0.5 nanoseconds and the 3 dB bandwidth of the power spectrum is approximately 3.2 GHz with maximum power density at 2 GHz.

In order to pass information over a UWB communication link, trains of pulses must be transmitted with some characteristic of a pulse or group of pulses varied in order to distinguish between "0" and "1." The time between consecutive pulses should be determined in a pseudo-random manner in order to smooth the energy spikes in the frequency spectrum. Reception of the transmitted pulse train is done by correlating the received signal with a similar sequence of pulses generated in the receiver. A large number of communication links

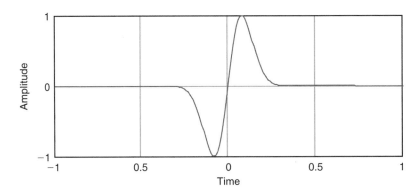

Figure 5.27: UWB monopulse

can be maintained simultaneously and independently by using different pseudo-random sequences for each link.

A pulse similar to that of Figure 5.27 can be generated by applying an impulse, or perhaps more conveniently a step-voltage or current, to a linear band limited network. Figure 5.29 is a simulation of a sequence of UWB pulses created by stimulating a bandpass filter with a pseudo-randomly spaced sequence of impulses. The figure also shows the power spectrum of that sequence. The network that creates the individual UWB pulses includes the transmitter antenna, the propagation channel, and the receiving antenna, whose characteristics, in terms of impulse response or amplitude and phase vs. frequency must be known and accounted for in designing the system.

There are several ways of representing a UWB pulse as "1" or "0." One method is to advance or retard the transmitted pulse with respect to the expected time of arrival of the pulse in the receiver according to the agreed pseudo-random time sequence. Another method is to send the pulse with or without inversion. In both cases the correlation of the received pulse with a "template" pulse generated in the receiver will result in a different polarity, depending on whether a "1" or a "0" was transmitted.

Detection of UWB bits is illustrated in Figure 5.30. A "1" monopulse is represented by a negative line followed by a positive line, and a "0" monopulse by the inverse—a positive line and a negative line. The synchronized sequence generated in the receiver is drawn on the

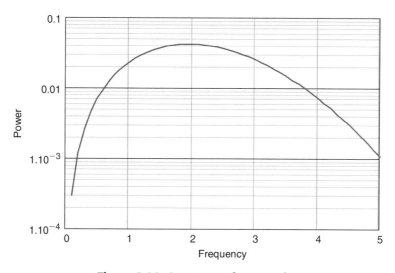

Figure 5.28: Spectrum of monopulse

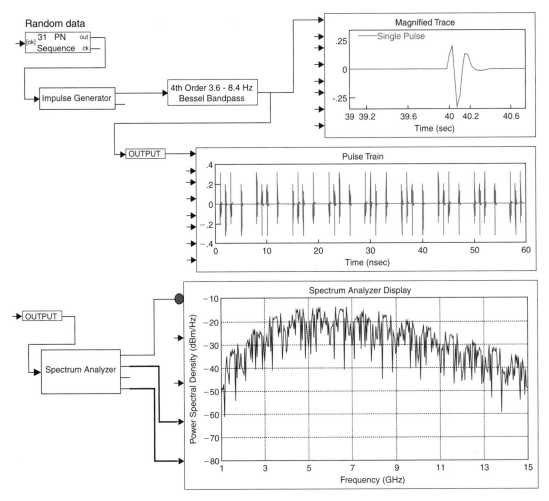

Figure 5.29: Simulated sequence of UWB pulses

second line and below it the result of the correlation operation $\int f(t) \cdot g(t)dt$ where $f(t)$ is the received signal and $g(t)$ is the locally generated sequence. By sampling this output at the end of each bit period and then resetting the correlator, the transmitted sequence is reconstructed in the receiver.

As mentioned above, an individual bit can be represented by more than one sequential monopulse. Doing so increases the processing gain by the number of monopulses per bit. Processing gain is also an inverse function of the pulse duty cycle. This is because, for

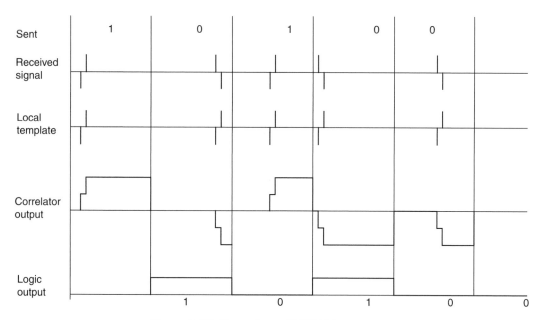

Figure 5.30: Detection of UWB bit sequence

constant average power, the power in the pulses contributing to each bit must be raised by (1/duty-cycle). By gating out the noise except during the interval of the expected incoming pulse, the signal-to-noise ratio will only be a function of the power in the pulse, regardless of the duty cycle. An example may make the explanation clearer. Let's say we are sending data at a rate of 10 Mbps. A UWB pulse in the transmitter is 200 picoseconds wide; 20 pulses represent one bit. The time between bits is $1/(10 * 10^6) = 100$ nanoseconds. So the time between pulses is $(100\,\text{ns})/20 = 5\,\text{ns}$. The duty cycle is $(200\,\text{ps})/(5\,\text{ns}) = 25$. Now the processing gain attributed to the number of UWB pulses per bit is $10\log(20) = 13\,\text{dB}$. That due to the duty cycle is $10\log(25) = 14\,\text{dB}$. Total processing gain is $13 + 14 = 27\,\text{dB}$.

A simplified block diagram of a UWB system is shown in Figure 5.31. A key to the generation of UWB pulses is the ability to create short impulse or step functions with rise times on the order of tens or at the most hundreds of picoseconds, and to detect the UWB pulses that result from their application. High speed integrated circuits can be employed or special circuit elements, such as tunnel diodes or step recovery diodes, can be incorporated.

Conditions for using UWB in Europe are also being considered by the European Union. Due to the fear of interference with vital wireless services in the wide bandwidths covered by UWB radiation, spectral density limits allowed by the FCC and presumably to be permitted

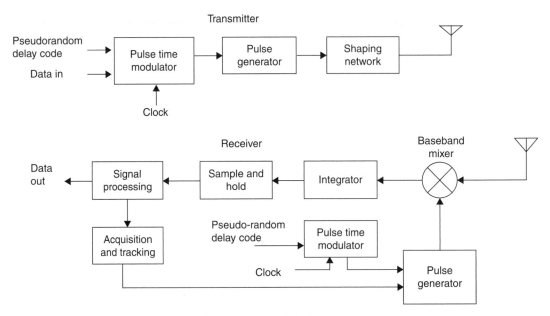

Figure 5.31: UWB simplified block diagram

in the European Union are relatively low, on the order of spurious radiation limits for conventional unlicensed transmissions. However, as we have seen, high-processing gains can be achieved with UWB and communication ranges on the order of tens or hundreds of meters at high-data rates can be expected. Also, the FCC has indicated its intention to monitor the effects of UWB transmissions on other services, once equipment has been put into service in significant quantities, and the agency may be expected to modify or make its limits more lenient if interference is found not to be a problem. In any case, the unique characteristics of UWB are attractive enough to make this technology an important part of the offerings for short-range wireless communication in the years to come.

5.6 Summary

The annually increasing volumes for Bluetooth and Wi-Fi products, stimulated in a large part by the acceptance of industrial standards by the major manufacturers, are causing prices to fall on complex integrated circuits as well as the basic RF components. This trend will open the way for the use of these parts in other short-range applications such as security and medical call systems. Use of sophisticated and proven two-way hardware and link protocols for these and other technically "low end" applications will open them up to much higher usage than they now command. A basic impediment to wireless use will still remain, however, and that is the problem of battery replacement. Reduced voltage and power consumption for integrated circuits will help, as will sophisticated wake-up protocols as are already built-in to Bluetooth. Range limitations may have to be dealt with by a greater use of repeaters than common in today's systems. Another area where advancements are affecting short-range radio is antennas. Both the use of higher frequencies and new designs are reducing antenna size and eliminating a visual reminder of the difference between wired and wireless devices.

The unconventional ultra-wideband technology, since its approval by the FCC, is opening up new civilian applications for short-range wireless, notably in the areas of distance measurement, concealed object location, and high precision positioning systems. Because of its high-interference immunity, and its property of not causing interference, it may successfully compete with and complement other technologies used for short-range radio applications such as personal communications systems, security sensors, and RFID tags.

In summary, advances in short-range radio communication developments in one area feeds its expansion in other areas. Overall, short-range radio will continue to play a major part in the ongoing communication revolution.

Bibliography

Petroff, Alan, & Withington, Paul. (2000). Time Modulated Ultra-Wideband (TM-UWB) Overview. *Presented at Wireless Symposium/Portable by Design*, February 25. San Jose, California. http://www.time-domain.com.

CHAPTER 6

RF Design Tools

Cheryl Ajluni

Chris Bowick's RF Circuit Design *has long been an engineering classic. When it was written 25 years ago, EDA tools were in their infancy. Verilog was just invented and VHDL was two years off. You had SPICE for analog simulation but for RF it was Smith charts, a calculator and prototypes. Cheryl Ajluni wrote the current chapter, which is a good introduction to the current design languages and tools available to assist RF designers.*

Not that there are a lot. While there's no shortage of EDA tools for digital designs—and an increasing number for analog/mixed-signal (AMS) ones—RF circuits continue to be a lot harder to model than, say, a memory controller or an audio amplifier. But being able to automate something as thorny as parasitic extraction could save a lot of time and headaches. There are tools that can do that, but not many.

Cheryl starts with an initial review of design languages that is useful, high-level and non-sectarian. There is no evaluation of C vs. SystemC vs. Catapult C vs. SystemVerilog or other contentious topics, though she does explain the features, benefits and to some extent the positioning of the various languages. A code sample shows how you can easily model a non-linear amplifier in VHDL-AMS and VHDL-FD, an extension that enables harmonic balance simulation for frequency domain analysis.

Modeling complex circuits is another kettle of fish. Going first down the RFIC design path, Ajluni takes you through the design flow from specification to full-chip verification. She then walks you through three RFIC design examples using (or referring to) a range of tools from Agilent, Cadence, Mentor Graphics, Applied Wave Research and even UMC, who supply the necessary bottom-up specs and models for their various foundry processes.

PCB design and packaging are both covered briefly before launching into a detailed case study: designing a full-blown 802.11A CMOS transceiver. This is the payoff for the chapter, and it wraps it up nicely.

If you're new to RF design, this chapter is a good place to start.

—**John Donovan**

Electronic design automation (EDA) is a category of automated design tools used to develop and manufacture the hardware component of a range of electronic systems including field programmable gate arrays (FPGAs)/programmable logic devices (PLDs), integrated circuits (ICs), printed circuit boards (PCBs), and systems-on-a-chip (SoCs). These systems often contain a mix of analog, digital and RF circuitry. While the analog and RF sections of the system make up a much smaller portion of the design than their digital counterpart, they are by far the most difficult to design, requiring substantially more time and expertise. In fact, in systems design, roughly 75% of the design time is spent on RF design. Of that 75%, 25% is spent on actual design work and 75% on interface, library and integration issues.

Given these facts, it might be natural to assume that RF design is today aided by a wealth of highly advanced EDA tools, flows and methodologies. This is not the case. In fact, EDA tools and methodologies for the digital realm remain far more advanced than those which are typically used for analog and RF design. Even so, automated design tools—whether those of the inexpensive, web-based variety or those which are costly and full-featured—can play an important role in enabling you to create fully optimized RF designs more productively than was previously possible with more manual techniques.

In this chapter, we will discuss the specific types of tools which can aid both the expert and novice design engineer with RF design tasks. But first, we begin by taking a look at exactly what constitutes an RFIC design flow and the design languages you will need to know to make that flow work.

6.1 Design Tool Basics

A wide range of EDA tools are used by RF engineers throughout the product lifecycle for the design, verification and test of RF circuits. They replace a process in which all design work and circuit layout was once done by hand in a sequential manner. The engineer would design a circuit and then hand it "over the wall" to another engineer who would verify the work. In turn, the verified design would then be handed off to an engineer who would lay out the circuit.

Design tools not only automate this process, but have transformed it from a sequential to a concurrent process as well. Today, via a methodology known as co-design, many different engineers can now work on various parts of a design at the same time. And the design process now begins at the system, as opposed to the circuit, level. The result is increased productivity and time to market, along with a reduction in costly, time-consuming errors. And, because EDA tools automate engineering best practices, acting as a repository of sorts for engineering expertise, even the novice engineer can be productive using them.

The most common types of EDA tools used today are:

- Schematic Capture tools—Captures a schematic representation of a design using components from a component library.

- Logic Synthesis tools—Translates high-level design languages into netlists.

- Place-and-Route (P&R) tools—Takes netlists and decides where to put each wire and transistor in the design layout.

- Simulation tools—Takes a description of a circuit (netlist) and simulates its behavior. These tools are critical to ensuring that you get your design right the first time.

- Verification tools—Verifies that a design will work as expected. Includes Design Rule Checkers (DRCs) and equivalence checkers which work to verify the design netlist and that the netlist matches the original design, respectively.

6.2 Design Languages

Design languages are used to describe, or model, the circuit. Much of today's RF design typically begins either by hand with a textual description or with an algorithmic description of a data transformation, such as a model in MATLAB or a C/C++ object class. These procedural languages are especially useful in developing individual untimed algorithms, but they do not work well as system-level design tools because they lack the built-in constructs necessary for time and concurrency.

Consider, for example, that in a mobile communication system, many algorithms operate at different rates in a base station. Algorithms also operate in the handset. Concurrently, RF energy bounces off objects such as a building or a tree to create a dynamically fading channel. Modeling the performance of an RF system realistically under these conditions requires the designer to capture the concurrency of all of this behavior, in addition to any significant physical properties. This is virtually an impossible task.

System-level design languages like SystemC or System Verilog offer a possible patch to this deficiency. SystemC, for example, provides transaction-level modeling constructs by adding a class library and a kernel to C++.

Whether the design starts life in the form of a system-level or procedural language, once its implementation is fleshed out, you will need to leverage a number of different easily accessible tools to assist in the hardware description language (HDL) coding process.

An HDL is a standard, text-based expression of the temporal behavior and/or spatial circuit structure of an electronic system. The two leading standard HDLs include Verilog and VHSIC Hardware Description Language (VHDL). Both have extensions (Verilog-AMS and VHDL-AMS) that allow them to work with analog and mixed-signal designs, including some RF. Generally, though, RF system design requires many more constructs than are present in these language extensions. Alternative language options exist that further extend the analog/mixed signal versions of Verilog and VHDL into the RF space, but because these additional constructs are vendor specific, they constrain the design database to a specific vendor.

It is important to note that whichever language you opt to use, the design description or model will eventually be used to perform mixed-level (e.g., analog, digital and RF) simulation. Therefore it may make sense to adopt a language that is broadly supported by a range of design tools and/or utilized by the rest of your design team. Many standard EDA tools offer support for a wide range of standard programming and HDLs.

What follows is a closer look at the languages in use today for RF design.

6.2.1 Verilog

Verilog is a hardware description language for describing electronic circuits and systems. It supports the design, verification and implementation of analog, digital, and mixed-signal circuits at various levels of abstraction and can be used for verification through simulation, for timing analysis, test analysis (e.g., testability analysis and fault grading) and logic synthesis.

6.2.2 Verilog-AMS

Verilog-AMS is an extension to Verilog that supports analog and mixed-signal modeling using both continuous-time and event-driven modeling semantics. It allows designers to use modules that encapsulate high-level behavioral descriptions as well as structural descriptions of systems and components.

6.2.3 Verilog-A

Verilog-A is an industry standard modeling language and is the continuous-time subset of Verilog-AMS. It is specifically designed for modeling behavior of analog components. The behavior of each module can be described mathematically in terms of its terminals and external parameters applied to the module. The structure of each component can be described in terms of interconnected sub-components.

```
HPEESOFSIM (*) 2003C.day Dec 14 2003 (built: 12/14/03 21:28:57)
Copyright Agilent Technologies, 1989-2003.

A verilog-A compiled model library (CML) cache has been created at
    '/users/bobl/hpeesof/agilent-model-cache/cml/0.99/hpux11'

Compiling Verilog-A file

'/rfdeExamples/VerilogA_Tutorial/sinRampSrc/veriloga/veriloga.va'

AGILENT-VACOMP (*) 2003C.day Dec 14 2003 (built: 12/15/03 01:49:33)
Tiburon Design Automation (R) Verilog-A Compiler Version 0.99.121203.
Copyright (C) Tiburon Design Automation, Inc. 2002-2003. All rights
reserved.

Loading Verilog-A module 'sinRampSrc' from
'/users/bobl/hpeesof/agilent-model-cache/cml/0.99/hpux11/veriloga_09106/1
ib.hpux11/veriloga.cml'.

TRAN Tran1[1] <input.ckt> time={0 s->100 ms)
Resource usage:
    Total stopwatch time: 23.21 seconds.
```

Figure 6.1: When a Verilog-A cell view is simulated for the first time, the Agilent RFDE Verilog-A Compiler compiles the associated Verilog-A code and caches the result so that it may be used directly in all subsequent simulations of that view.

As an example of Verilog A, consider the Advanced Design System (ADS) and RF Design Environment (RFDE) solutions from Agilent Technologies (www.agilent/com/find/eesof). Both feature a Verilog-A solution—Verilog-A Compiler—which is suitable for RF circuit and system design.

Verilog-A Compiler is based on technology from Tiburon Design Automation and enables simulation speed that is close to built-in C models. It can be used to create behavioral models of RF blocks such as mixers, amplifiers, etc., as well as transistor models for things like MOSFETs, BJTs and HBTs. Many transistor model examples are provided with the product, and they can be used as the basis for more complete device models.

A typical simulation status showing the cache creation message and a compilation/load message is illustrated in Figure 6.1.

6.2.4 SystemVerilog

SystemVerilog is considered the industry's first unified hardware description and verification language (HDVL) standard (IEEE 1800), and is a major extension of the established IEEE 1364Verilog language. Based on extensions to Verilog, it was developed originally by Accellera to dramatically improve productivity in the design of large gate-count, intellectual property (IP)-based, bus-intensive chips. SystemVerilog is targeted primarily at the chip implementation and verification flow, with powerful links to the system-level design flow.

6.2.5 VHDL

VHDL is a fairly general-purpose language, although it requires a simulator on which to run the code. It can read and write files on the host computer, so a VHDL program can be written that generates another VHDL program to be incorporated in the design being developed. As a result, it is possible to use VHDL to write a testbench that verifies the functionality of the design using files on the host computer to define stimuli, interacts with the user, and compares results with those expected.

6.2.6 VHDL-AMS

VHDL-Analog Mixed Signal (VHDL-AMS), the IEEE-endorsed standard modeling language (IEEE 1076.1), was created to provide a general-purpose, easily exchangeable and open language for modern analog-mixed-signal designs. Models can be exchanged between all simulation tools that adhere to the VHDL-AMS standard. The example in Figure 6.2 illustrates the architecture basics of an amplifier in VHDL-AMS.

6.2.7 VHDL-AMS/FD

There are also extensions to VHDL-AMS that are specifically geared toward high-frequency design (Figure 6.3). VHDL-AMS/ Frequency Domain (FD) is an extension language that supports harmonic balance simulation for frequency domain analysis. It provides the RF designer with the capability to describe complex microwave and RF devices (e.g., wireless systems) from architecture to transistors. The RICON Harmonic Balance simulator is currently the only VHDL-AMS simulator to use the VHDL-AMS/FD extensions.

6.2.8 VHDL-RF/MW

VHDL-RF/MW is an extension to VHDL-AMS that supports design at radio and microwave frequencies. Extensions facilitate non-lumped terminals, finite-element modeling, parasitics, and frequency-domain modeling. VHDL-RF/MW is supported in the FTL Systems' Auriga modeling and verification solution which spans system level, register-transfer-level (RTL), analog/mixed signal, radio-frequency down to parasitic/microwave detail.

6.2.9 C/C++

C is a general-purpose, procedural, imperative computer programming language. C++ is an extension of C that provides object-oriented functionality with C-like syntax. It adds greater

```
-- This model contains simultaneous conditional use statements, but
-- the conditions are such that all quantities are continuous over the
-- switch between equations; thus no "break" is required.
--
library IEEE;
architecture basic of amplifier is
    terminal Cout: electrical;
    constant c1: real := imax/(sr*1.0e6);
    constant gmnom: real := 2.0 * IEEE.math_real.math_pi * ugf * c1;
    constant r1: real := gain/gmnom;
    constant dvmax: real := imax/gmnom;
--
-- The reader will find that diagraming these quantities as the edges of
-- a directed graph on the terminals will aid in understanding the
-- design.
--
    quantity Iinput through Vinput across vin to vinb;
    quantity Ivin through refer to vin;
    quantity Ivinb through refer to vinb;
    quantity Icout through Vcout above refer to cout;
    quantity Icout_vout through Vcout_vout above cout to vout;
    quantity Vvout_vp above vout to vp;
    !
quantity Vvout_vn above vout to vn;
--
-- These non-branch quantities are used as temporaries in building up
-- the equation for Icout
--
    quantity gain_current, dp_current, outstage_current: current;
begin
--
-- local terminal Cout
    Icout == gain_current + dp_current - outstage_current;
    Icout_vout == Vcout_vout/r0;
--
-- input stage.
    Iinput == Vinput/rin;
    Ivin == iin;
    Ivinb == iin;
--
-- gain stage.
-- The following relationship must hold between gmnon,dvmax and imax
-- to prevent a discontinuity in gain_current
    assert abs(gmnon*dvmax) = abs(imax);
    if Vinput > dvmax use
        gain_current == imax;
    elsif Vinput < -dvmax use
        gain_current == -imax;
    else
        gain_current == gmnom * Vinput;
    end use;
--
-- dominant pole.
    dp_current == c1 * Vcout'dot + Vcout/r1;
--
```

Figure 6.2: An amplifier, as described in VHDL-AMS. This code example was excerpted from a conference paper entitled "Mixed-Mode Simulation," by Kenneth Bakalar from Compass Design Automation, now Synopsys.

```
-- output stage limiting.
-- outstage_current is zero when Vvout_vp=-so!
ft or Vvout_vn=+soft
-- so there is no discontinuity at the transit

ion points
    if Vvout_vp > -soft use
        outstage_current == gmnom * (Vvout_vp+soft);
    elsif Vvout_vn < soft use
        outstage_current == gmnom * (Vvout_vn-soft);
    else
        outstage_current == 0.0;
    end use;
end basic;
```

Figure 6.2: *(Continued).*

```
entity nonlinear_amp is
generic (k1, k2, k3: real); port (quantity a_in, a_out : real);
end entity amp;
architecture eq of amp is
begin
a_in*k1+a_in**2*k2+a_in**3*k3==a_out; - nonlinear effects included
end architecture;
...
The example shows that the entity "nonlinear_amp" contains the parameters k1, k2, k3
The test bench for the amplifier is the following:
# amplifier jobs: - comment line
res1=HBparametric([3.e6,5,0.01,1000])# - performs HB
print "================== results ================="
print res1; # - print res1 (complex structure)
print "================== tout.H1 =================".
print res1.getX("tout'reference",[1]) # of "tout'reference" variable
# - gets and prints only spectrum
# of "tout'reference" variable at first harmonic
```

Figure 6.3: Code for defining an amplifier using VHDL-FD. This code example was excerpted from "VHDL-AMS Extensions Enable RF Harmonic Balance Simulation," originally published in *High Frequency Electronics* magazine in March 2004 and written by Mark Rencher, Ridgetop Group.

typing strength, scoping and other tools useful in object-oriented programming, and permits generic programming via templates to simplify high-level programming and offer a better approach to large-scale programming. C++ is also a larger language with more features and complexity than C, but C++ can improve productivity with its greater number of features. A list of features that C++ supports and C does not is shown in Table 6.1.

Table 6.1: Features supported by C++ and not by C

Classes	Inline functions
Member functions	Default arguments
Constructors and destructors	Function overloading
Derived classes	Namespaces
Virtual functions	Exception handling
Abstract classes	Run-time type identification
Access control (public, private, protected)	// comments
Friend functions	True const
Pointers to members	Declarations as statements
Static members	Automatically typedef struct tags
Mutable members	Type safe linkage
Operator overloading	New and delete
References	Bool keyword
Templates	Safer and more robust casting

6.2.10 SystemC

SystemC provides hardware-oriented contructs within the context of C++ as a class library implemented in standard C++. Its use spans design and verification from concept to implementation in hardware and software. The language provides an interoperable modeling platform, which enables the development and exchange of very fast system-level C++ models. It also provides a stable platform for development of system-level tools.

6.2.11 MATLAB/RF Toolbox/Simulink

MATLAB is a high-level programming language and interactive, integrated environment, developed by The MathWorks (www.mathworks.com/products/matlab/). It enables the engineer to perform computationally intensive tasks faster than with traditional programming languages such as C and C++. It is used for numerical computations, symbolic computations and scientific visualizations and runs in interpreted, as opposed to compiled mode.

RF toolbox is a MATLAB function and class library that RF engineers use for designing, modeling, analyzing, and visualizing networks of RF components (Figure 6.4). It works within the Simulink environment and offers users a library of blocks to model the behavior of RF amplifiers, mixers, filters, and transmission lines.

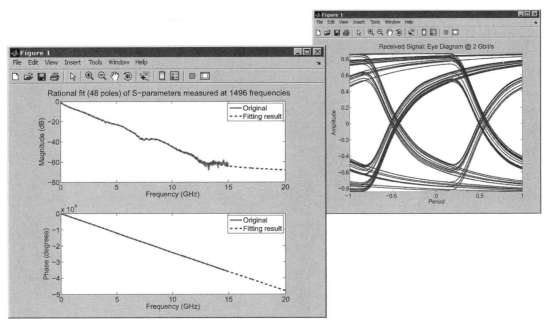

Figure 6.4: RF Toolbox from The MathWorks enables engineers to design, model, analyze, and visualize networks of radio frequency components.

Simulink is a block library tool for modeling, simulating and analyzing dynamic systems. It works with MATLAB.

6.2.12 SPICE

SPICE is a powerful general-purpose analog circuit simulator, developed at the University of California, Berkeley. It is used to verify circuit designs and to predict the circuit behavior. It falls into the category of an equivalent-circuit model software package. The SPICE simulator was long considered the de facto industrial standard for computer-aided circuit analysis. However, the limitations of SPICE have forced the creation of new languages like Verilog and VHDL-AMS.

6.3 RFIC Design Flow

The process of chip design from concept to production is called a design flow. A design methodology is a set of procedures that accompany a design flow to achieve a particular outcome. For example, a design tool vendor specializing in locating and identifying signal

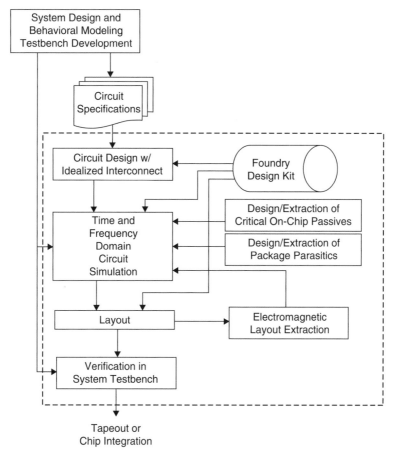

Figure 6.5: This flow chart, courtesy of UMC, depicts typical RFIC design and verification flow.

integrity issues may develop a methodology that helps ensure any potential signal integrity issues are eliminated early in the design flow when they are easier to find and cheaper to fix. Note that most design tool vendors, especially those known for offering solutions that span the entire product lifecycle (e.g., from R&D to production), typically offer their own individual flows and methodologies.

A typical RFIC design flow is pictured in Figure 6.5. Recall that while digital designers have grown accustomed to unified design flows, RF designers must often piece together tools from various vendors to develop a unified flow for their designs. EDA vendors like Agilent Technologies, Ansoft, Applied Wave Research (AWR), Cadence Design Systems, and Mentor Graphics are now working to rectify this situation via the development of RFIC flows

Design flow

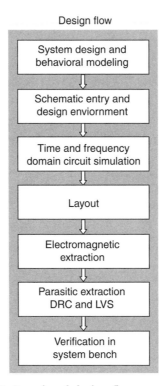

Figure 6.6: Functional design flow representation

developed around their key RF design tools and platforms. Therefore, while the flow depicted in Figure 6.5 may be typical, it is by no means the only possible RFIC flow available today.

A functional representation of this flow is provided in Figure 6.6. The key steps that comprise this flow are system design, circuit design (e.g., design capture and simulation), circuit layout, parasitic extraction, and full-chip verification.

6.3.1 System Design

During the system-level design phase, an RF specification in the form of a behavioral model is created using either Matlab, C or some other system-level design language. A behavioral model is essentially a model that describes what the system does. To create the model, the engineer begins with perfect RF components and then adjusts the performance specifications of the components until performance degradation occurs. The goal of this process is to determine what level of noise, nonlinearity, and frequency domain distortions the system can tolerate.

A behavioral testbench is also developed during this phase. It will serve as the framework for more complex mixed-level (e.g., analog, digital and RF) simulations, where blocks can be inserted at the transistor level and verified in a system context. The behavioral model and testbench are often referred to as an executable specification and are used together to validate the specification. Once the specification has passed verification (via any number of functional verification tools), it is passed to the circuit designer where the detailed design, RTL or board-level schematics, are created and compared to the original behavioral description. When they match, the design is deemed "functionally" complete.

Note that the engineer might also utilize an advanced architectural planning tool to help determine the design's specification (Figure 6.7). Such tools work by allowing the engineer to specify the correct system architecture. They then generate suitable specifications for each of the underlying components in complex communications designs. As a result, they help designers reduce time-to-market by eliminating iterations and rework, and cut system costs by ensuring that components are not over-specified and thus unnecessarily expensive.

6.3.2 Circuit Design

During the circuit design phase, the verified executable specification is further refined to incorporate results generated by the engineer while creating the actual circuit design. The process normally begins with the conversion of the specification into a block diagram of the various functions that the circuit must perform. The various blocks are considered in detail still—at an abstract stage—but with much more focus on the details of the electrical functions to be provided. The block diagram is then synthesized into a netlist for simulation. This will allow the engineer to find mistakes early in the design cycle before a physical device is fabricated.

Circuit simulation is performed in the time and frequency domains to characterize critical performance metrics. The choice of domain will depend on a number of factors including: the type of circuit and its size, the type of simulation and desired output, the designer's comfort level, and any personal preference the designer may have (Figure 6.8). As circuits are completed at block level, they are verified within the top-level context with behavioral stimulus and descriptions for the surrounding chip. If the engineer is unable to achieve the specification for a certain component, the system-level model and executable specification can be modified accordingly.

There are two important things to note during the circuit-design phase. First, it is common during this stage to employ the use of modeling tools (e.g., for spiral inductors). This task will be discussed in detail later in the chapter. It is worth noting, however, that these tools allow

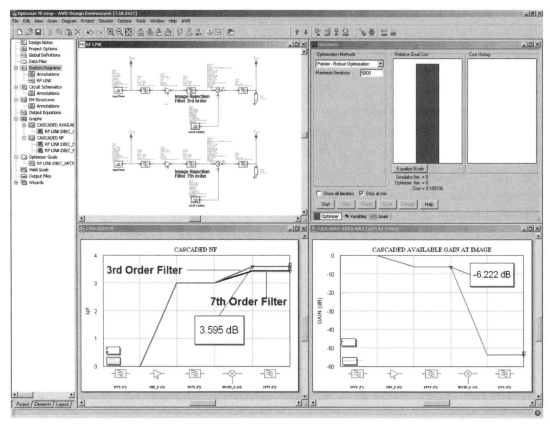

Figure 6.7: The Visual System Simulator (VSS) from Applied Wave Research is a comprehensive software suite for the design of complete, end-to-end communications systems. The tool's RFA RF Budget Analysis module, shown here, enables designers to find potential pitfalls early in the design process, at the system-level design phase, thereby saving significant design cycle time and speeding products to market. Using it, designers can make traditional RF cascaded measurements such as gain, noise figure, and third-order intercept, inclusive of image noise, along a communication link.

you to determine what is and is not feasible to achieve with your circuit. The information that you obtain from this exercise is then fed back into earlier stages of the design process to verify system-level performance.

Secondly, prior to the start of the circuit design phase, the engineer will select a process technology (e.g., 0.18 or 0.25 micron design geometries). Once this selection is made, a process design kit (PDK) can be obtained from the appropriate foundry or IC fabrication

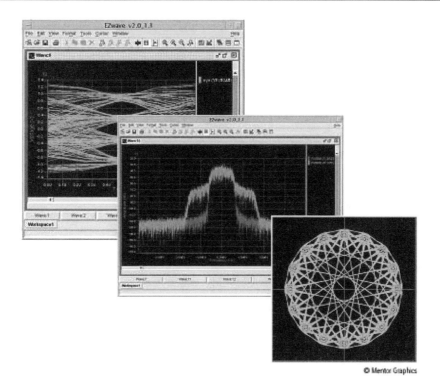

© Mentor Graphics

Figure 6.8: The Eldo RF simulator from Mentor Graphics provides a set of dedicated algorithms to accurately and efficiently handle the multi-GHz signals in modern wireless communication applications. It features steady-state analysis of RFIC circuits excited with periodic (single-tone) or quasi-periodic (multi-tone) sources as well as a complete RF toolbox, including Smith Chart diagrams, gain and stability circles, and minimum noise figure. Eldo RF is part of the Mentor IC design flow, which includes Design Architect IC and IC Station for front-end and back-end analog and mixed-signal design.

house and installed within your design environment. The PDK helps jump start the design process by providing you with all of the foundry-specific process models and data that you will need for circuit design. The PDK is provided in a file format that is compatible with the design environment (e.g., IC design flow and tools) you are using.

6.3.3 Circuit Layout

The next step in the process is to perform circuit layout. Here, the individual circuit components which you have chosen to carry out each function in the system, are physically laid out and the electrical connections of each component are decided. Automated design-rule-driven and connectivity-driven layout tools may be used judiciously, especially to take advantage of direct ties to schematic and design-rule checking. Critical analog blocks are generally manually routed using a full custom approach to ensure that highly sensitive analog circuitry meets specifications.

6.3.4 Parasitic Extraction

The next step in the design process is to extract parasitics. This is a crucial step as high-speed requirements make RF circuits extremely sensitive to the effects of parasitics, including parasitic inductance, passive component modeling, as well as signal integrity issues. Once extracted, these effects are added to the circuit simulations.

For sensitive blocks like VCOs and critical radio blocks, full-wave three-dimensional electromagnetic (EM) simulation allows the extraction of the full layout at the block level. This rigorous method simulates all high-frequency layout effects including on-chip inductors, interconnect, coupling between on-chip passives and to other interconnect structures, and substrate coupling. No assumptions are made regarding parasitics or coupling. The resulting accurate models of these effects are used to replace the models that were created earlier in the design process.

Note that net-based RLC extractors also have their place in the RFIC flow, but they require designer input to manage which parasitic effects to include in the extraction. Unfortunately, it is not always clear which parasitic effects are most critical in the circuit context. Rigorous EM extraction of the entire block should be used to remove any doubt in the process.

6.3.5 Full-Chip Verification

The final step in the design process, prior to tape-out or additional chip integration, is to perform full-chip verification in a system (e.g., behavioral) testbench. The verification can include transistor-level circuits for multiple circuit blocks with incorporation of all extracted parasitics. The system should allow designers to select the particular level of abstraction for individual circuit blocks in order to make reasonable trade-offs between accuracy and simulation run time.

6.4 RFIC Design Flow Example

What follows is an example of a wireless RFIC flow that is part of the Cadence Design Systems RF Design Methodology Kit which is based on the Advanced Custom Design Methodology (ACD).

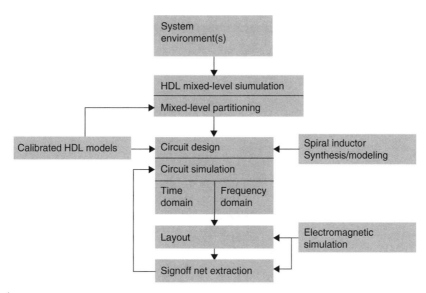

Figure 6.9: Wireless RFIC flow utilized as part of the Cadence Design Systems RF Design Methodology Kit

Figure 6.9 depicts the wireless RFIC flow. In this flow, the design collateral from the system design process is used as the first and highest abstraction level. System-level descriptions form the executable testbench for the top-level chip. Models of the surrounding system are combined with a high-level model of the chip, producing an executable specification. System requirements serve as the first specification to drive the chip-level requirements, and ultimately turn into repeatable testbenches and regression simulations.

Once a high-level executable specification is achieved, the design process continues by identifying areas of concern in the design. Plans are then developed for how each area of concern will be verified. The plans specify how the tests are preformed and which blocks are at the transistor level during the test. Resist the temptation to specify and write models that are more complicated than necessary. Start with simple models and only model additional effects as needed.

6.4.1 HDL Multi-Level Simulation

This process starts with HDL modeling (including all RF blocks and any analog content and/or digital blocks) for the entire RFIC being added to the system-level testbench. First a behavioral model of the full chip within a top-level testbench is created. Initially it is used to verify the partitioning, block functionality and ideal performance characteristics of the IC. Then, it serves as the basis to facilitate mixed-level simulations, where blocks can be inserted at the transistor level and verified in a top-level context. This behavioral model/testbench also serves as the regression template, allowing for continuous verification as blocks mature.

In the full simulation environment, several views of the same circuit will exist including, for example, a behavioral view, a pre-layout transistor-level view, and several views of parasitic information. As blocks mature, it may be necessary to add more transistor-level information to test RF/analog and RF/digital interfaces. This will require the use of a mixed-signal simulator capable of handling analog, digital, and RF descriptions and mixed behavioral-level with transistor-level abstractions. You will need to pick the appropriate views of each block and manage the runtime versus accuracy tradeoffs through simulation options by—as an example—sending the transistors to a FastSPICE simulator or keeping the transistors in a full SPICE mode.

6.4.2 Block Circuit Design

During this stage a preliminary circuit design is created that allows for early circuit exploration and a first-cut look at performance specifications. This early exploration leads to a top-level floorplan, which for RFIC is sensitive to noise concerns and block-level interconnect. Passive components (e.g., spiral inductors) are then synthesized and an initial placement of them on the chip is completed.

Note that this approach enables two key activities: creating early models for spiral inductors that can be used in simulation before the block-level layouts are complete, and allowing for an initial analysis of mutual inductance between the spirals.

Next, simulation is performed. A single PDK and associated environment allows for a smooth determination and selection of the simulation algorithm desired. Results are displayed through an appropriate display for the simulation type selected. As circuits are completed at block level, they are verified within the top-level context with behavioral stimulus and descriptions for the surrounding chip.

6.4.3 Physical Implementation

Layout automation (e.g., automated routing, connectivity-driven layout, design-rule-driven layout, and placement) is used to implement the circuit. Analog-capable routers can help with differential pairs and shielding wires, and allow for manual constraints per line. Highly sensitive circuitry requires a manual approach.

6.4.4 Parasitic Extraction

Once the layout is complete, EM simulation is used to provide highly accurate models for passive components. Net-based parasitic extraction becomes a key element of the process as layouts emerge. Less sensitive interconnects may require RC only, whereas more sensitive lines may require RLC. For lines with spirals attached, these can be extracted fully with RLC plus the associated inductor component, even with substrate effects added for those lines that are the most sensitive. Lines that contain a "full" extraction can be mixed and matched with the component models for passive components that were created earlier.

Also during this stage, designers will check to see whether noisy circuits (such as digital logic and perhaps PLLs) are affecting the highly sensitive RF circuits. If so, they can either modify the floorplan accordingly or add guardbands around the noisy circuitry.

Note that it is often impractical to both simulate the entire design at transistor level and include all the parasitic information. One approach is to extract calibrated behavioral models using the extracted view of the design blocks. This will not capture the effects of the parasitics on interconnect between blocks; therefore, hierarchical extraction capabilities that extract only parasitics of interconnect between design blocks is required.

6.4.5 Calibrated Models

As blocks are completed, the initial behavioral models can be back annotated for key circuit performance parameters to provide more accurate HDL-level simulation. Verification of a block by mixed-level simulation, therefore, becomes a three-step process. First, the proposed block functionality is verified by including an idealized model of the block in system-level simulations. Next, the functionality of the block as implemented is verified by replacing the idealized model with the netlist of the block. Finally, the netlist of the block is replaced by an extracted model. By comparing the results achieved from simulations that involve the netlist and extracted models, the functionality and accuracy of the extracted model can be verified. From then on, mixed-level simulations of other blocks are made more representative by using the verified extracted model of the block rather than the idealized model.

6.5 Simulation Example 1

To better understand how to verify that the blocks in an RFIC (e.g., amplifiers, mixers, frequency dividers, and baseband chain) are wired together correctly, consider the example of a receiver with a wireless local area network (WLAN) 802.11b signal as shown in Figure 6.10. This example utilizes the RF Design Environment (RFDE) from Agilent Technologies.

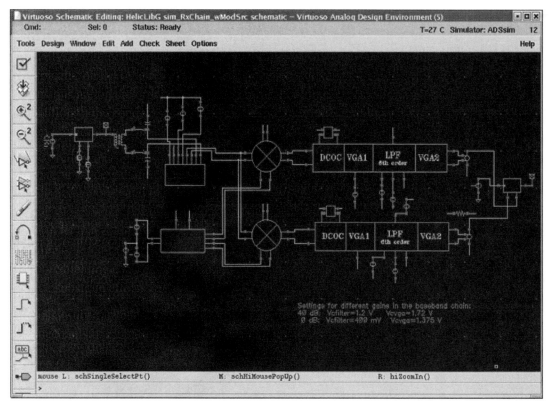

Figure 6.10: This schematic, as viewed using Agilent Technologies' RFDE software design platform for large-scale RF/mixed-signal IC design, shows a direct-conversion receiver.

For this example, the direct-conversion receiver consists of an LNA, I and Q mixers and baseband receive chains, and a frequency divider to generate quadrature LO signals. In this example, the task is to supply a WLAN 802.11b input signal to the LNA, and verify that the I and Q baseband output signals still track the input modulation, at various points along the baseband chain (outputs from the mixers, outputs from the DC offset cancellation circuit, outputs from the VGAs and from the tunable filters).

The simulation uses time-domain baseband data sources that are the I and Q data of a WLAN 802.11b signal. These data sources modulate a sinusoid via an ideal modulator to generate the test signal at the input. If you have the I and Q time-domain data waveforms for some other modulation format, you may use this technique along with the Cadence Envelope Follower simulator or the Agilent Circuit Envelope simulator to generate a modulated test signal.

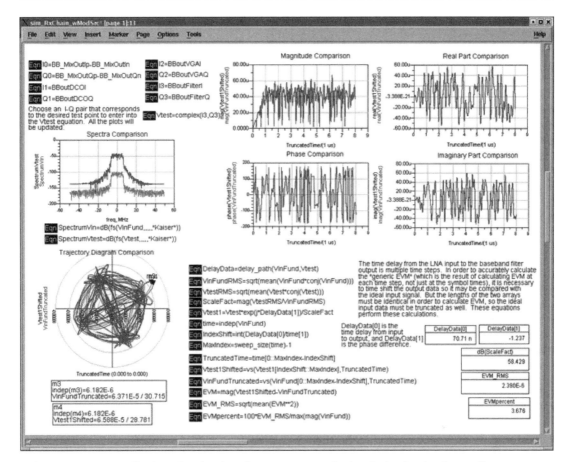

Figure 6.11: Data display of a simulated WLAN 802.11b signal obtained using Agilent's RFDE software design platform. Note that the data display also calculates the "generic" EVM, which is the result of including every time point (not just the symbol samples) in the EVM calculation. This data may also be displayed as part of the Cadence simulation environment.

The simulation time required depends linearly on the desired stop time, with about 10 minutes being the minimum to get useful information. The circuit has 2547 devices, 1377 of which are nonlinear.

The data display in Figure 6.11 shows comparisons between the input signal and the signals at various points in the receive chain. By changing the Vtest equation, you can select the test point to be the output of the mixer, the output of the DC offset cancellation circuit, the output of the first variable gain amplifier, or the output of the baseband filter. The plots compare the spectra, magnitudes, phases, real and imaginary parts, and the trajectories of the input and Vtest signals.

By changing the Vtest equation, you can see if there is a block in the receiver chain that significantly degrades the EVM. Also, you can make changes to various blocks to see if the EVM can be improved.

A good strategy for minimizing BER is to first run relatively quick simulations like this to test the EVM. When the EVM is minimized, the BER should be minimized as well, and these EVM simulations are much faster than simulating BER. Using this type of simulation, you can easily see whether your receiver or transmitter is wired up correctly and which block is causing degradation.

6.6 Simulation Example 2

The high-frequency nature of RFIC circuits makes them especially vulnerable to circuit impairments like compression, noise, distortion and phase noise, as well as physical parasitics that include interconnect impedance and coupling. In order to handle these issues properly, you will require a simulator with the ability to simulate large circuits with extracted parasitics and high nonlinear designs. It should be capable of simulating signals at whatever frequency you want and running EM simulations on arbitrary layout structures, which is more accurate than using analytical models. It should also be able to run simulations that include post-layout extractions. Any limitation on the type, range, or capacity of simulation tools can impose major limits on accuracy, flexibility and design exploration.

As an example of the flexibility required for RFIC simulation, consider Figure 6.12. The graphic depicts the spectrum for a direct downconversion of either a direct-conversion receiver or transmitter. Two input signals are shown near the local oscillator (LO) frequency, with downconversion to baseband terms at the output. The simulation can be run using either the Agilent Harmonic Balance or Circuit Envelope simulators available from either Cadence or Agilent via the Cadence SpectreRF or Agilent Advanced Design System (ADS) solutions. Both Cadence and Agilent provide these solutions.

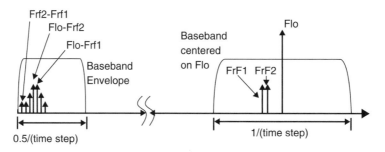

Figure 6.12: Spectrum for direct downconversion

Using the Harmonic Balance solution, this simulation would require three large-signal tones, one for the LO and one for each RF signal. A two-tone simulation can also be used if the frequency difference between Flo and Frf1 is an integer multiple of the frequency difference between Frf1 and Frf2.

With Circuit Envelope, only one large signal tone, for the LO, is required. Circuit Envelope simulation is a hybrid time- and frequency-domain simulation technique. Signals that are within the envelope bandwidth that is centered on each large-signal analysis tone are generated without requiring any additional large-signal analysis tones. The envelope bandwidth is equal to 1/(simulation time step). Tones that are within 0.5/(simulation time step) above 0 Hz are also generated.

The large-signal analysis frequencies in this Circuit Envelope simulation are the LO frequency and its harmonics. The LO signal can be generated by an oscillator and/or a frequency divider, in which case its frequency is determined automatically by the simulator while solving the oscillator and frequency divider.

Figure 6.13 displays the Circuit Envelope simulation results. The equations show how a particular spectral tone can be plotted as a function of the input signal amplitude. The design variables Flo, Frf1, and Frf2 specify the spectral tone. The input signal amplitude is VRFamp_dB.

Note that, in this example, the LO and its harmonics had to be simulated along with baseband signals near DC due to the large difference in signal frequencies. While a time-domain simulator cannot simulate such circuits efficiently, frequency-domain simulators like Harmonic Balance and Circuit Envelope are well suited for circuits of this type.

6.7 Modeling

As opposed to system-level behavioral modeling, which describes what the system does, modeling of active and passive on-chip elements during circuit design is done to improve the fidelity of the simulation. Especially problematic is the process of modeling spiral inductors on RFIC circuits, which are critical for filtering and tuning purposes (Figure 6.14). When modeling these elements it is vital that the Q of the inductor and the resonance be predicted accurately. Because of their complex structures, this is often an extremely difficult, nontrivial task (Figure 6.15). Note that the Cadence SpectreRF Simulator includes a Virtuoso Passive Component Modeler (VPCM) capability which enables synthesis, verification and modeling of spiral inductors and transformers. It is a complete flow, integrated in the Virtuoso Schematic editor and layout suite, and is compatible with Assura. It produces a complete PDK component with schematic, symbol, pcell layout, S-parameter file and Spectre models.

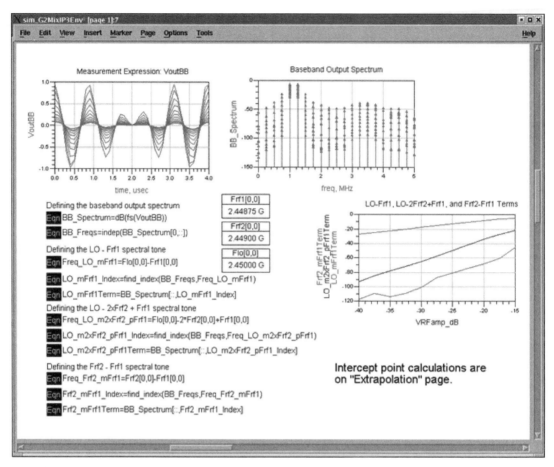

Figure 6.13: Shown here are the results of simulating a downconverting mixer using the Circuit Envelope simulator as part of Agilent's Advanced Design System (ADS)/RFDE. The RF tones are at 2.44875 and 2.44900 GHz, while the LO is at 2.45 GHz. The simulation required about 4.5 minutes.

Note that a Foundry Design Kit (FDK) will provide many of the passive models (e.g., metal-insulator-metal or metal-oxide-metal capacitors) needed for designing RFICs. The RF Design Methodology Kit from Cadence can be used to help generate a design's layout and an electrical model. As long as design rules are followed and parameter ranges are not exceeded, these models are highly accurate. Because most foundries provide only corner-case device models for device modeling, these models may not be well suited for RF designs. Ideally, a model library should include the following: corner cases, statistical models, digital/analog

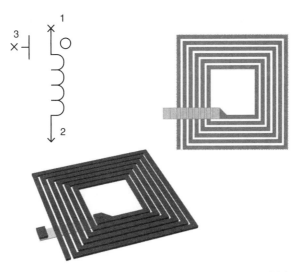

Figure 6.14: Spiral filters are complicated structures, with non-zero thickness lines, underpasses, and more than one spiral metal layer possible.

Graphic courtesy of Applied Wave Research

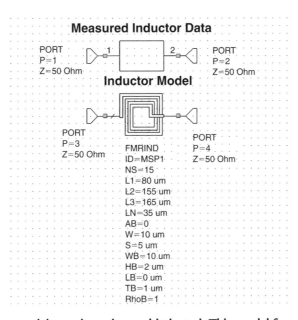

Figure 6.15: Inductor models can be quite sophisticated. This model for a single layer spiral has thirteen physical parameters in order to match the experimental data accurately.

Graphic courtesy of Applied Wave Research

mismatch models, pad models with RF electrostatic discharge, flicker-noise models, and substrate-resistance models. The library also should include well-proximity and shallow trench isolation stress effects.

6.7.1 Modeling Issues

If a suitable FDK is not available, the task of creating the model falls to the engineer. Some of the issues that will need to be considered when developing a spiral inductor model are:

- The model must have some form of frequency-dependent resistance to compensate for conductor losses.

Normally, spiral inductors are placed on the top layer metal. The metal thickness and width are typically on the order of a few microns in each dimension. At GHz frequencies, these dimensions are on the order of a skin depth for aluminum or copper, forcing the current distribution in the conductors to change (e.g., crowding toward the surface of the line). This current crowding increases the resistance in the inductor and decreases the Q.

- Inductance effects must be considered.

Current crowding affects the inductance as a function of frequency. As it occurs at higher frequencies, the internal inductance will go down, as the current and flux are being excluded from the interior of the lines. Therefore, there is a frequency dependent inductance effect, which also affects the Q.

- Shielding can be used to isolate the inductor from the substrate and minimize substrate effects.

The lossy substrate can contribute substantially to reduction of the Q by providing a resistive loss mechanism. Shielding placed below the inductor to isolate it from the substrate can mitigate the substrate's effect on Q.

- Make an assumption regarding location of the ground return.

Inductance is only uniquely defined when there is a loop of current. Unfortunately, the modeler rarely has the luxury of knowing exactly where the return current is going to be. Therefore, certain assumptions will have to be made regarding where the return current is, e.g., on the substrate or on an intervening power plane.

- Capacitance can be difficult to predict.

A multi-turn spiral has a large amount of capacitive coupling to itself, which does not change much with frequency. Charge will stay on the surface of a "good" conductor. Aluminum, for example, is a good conductor well past 100 GHz. However, it is often difficult to predict the capacitance, as the lines are three-dimensional structures. This is where experience and intuition can be important.

- Be careful when using the model in SPICE.

Resistance and inductance will change with frequency. As a result, it may be tempting to make a model with frequency-dependent resistance and inductance. When the model is exported to SPICE, however, undesirable behavior can occur. The response can become completely nonphysical.

- Include underpass effect in the model.

The signal must be brought out from the center port of the spiral inductor. Typically this is accomplished by connecting to a line on a lower metal layer, creating an underpass. It is important that the effect of the underpass be included in the model.

Electromagnetic simulation software can be used to obtain simulated data for inductor modeling and design. Several products are commercially available that use a variety of methods. They range from general-purpose simulators to those that are specifically customized for spiral inductor modeling in RFICs.

6.8 PCB Design

The printed circuit board (PCB) is used to mechanically support and electrically connect electronic components using conductive pathways, or traces, that are etched from copper sheets laminated onto a nonconductive substrate. In high-volume production, the PCB is inexpensive, rugged and reliable.

6.8.1 The Flow

A typical PCB functional design flow is illustrated in Figure 6.16. In the system specification phase all design-specific requirements are gathered and a specification is created. Next, a schematic representation of the specification is created or captured, using a range of schematic capture tools. Components for the schematic are chosen by the engineer based on a wide range of criteria, which might include such things as price, support and availability. A netlist is then automatically generated, which will serve as input to the PCB Layout tool.

Schematic capture tools allow you to take components from a component library and place them on the schematic. You will need to edit them as required to ensure that the component package details and silkscreen legend to be printed on the PCB are as per requirements.

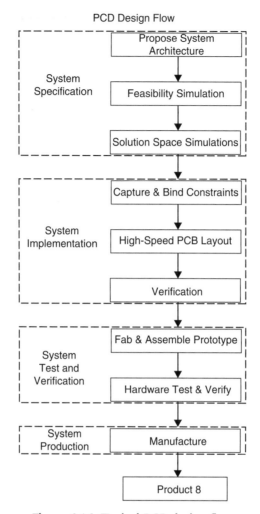

Figure 6.16: Typical PCB design flow

If a component is not in the library, the component is defined and its definition and details are added to the component library. Existing library components may also need to be further edited to reflect the actual circuit diagram, a step which requires that the pins of a particular component have to be moved around as per desired circuit schematics. Once all the parts have been placed on the schematics, the required connections between pins of various components are drawn.

Next comes the PCB layout phase in which the schematic representation is laid out via the use of a PCB layout design tool. Considerations during this design phase include such things as design for manufacturability (DFM), signal integrity and EMI/EMC. These considerations will generally impose certain restrictions on the PCB layout design and should be included with your basic design rules (e.g., the rules that you follow for layout pertaining to things like track width, track spacing, pad sizes, via sizes, and routing types). Verification that the layout will provide the performance and function documented in the specification is then necessary.

During layout, components that form the schematic are taken from the parts library or libraries and placed on the PCB layer. This process is aided by the use of PCB placement and routing (P&R) tools. All component placement constraints are taken into account. For example, heat dissipation considerations may dictate that some parts should be a certain minimum distance away from others, while signals that need to be limited in length may dictate close placement of certain parts that involve those signals.

Following placement, routing takes place. Traditionally, this process would have involved a combination of both manual and automatic routing techniques. Today, though, with 80–90% or more of signals considered critical it is no longer feasible to manually route critical nets. Instead, today's auto routers are able to route "correct by design" signals, following all the signal constraints for each signal. Modern auto routers also route signals in a manufacturing-optimized manner so that manual cleanup is not necessary (Figure 6.17). While some engineers may choose to spend time manually cleaning up their designs, this process is just not "do-able" for very large designs, as it would take several weeks.

After routing, the engineer must perform a DRC check, whereby the PCB layout is tested against the netlist and the set of previously specified design rules. Any violations reported are corrected. After DRC, a final layout cleanup is done.

Note that radio transmitters and radio receivers are especially difficult to design. PCB designers must therefore minimize parasitic effects due to layout of components, or take them into account with a general model and use simulation software such as SPICE. Fortunately, many practical circuits can be laid out using a much simpler lumped element model.

At this stage, a final verification is done to ensure that the design will perform as expected. If any error is found then the PCB design process must be iterated (e.g., from design specification, layout, etc.) until such a time as the prototype passes verification. Once the design has been fully verified, it can be produced on a volume scale for delivery to the customer.

If desired, the engineer can create a prototype for the purpose of final verification, using the design files generated during PCB layout (e.g., ODB++ formatted manufacturing data). More often than not, though, it has become common to skip the prototype altogether. This

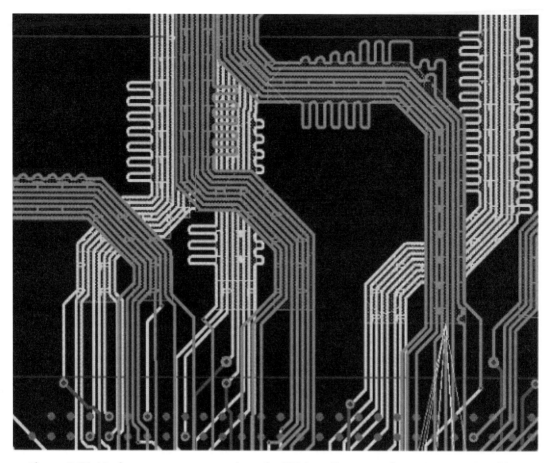

Figure 6.17: Modern auto routers, such as the RE AutoRouter option to Mentor Graphics'
Board Station RE PCB placement-and-routing environment, mimic the manual strategies
used by a designer.

process can be very time-consuming. More importantly, today's verification tools now make it
possible to get the design right the first time, without having to build a prototype.

6.8.2 PCB Design Tools

At every stage throughout this process, design tools can be used to assist the RF engineer.
These tools play a special role in ensuring design success, especially given the high-frequency
circuit impairments in today's complex analog and RFICs, such as compression, noise,
distortion and phase noise, as well as the physical parasitics like interconnect impedance and

coupling, which make achieving design closure between RFIC's system and circuit, electrical and physical, and design and test activities difficult at best.

To aid RF designers, design tool vendors now offer not just point tools, but PCB design flows that are unified with both the IC and package domains. This allows the RF engineer to perform early analysis, I/O buffer planning, optimization and implementation across all three implementation domains.

6.9 Packaging

A package is used to house the component, system or subsystem being designed. RF packages are substantially different from digital IC packages. They operate at much higher frequencies and usually have far fewer I/O's. Consequently, greater care must be taken when doing the layout of individual signal interconnects and transmission line junctions.

Additionally, in the GHz range, the effects of the physical properties of interconnect on electrical signals (e.g., dispersion, radiation losses, resonance and skin effects and even the surface roughness of a conductor) can have a significant impact on RF performance. In fact, the electrical parasitics arising from the packaged product have become such a key problem for chip designers that it has now brought packaging into the spotlight as a system-level concern. These effects, therefore, must be carefully considered when designing RF packaging. Electromagnetic modeling and three-dimensional EM models often play a critical role in the RF package design process.

6.9.1 Options

There are a variety of packaging options and materials to choose from. Which one you choose will depend in part on your end application. For example, if the RF system in question is for a wireless mobile product, then you will need to trade off your packaging choices against customer-driven factors like cost, performance and size.

Packaging options range from standard single-die packages to System-in-Package (SiP). SiP technology builds on the innovative array interconnect of ball grid array (BGA). It allows multiple die with complementary device technologies to be combined in a single package. Passive devices may also be included in the package to deliver highly functional integration for digital and radio frequency applications.

Another packaging option, RF Stacked Die System-in-Package, is typically used in designs where the X-Y size constraint is the critical requirement, such as for wireless communication applications. It allows for the combination of die (e.g., 2 to 4 separate die) from different fabrication processes into a single package. Board area savings are realized by stacking the die vertically versus a side-by-side approach.

RF SiP is supported by design tools like the RF SiP Methodology Kit from Cadence Design Systems. The Kit accelerates the application of advanced EDA technologies to SiP designs for RF/wireless applications. It provides methodologies that maximize design productivity and predictability for customers leveraging the advantages of SiP implementation. An integrated set of SiP design products, built around proven methodologies, enables complete front-to-back SiP design and implementation, which you can leverage for techniques and methods to apply to your own design. The Kit is comprised of a complete documented step-by-step methodology and flow for designing a real, multi chip IC package. It includes all the design data, flow guides, and tutorials to jumpstart customers on a multi-chip RF SiP Module design methodology, including package layout. This methodology is also integrated with Cadence's Virtuoso environment.

RF packages can be designed in a wide range of materials (e.g., ceramics or advanced low dielectric materials). Two of the more common options in use are ceramics and laminates.

6.9.1.1 Ceramics

Ceramics are inherently robust and allow for hermetic packaging—a feature that makes them especially useful in applications requiring high reliability over a long period of time or for those that will be used in hostile environments such as in space or on the battlefield. Because there is such a wide range of ceramics technologies to choose from, the use of this material in RF packaging is quite common today.

Fabrication process options include thin film or thick film. Thin film technology offers the smallest feature size and the highest resolution. In contrast, thick film technology allows the printing of resistors and inductors directly onto the circuit. This technology is particularly useful in RF packaging where normal lumped components can be difficult to apply due to their inherent parasitic properties.

Low temperature co-fired ceramics (LTCC) is a multilayer ceramic substrate technology. It is especially well suited for RF applications and products where a high integration level and/or high reliability is needed.

Ceramic material options range from a standard material like high temperature co-fired ceramic (HTCC) alumina to more specialized options, like aluminum nitride with its excellent thermal properties or HiTCE ceramics, which have thermal expansion coefficients that are better matched to the expansion coefficient of PCB materials. Ceramic materials are also available with a variety of electrical properties. Lower dielectric constant ceramics, for example, allow for larger feature sizes and reduced propagation delays.

6.9.1.2 Laminates

A laminate is a material that is made up of bonded layers. A hybrid laminate is a PCB like structure, normally with a FR4 material core, that uses advanced materials on selected layers in order to meet rigorous RF performance requirements. These types of package substrates are particularly useful for RF systems where signals in the GHz frequency range may be routed on the substrate surface between various active circuit elements. They have become increasingly important in RF packaging due to their low cost.

6.9.2 Design Solutions

Given the possible trade-offs and packaging options RF engineers now have at their disposal, it is clear to see that designing an RF package requires a broad knowledge of material properties and assembly processes, as well as a thorough understanding of high-frequency signal behavior. Making the right choice can either make or break the success of your product. Therefore, the choice of packaging should never be an afterthought. You must implement a packaging strategy during the design phase in order to ensure that you will meet all your requirements for reliability, manufacturability, RF performance, size and cost.

Various EDA tools are now available to help you design an RF package. In addition to the EM modeling tools previously mentioned, there are also advanced packaging kits, such as the RF SiP Methodology Kit from Cadence Design Systems, which offer guidance on different design techniques (e.g., for SiP). They cover RF components, together with packaging strategies, to help you meet requirements for reliability, manufacturability, RF performance, size and cost. Kits also explain design rules for different types of packaging such as laminate and LTCC modules, as well as the trade-offs between them, and present working guidelines for IC partitioning decisions early in the design phase.

6.10 Case Study

Now that you have a better understanding of EDA tools, it is time to take a closer look at how they can be utilized to streamline the design process in the real world. The following case study examines the design and analysis of an IEEE 802.11a (wireless local area network) RF complementary metal-oxide semiconductor (CMOS) transceiver using the Analog Office software solution and design flow illustrated in Figure 6.18. The design employs

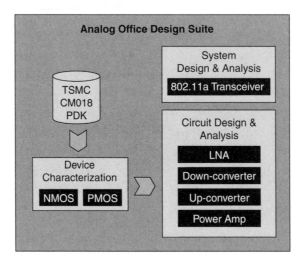

Figure 6.18: The 802.11a design flow within the Analog Office design environment from
Applied Ware Research

a conventional configuration widely used in the silicon RFIC community and 0.18-μm CMOS
technology from Taiwan Semiconductor Manufacturing Company. Note that other design
tools—such as the Genesys software environment from Agilent Technologies—may also be
used for this design process.

6.10.1 System-Level Transceiver Design

The system-level diagram of the 802.11a transceiver, shown in Figure 6.19, is captured
using pre-built behavioral blocks in the Visual System Simulator (VSS) software. The
design employs a homodyne architecture, i.e., direct conversion of the RF signal in 5.15
to 5.825 GHz band to baseband. The transceiver consists of a baseband and RF circuits.
The transmitter baseband circuits include a 64 quadrature amplitude modulation (QAM)
modulator and orthogonal frequency division multiplexing (OFDM) modulator. Receiver
baseband circuits include a 64 QAM demodulator and OFDM demodulator. The transmitter
RF circuits are comprised of up-converter mixers and power amplifiers, while the receiver RF
circuits include up-converter mixers and low-noise receiver amplifiers. The local oscillator
frequency is set to RF frequency as per the homodyne architecture.

The goal of the system-level design is to determine RFIC specifications from system-
level specifications. During system simulation, the noise figure and nonlinearity of the

**802.11a/g, 54 Mb/s 64-QAM Signal Generator
and BER measurement**

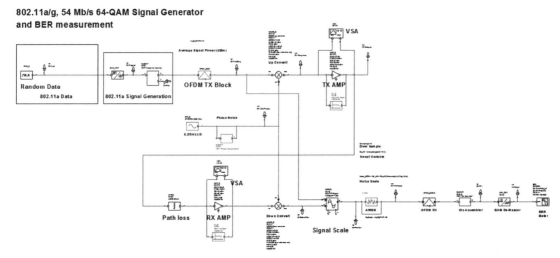

Figure 6.19: The 802.11a transceiver system-level diagram with behavioral blocks representing major functions of the entire RF transceiver chain, including the channel characteristics

power amplifier (PA), LNA, and up-and down-converter mixers and phase noise of the local oscillator (LO), are varied and system packet error rate (PER) is monitored against specifications.

Prior to the start of the circuit design phase, the TSMC 0.18-μm RF CMOS process is selected. A process design kit (PDK) is then developed and installed within the Analog Office design environment.

6.10.2 Circuit-Level Receiver Design

The homodyne receiver portion of the IEEE 802.11a transceiver consists of two main RF circuits: an LNA and a down-converter. A modulated RF signal in the 5.150 to 5.825 GHz band is the input to the LNA. The LNA must provide a sufficient gain to minimize the impact of mixer noise on the overall noise figure of the receiver. A single-ended, 50 Ω input source is assumed for the LNA, which amplifies the input signal by 20 dB, with a noise figure of 1.2 dB.

The down-converter directly converts the amplified RF signal to DC. It consists of two mixers with LO signals offset by 90° between the I and Q mixers. The mixer has a conversion gain of 6 dB, and noise figure of 5 dB.

6.10.3 LNA Design

The LNA selected is a differential cascode common source amplifier. Note that in this example, noise figure, power consumption, input impedance, gain, and linearity are considered together in designing the LNA.

6.10.4 Device Characterization

The design optimizes the noise figure of the LNA, while taking power consumption into account. The optimum device width is about 160 μm (NR = 64). DC and AC characteristics of a device with a width of 160 μm are simulated, as shown in Figure 6.20. The DC simulation displays the drain current (I_{ds}) with various bias settings of drain to source voltage

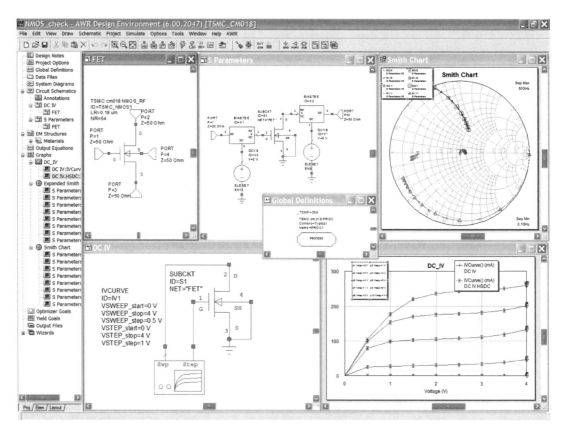

Figure 6.20: The NMOS transistor is characterized before use in the circuit to ensure accurate and adequate performance.

(V_{ds}), and gate to source voltage (V_{gs}) of the negative-channel metal-oxide semiconductor (NMOS) device. A set-up for measuring I_{ds} vs. V_{ds} for various settings of V_{gs} is also shown. A set-up for measuring I_{ds} vs. V_{gs} can be used to extract g_m.

AC simulation displays f_T and Y- and S-parameters. A set-up for measuring a gain of a transistor versus frequency can be used to observe f_T of the transistor at a particular bias setting as the gain goes to 1. For the bias setting of interest, the f_T of the device is about 44 GHz. It is also used to monitor the Y-parameters. A setup for measuring S-parameters of the transistor at a particular bias setting is shown.

6.10.5 Circuit Design

Once the device characteristics are simulated, a bias current of 14 mA is selected for the differential pair, as shown in Figure 6.21. An estimated noise figure within the power

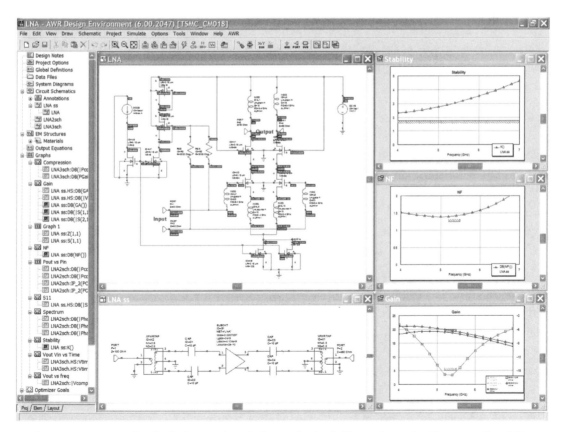

Figure 6.21: LNA circuit design and analysis results including gain, noise figure, and stability

constraint is about 1.5 to 1.9 dB. A noise simulation shows a noise figure of 1.2 dB in the frequency range of 4 to 6 GHz and 1.1 dB at 5.2 GHz.

Next, input impedance is matched to 100 Ω differential impedance. Since input impedance of an NMOS device is mostly capacitive, a real term of the impedance must be provided. In order to minimize the noise, an inductive source degeneration method is used to present a real term in the frequency of interest. The capacitive term of input impedance is "resonated out" by a series inductor at the input. A high Q of the series inductor is desirable to minimize the noise figure. S11 is −20 to −30 dB in the frequency of interest.

Note that output impedance is not matched to 100 Ω differential because the LNA is intended to be integrated with an I/Q demodulator. An inductor is used as a load at the drain to provide more headroom, to resonate out the capacitance seen at the output load, and to provide sufficient gain. The Q of the load inductors at the drain must be low enough to cover the entire band of 5.15 to 5.85 GHz. The gain of the LNA is 20 dB over the band of interest.

A stability factor, K, of the LNA is greater than 1 to ensure the stability over the band of interest and other frequencies. To prevent a negative resistance and thus avoid instability, it is important that the differential pair not see a capacitance at the source. In order to take advantage of automatic measurements of impedance, S-parameter, gain, noise, 1-dB compression point (and IP3), and intermodulation components, a balun is used to convert the differential inputs of the LNA input single-ended.

The LNA circuit is duplicated with differential input signal sources for different simulations. One circuit is used to simulate gain, noise, S11, and Z_{in}, and stability factor K. Another circuit with two tone frequencies and a power swept at the input port is used to simulate input IP3 and 1-dB compression point. The 1-dB compression point is −10 dBm and IP3 is >10 dBm.

Another circuit with a single tone frequency and a power at the input port is used to simulate the input and output voltages in time domain.

6.10.6 Down-Converter Circuit Design

A down-converter consists of two mixers, converting from RF to DC. Figure 6.22 shows an example of an RF mixer design using a Gilbert cell configuration. One mixer circuit is used for the simulation. The simulation shows a noise figure of only 5 dB, so it may not be including all the noise sources. The bias and the load resistor determine the conversion gain. The conversion gain is about 6 dB as shown in S21 of a large signal S-parameter simulation and noise gain simulation.

The source degeneration inductors at the source of the RF differential pair improve the linearity of the mixer. The 1-dB compression point is about −10 dBm of input power at the mixer RF port.

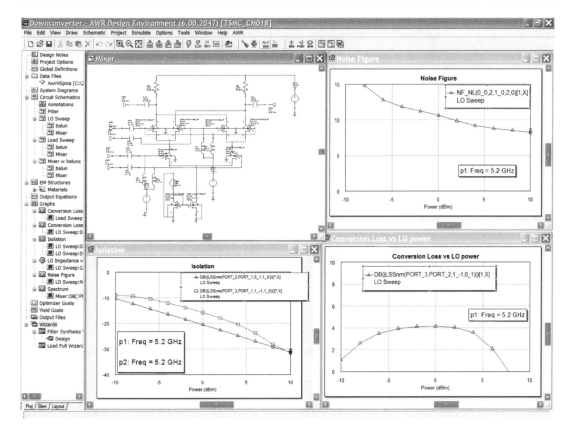

Figure 6.22: The down-converter mixer design circuit schematic with bias voltages annotated and several simulation results including isolation, conversion loss vs. LO power and noise figure

The isolation of LO-IF ports and LO-RF ports is simulated by mismatching the RF driver differential pair and LO switching pairs. The LO signal is monitored at the IF and RF ports separately, using a large signal S-parameter (LSS) NM measurement.

Another identical mixer circuit with a single RF input power level is used to simulate the intermodulation components of the mixer.

6.10.7 Transmitter Circuit Design

The direct conversion transmitter consists of two main RF circuits: an up-converter and a PA. The PA in the design has an output power of 40 mW and outputs a modulated RF signal in the 5.18 to 5.26 GHz band.

6.10.8 Up-Converter Design

The up-converter consists of two double-balanced active mixers for I and Q modulation. The lower differential pair takes baseband signals and the upper switching pairs take differential local oscillator signals tuned to the desired RF output signal. The local oscillator frequency should be varied from 5.18 to 5.26 GHz. For simulation purposes, the input center frequency is 50 MHz. If the outputs of I and Q paths are summed, the lower sideband will be selected as the output. If the outputs of I and Q paths are subtracted, the upper sideband will be selected as the output. The linearity requirement is set high to ensure no degradation of linearity in the overall transmitter for OFDM and 64 QAM modulation. The 1-dB compression point is set 10 dB higher than the operating power level. For instance, the input 1-dB compression point is 8 dBm and the input power level in use is −2 dBm or less. An RF system simulation using VSS software with a behavioral model for up-converter showed 8 dB back-off from the 1-dB compression point to guarantee no BER degradation, assuming 4 to 5 dB back-off from the 1-dB compression point for PA operation.

6.10.9 Mixer Design

The bias and size of the transistors are selected to provide the necessary linearity and gain. The mixer has a conversion loss of about 5 dB, as shown by the LSS_{21} measurement if the differential LO power level is 10 dBm. The noise figure is about 1.5 dB, which means it is not simulating correctly. The large signal impedance is monitored at the LO port for matching purposes.

It is important to provide a proper isolation between the LO and the output port. The LSS measurement monitors the LO signal level at the output port. The 1-dB compression point at the input is 8 dBm as shown by sweeping the baseband input power as the RF output is monitored. 8 dBm of a differential input power is referenced to a $100\,\Omega$ input impedance load. Since the input of the up-converter is at baseband frequency, the input signal should have an option to sweep voltages rather than power. Two-tone simulations should show the intermodulation and harmonic components. A two-tone harmonic balance source with a fixed or swept power is applied to the input port while the intermodulation components are monitored at the output port. Small signal impedance and S-parameters as well as large signal impedance are monitored for the matching. Figure 6.23 shows the various measurements mentioned above.

6.10.10 PA Design

The PA has a single-ended, common-source configuration with two gain stages. The target frequency band is 5.18–5.26 GHz. Although the average output power in this frequency band

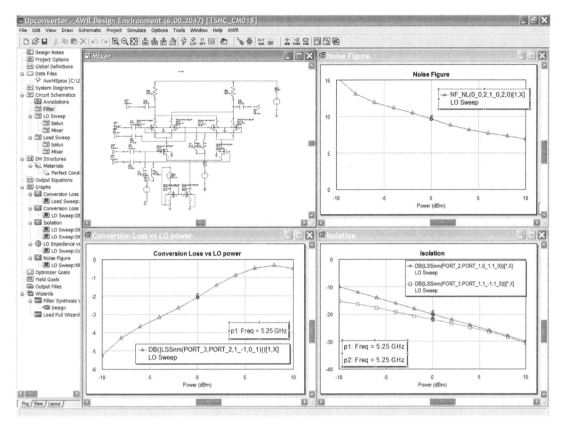

Figure 6.23: Up-converter mixer design and analysis results, including noise figure, conversion loss vs. LO power, and isolation

is 40 mW (16 dBm), OFDM and 64 QAM modulation signaling raises the required output peak power to 21 dBm. An RF system simulation using VSS software with a behavioral model for the PA also shows that 21 dBm is the required 1-dB compression point at the output. It is challenging to design a PA to deliver the necessary output power with a required linearity in CMOS technology at a low power supply voltage. Therefore, the bulk of the effort in this design is focused on exploring the design space to achieve optimum linearity.

6.10.11 PA Device Characterization

The output device size is selected by sweeping the current vs. bias and device size, given the trade-off between power gain, linearity and efficiency. Allowing for loss, the output power is designed for 22 to 23 dBm. Although the output resistance of $22\,\Omega$ is required to deliver 23 dBm with a 3 V supply voltage, a larger device of $960\,\mu\text{m}/0.18\,\mu\text{m}$ (thus smaller output resistance) is selected to achieve a better linearity and power gain. The gate is biased with a trade-off between the linearity and efficiency. The emphasis, however, is placed on the linearity, and therefore, it is biased as a class A amplifier (with $V_{gs} - V_{th} > 0.1\,\text{V}$). The device size of the first stage is $480\,\mu\text{m}/0.18\,\mu\text{m}$.

6.10.12 PA Circuit Design

Given the output power and linearity requirement, the transistors are sized to provide the necessary load and gain. Each stage is first designed separately and then combined to resimulate the overall performance. Since a PA operates at large input and output levels, it is important to simulate it in a large signal domain. For instance, gain contours with an optimal load can be displayed in a Smith Chart. In Figure 6.24, four gain contours with maximum

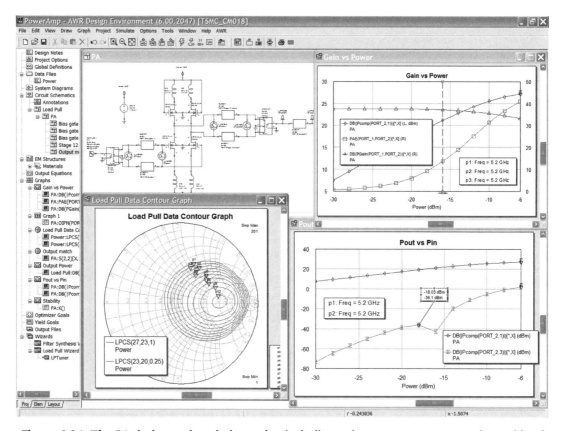

Figure 6.24: The PA design and analysis results, including gain vs. power, pout vs. pin, and load pull data contour

incremental power gains of 20.8 dB and 0.1 dB are displayed along with S11 and S22. The input and output matching should maximize the gain. The contour plot shows that they are optimally matched. S11 and S22 are better than −25 dB over the frequency band of interest.

Prior to matching to 50 Ω input and output ports, the circuit must be stabilized. Since the input impedance of the circuit is negative, a small inductor at the emitter to ground (simulating the bond wire connection) should help without degrading the noise figure. Additionally, a small resistor at the gate can provide stability over a wider band. The K value is greater than 2 at the frequency band of interest.

A proper matching of input and output delivers an optimal power gain. A matching network consists of a DC blocking capacitor to prevent the loading of circuits and an impedance transformation. A high pass L-match is used. The impedance at the input and output is transformed up to 50 Ω. Since the input and output impedance changes with the power level, the matching should be monitored at a varying power level. The software displays large-signal impedance (Z-comp) as the input power level is swept.

The power-added efficiency is displayed by selecting "PAE". It is about 27% at the 1-dB compression point, as the PA is designed for high linearity.

Since the linearity is critical, it is important to simulate a 1-dB compression point and IM3 and IP3. The software simulates them by applying a single-tone or two-tone harmonic balance source. Two-tone harmonic balance simulation shows a large signal S21 (LSS21) of 21 dB with an input 1-dB compression point at 3 dBm (output at 22.6 dBm). The output IP3 is about 40 dBm and IM3 is about 33 dB down from the output power level of 18 dBm. The 1-dB compression point at the output of 22.6 dBm allows for better than 6 dB of peak-to-average ratio. The design target was 5 dB.

The resulting 802.11a RF CMOS transceiver is an optimum design. The LNA achieves adequate gain, low noise, and IP3, and input match and stability are also adequate. The mixer achieves adequate conversion gain and isolation, low noise figure, and IP3.

6.11 Summary

EDA solutions for the RF industry are certainly far from being as mature and complete as those available to digital designers. Nonetheless, they can be used to more accurately and quickly design RFICs and PCBs. A number of design flows and tools now support these efforts, regardless of how simple or complex the design may be.

CHAPTER 7

On Memory Systems and Their Design

Bruce Jacob
Spencer W. Ng
David T. Wang

Memory matters. As Bruce Jacob points out, the increasing gap between processor and memory speeds has made memory organization and architecture a critical part of portable design. In fact, to Jacob, it's the most important part of the design. With tongue only slightly in cheek, he claims "little else matters right now, so stop wasting time and resources on other facets of the design."

He has a point. Changes to the memory hierarchy can indeed, as he claims, often improve execution time by 2–10✕, in contrast to changes to processor parameters, which are more likely to make a difference of 2–10%. They can also have a similar impact on the energy profile of portable designs, which are ultimately the barrier between what handset buyers, for example, want and what you can deliver.

As someone who learned to program on a DEC 11/70, I hated the segmented architecture of Intel's 8086, with all the hoops it made you jump through so you could work in a tiny memory space. A flat memory architecture is still alluring, but it's not a reasonable option. You need a memory hierarchy that can keep up with the processor, have the lowest possible power profile and a cost-per-bit approaching that of the cheapest component. This generally translates to using SRAM for program cache, DRAM as an operating store and a hard disk for permanent storage. With distributed memory, you then get to deal with bus latency and contention issues, which is enough to make you nostalgic for a large, sequential address space.

Jacob takes a holistic approach to memory system design. He claims that modular design—where separate teams work on different subsystems—is a dangerously outmoded concept. That approach assumes there is no interaction between individual modules and the system architecture, when in fact interconnect physics, caching and scheduling policies can have significant interactive effects throughout the memory system. He makes the point repeatedly and forcefully:

This is extremely important, so it bears repeating: the bulk of lost performance is not due to the number of CPU pipeline stages or functional units or choice of branch prediction algorithms or even CPU clock speed; the bulk of lost performance is due to poor configuration of system-level parameters such as bus widths, granularity of access, scheduling policies, queue organizations, etc.

Jacob's book Memory Systems: Cache, DRAM, Disk *is the first systematic treatment of the whole memory hierarchy in a very long time. While he considers the following chapter an overview, it has both the breadth and enough depth to make it a very worthwhile read for any system designer.*

—**John Donovan**

Memory is essential to the operation of a computer system, and nothing is more important to the development of the modern memory system than the concept of the memory hierarchy. While a flat memory system built of a single technology is attractive for its simplicity, a well-implemented hierarchy allows a memory system to approach simultaneously the performance of the fastest component, the cost per bit of the cheapest component, and the energy consumption of the most energy-efficient component.

For years, the use of a memory hierarchy has been very convenient, in that it has simplified the process of designing memory systems. The use of a hierarchy allowed designers to treat system design as a modularized process—to treat the memory system as an abstraction and to optimize individual subsystems (caches, DRAMs [dynamic RAM], disks) in isolation.

However, we are finding that treating the hierarchy in this way—as a set of disparate subsystems that interact only through well-defined functional interfaces and that can be optimized in isolation—no longer suffices for the design of modern memory systems. One trend becoming apparent is that many of the underlying implementation issues are becoming significant. These include the physics of device and interconnect scaling, the choice of signaling protocols and topologies to ensure signal integrity, design parameters such as granularity of access and support for concurrency, and communication-related issues such as scheduling algorithms and queueing. These low-level details have begun to affect the higher level design process quite dramatically, whereas they were considered transparent only a design-generation ago. Cache architectures are appearing that play to the limitations imposed by interconnect physics in deep submicron processes; modern DRAM design is driven by circuit-level limitations that create system-level headaches; and modern disk performance is dominated by the on-board caching and scheduling policies. This is a nontrivial environment in which to attempt optimal design.

This trend will undoubtedly become more important as time goes on, and even now it has tremendous impact on design results. As hierarchies and their components grow more complex, *systemic* behaviors—those arising from the complex interaction of the memory system's parts—have begun to dominate. The real loss of performance is not seen in the CPU or caches or DRAM devices or disk assemblies themselves, but in the subtle interactions between these subsystems and in the manner in which these subsystems are connected. Consequently, it is becoming increasingly foolhardy to attempt system-level optimization by designing/optimizing each of the parts in isolation (which, unfortunately, is often the approach taken in modern computer design). No longer can a designer remain oblivious to issues "outside the scope" and focus solely on designing a subsystem. It has now become the case that a memory-systems designer, wishing to build a properly behaved memory hierarchy, must be intimately familiar with issues involved at all levels of an implementation, from cache to DRAM to disk. Thus, we wrote this book.

7.1 Memory Systems

A memory hierarchy is designed to provide multiple functions that are seemingly mutually exclusive. We start at random-access memory (RAM): all microprocessors (and computer systems in general) expect a random-access memory out of which they operate. This is fundamental to the structure of modern software, built upon the von Neumann model in which code and data are essentially the same and reside in the same place (i.e., memory). All requests, whether for instructions or for data, go to this random-access memory. At any given moment, any particular datum in memory may be needed; there is no requirement that data reside next to the code that manipulates it, and there is no requirement that two instructions executed one after the other need to be adjacent in memory. Thus, the memory system must be able to handle randomly addressed[1] requests in a manner that favors no particular request. For instance, using a tape drive for this primary memory is unacceptable for performance reasons, though it might be acceptable in the Turing-machine sense.

Where does the mutually exclusive part come in? As we said, all microprocessors are built to expect a random-access memory out of which they can operate. Moreover, this memory must be *fast*, matching the machine's processing speed; otherwise, the machine will spend most of its time tapping its foot and staring at its watch. In addition, modern software is written to expect gigabytes of storage for data, and the modern consumer expects this storage to be cheap. How many memory technologies provide both tremendous speed and tremendous

---

[1] Though "random" addressing is the commonly used term, authors actually mean *arbitrarily* addressed requests because, in most memory systems, a *randomly* addressed sequence is one of the most efficiently handled events.

storage capacity at a low price? Modern processors execute instructions both out of order and speculatively—put simply, they execute instructions that, in some cases, are not meant to get executed—and system software is typically built to expect that certain changes to memory are permanent. How many memory technologies provide nonvolatility and an *undo* operation?

While it might be elegant to provide all of these competing demands with a single technology (say, for example, a gigantic battery-backed SRAM [static RAM]), and though there is no engineering problem that cannot be solved (if ever in doubt about this, simply query a room full of engineers), the reality is that building a full memory system out of such a technology would be prohibitively expensive today.[2] The good news is that it is not necessary. Specialization and division of labor make possible all of these competing goals simultaneously. Modern memory systems often have a terabyte of storage on the desktop and provide instruction-fetch and data-access bandwidths of 128 GB/s or more. Nearly all of the storage in the system is nonvolatile, and speculative execution on the part of the microprocessor is supported. All of this can be found in a memory system that has an average cost of roughly 1/100,000,000 pennies per bit of storage.

The reason all of this is possible is because of a phenomenon called *locality of reference* [Belady 1966, Denning 1970]. This is an observed behavior that computer applications tend to exhibit and that, when exploited properly, allows a small memory to serve in place of a larger one.

7.1.1 Locality of Reference Breeds the Memory Hierarchy

We think linearly (in steps), and so we program the computer to solve problems by working in steps. The practical implications of this are that a computer's use of the memory system tends to be nonrandom and highly predictable. Thus is born the concept of *locality of reference*, so named because memory references tend to be localized in time and space:

- If you use something once, you are likely to use it again.
- If you use something once, you are likely to use its neighbor.

The first of these principles is called *temporal locality;* the second is called *spatial locality*. For now it suffices to say that one can exploit the locality principle and render a single-level

---

[2] Even Cray machines, which were famous for using SRAM as their main memory, today are built upon DRAM for their main memory.

memory system, which we just said was expensive, unnecessary. If a computer's use of the memory system, given a small time window, is both predictable and limited in spatial extent, then it stands to reason that a program does not need all of its data immediately accessible. A program would perform nearly as well if it had, for instance, a *two-level* store, in which the first level provides immediate access to a subset of the program's data, the second level holds the remainder of the data but is slower and therefore cheaper, and some appropriate heuristic is used to manage the movement of data back and forth between the levels, thereby ensuring that the most-needed data is usually in the first-level store.

This generalizes to the *memory hierarchy*: multiple levels of storage, each optimized for its assigned task. By choosing these levels wisely a designer can produce a system that has the best of all worlds: performance approaching that of the fastest component, cost per bit approaching that of the cheapest component, and energy consumption per access approaching that of the least power-hungry component.

The modern hierarchy is comprised of the following components, each performing a particular function or filling a functional niche within the system:

Cache (SRAM): Cache provides access to program instructions and data that has very low latency (e.g., 1/4 nanosecond per access) and very high bandwidth (e.g., a 16-byte instruction block and a 16-byte data block per cycle => 32 bytes per 1/4 nanosecond, or 128 bytes per nanosecond, or 128 GB/s). It is also important to note that cache, on a per-access basis, also has relatively low energy requirements compared to other technologies.

DRAM: DRAM provides a random-access storage that is relatively large, relatively fast, and relatively cheap. It is large and cheap compared to cache, and it is fast compared to disk. Its main strength is that it is just fast enough and just cheap enough to act as an operating store.

Disk: Disk provides permanent storage at an ultra-low cost per bit. As mentioned, nearly all computer systems expect *some* data to be modifiable yet permanent, so the memory system must have, at some level, a permanent store. Disk's advantage is its very reasonable cost (currently less than 50¢ per gigabyte), which is low enough for users to buy enough of it to store thousands of songs, video clips, photos, and other memory hogs that users are wont to accumulate in their accounts (authors included).

Table 7.1 lists some rough order-of-magnitude comparisons for access time and energy consumption per access.

Why is it not feasible to build a flat memory system out of these technologies? Cache is far too expensive to be used as permanent storage, and its cost to store a single album's worth

Table 7.1: Cost-performance for various memory technologies

Technology	Bytes per Access (typ)	Latency per Access	Cost per Megabyte[a]	Energy per Access
On-chip Cache	10	100 of picoseconds	$1–100	1 nJ
Off-chip Cache	100	Nanoseconds	$1–10	10–100 nJ
DRAM	1000 (internally fetched)	10–100 nanoseconds	$0.1	1–100 nJ (per device)
Disk	1000	Milliseconds	$0.001	100–1000 mJ

[a] Cost of semiconductor memory is extremely variable, dependent much more on economic factors and sales volume than on manufacturing issues. In particular, on-chip caches (i.e., those integrated with a microprocessor core) can take up half of the die area, in which case their "cost" would be half of the selling price of that microprocessor. Depending on the market (e.g., embedded versus high end) and sales volume, microprocessor costs cover an enormous range of prices, from pennies per square millimeter to several dollars per square millimeter.

of audio would exceed that of the original music CD by several orders of magnitude. Disk is far too slow to be used as an operating store, and its average seek time for random accesses is measured in milliseconds. Of the three, DRAM is the closest to providing a flat memory system. DRAM is sufficiently fast enough that, without the support of a cache front-end, it can act as an operating store for many embedded systems, and with battery back-up it can be made to function as a permanent store. However, DRAM alone is not cheap enough to serve the needs of human users, who often want nearly a terabyte of permanent storage, and, even with random access times in the tens of nanoseconds, DRAM is not quite fast enough to serve as the only memory for modern general-purpose microprocessors, which would prefer a new block of instructions every fraction of a nanosecond.

So far, no technology has appeared that provides every desired characteristic: low cost, nonvolatility, high bandwidth, low latency, etc. So instead we build a system in which each component is designed to offer one or more characteristics, and we manage the operation of the system so that the poorer characteristics of the various technologies are "hidden." For example, if most of the memory references made by the microprocessor are handled by the cache and/or DRAM subsystems, then the disk will be used only rarely, and, therefore, its extremely long latency will contribute very little to the average access time. If most of the data resides in the disk subsystem, and very little of it is needed at any given moment in time, then the cache and DRAM subsystems will not need much storage, and, therefore, their higher costs per bit will contribute very little to the average cost of the system. If done right, a memory system has an average cost approaching that of bottommost layer and an average access time and bandwidth approaching that of topmost layer.

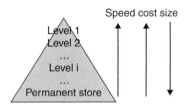

Figure 7.1: A memory hierachy

The memory hierarchy is usually pictured as a pyramid, as shown in Figure 7.1. The higher levels in the hierarchy have better performance characteristics than the lower levels in the hierarchy; the higher levels have a higher cost per bit than the lower levels; and the system uses fewer bits of storage in the higher levels than found in the lower levels.

Though modern memory systems are comprised of SRAM, DRAM, and disk, these are simply technologies chosen to serve particular needs of the system, namely permanent store, operating store, and a fast store. Any technology set would suffice if it (a) provides permanent and operating stores and (b) satisfies the given computer system's performance, cost, and power requirements.

7.1.1.1 Permanent Store

The system's permanent store is where everything lives … meaning it is home to data that can be modified (potentially), but whose modifications must be remembered across invocations of the system (power-ups and power-downs). In general-purpose systems, this data typically includes the operating system's files, such as boot program, OS (operating system) executable, libraries, utilities, applications, etc., and the users' files, such as graphics, word-processing documents, spreadsheets, digital photographs, digital audio and video, email, etc. In embedded systems, this data typically includes the system's executable image and any installation-specific configuration information that it requires. Some embedded systems also maintain in permanent store the state of any partially completed transactions to withstand worst-case scenarios such as the system going down before the transaction is finished (e.g., financial transactions).

These all represent data that should not disappear when the machine shuts down, such as a user's saved email messages, the operating system's code and configuration information, and applications and their saved documents. Thus, the storage must be *nonvolatile*, which in this context means not susceptible to power outages. Storage technologies chosen for permanent store include magnetic disk, flash memory, and even EEPROM (electrically erasable programmable read-only memory), of which flash memory is a special type. Other forms of programmable ROM (read-only memory) such as ROM, PROM (programmable ROM), or EPROM (erasable programmable ROM) are suitable for nonwritable permanent information such as the executable image of an embedded system or a general-purpose system's boot code and BIOS.[3] Numerous exotic nonvolatile technologies are in development, including magnetic RAM (MRAM), FeRAM (ferroelectric RAM), and phase-change RAM (PCRAM).

In most systems, the cost per bit of this technology is a very important consideration. In general-purpose systems, this is the case because these systems tend to have an enormous amount of permanent storage. A desktop can easily have more than 500 GB of permanent store, and a departmental server can have one hundred times that amount. The enormous number of bits in these systems translates even modest cost-per-bit increases into significant dollar amounts. In embedded systems, the cost per bit is important because of the significant number of units shipped. Embedded systems are often consumer devices that are manufactured and sold in vast quantities, e.g., cell phones, digital cameras, MP3 players, programmable thermostats, and disk drives. Each embedded system might not require more than a handful of megabytes of storage, yet a tiny 1¢ increase in the cost per megabyte of memory can translate to a $100,000 increase in cost per million units manufactured.

7.1.1.2 Operating (Random-Access) Store

As mentioned earlier, a typical microprocessor expects a new instruction or set of instructions on every clock cycle, and it can perform a data-read or data-write every clock cycle. Because the addresses of these instructions and data need not be sequential (or, in fact, related in any detectable way), the memory system must be able to handle *random access*—it must be able to provide instant access to any datum in the memory system.

The machine's operating store is the level of memory that provides random access at the microprocessor's data granularity. It is the storage level out of which the microprocessor could conceivably operate, i.e., it is the storage level that can provide random access to its storage, one data word at a time. This storage level is typically called "main memory." Disks cannot serve as main memory or operating store and cannot provide random access for two reasons: instant access is provided for only the data underneath the disk's head at any given moment, and the granularity of access is not what a typical processor requires. Disks are block-oriented devices, which means they read and write data only in large chunks; the typical granularity is 512 B. Processors, in contrast, typically operate at the granularity of 4 B

---

[3] BIOS = basic input/output system, the code that provides to software low-level access to much of the hardware.

or 8 B data words. To use a disk, a microprocessor must have additional buffering memory out of which it can read one instruction at a time and read or write one datum at a time. This buffering memory would become the *de facto* operating store of the system.

Flash memory and EEPROM (as well as the exotic nonvolatile technologies mentioned earlier) are potentially viable as an operating store for systems that have small permanent-storage needs, and the nonvolatility of these technologies provides them with a distinct advantage. However, not all are set up as an ideal operating store; for example, flash memory supports word-sized reads but supports only block-sized writes. If this type of issue can be handled in a manner that is transparent to the processor (e.g., in this case through additional data buffering), then the memory technology can still serve as a reasonable hybrid operating store.

Though the nonvolatile technologies seem positioned perfectly to serve as operating store in all manner of devices and systems, DRAM is the most commonly used technology. Note that the only requirement of a memory system's operating store is that it provide random access with a small access granularity. Nonvolatility is not a requirement, so long as it is provided by another level in the hierarchy. DRAM is a popular choice for operating store for several reasons: DRAM is faster than the various nonvolatile technologies (in some cases *much* faster); DRAM supports an unlimited number of writes, whereas some nonvolatile technologies start to fail after being erased and rewritten too many times (in some technologies, as few as 1–10,000 erase/write cycles); and DRAM processes are very similar to those used to build logic devices. DRAM can be fabricated using similar materials and (relatively) similar silicon-based process technologies as most microprocessors, whereas many of the various nonvolatile technologies require new materials and (relatively) different process technologies.

7.1.1.3 Fast (and Relatively Low-Power) Store

If these storage technologies provide such reasonable operating store, why, then, do modern systems use cache? Cache is inserted between the processor and the main memory system whenever the access behavior of the main memory is not sufficient for the needs or goals of the system. Typical figures of merit include performance and energy consumption (or power dissipation). If the performance when operating out of main memory is insufficient, cache is interposed between the processor and main memory to decrease the average access time for data. Similarly, if the energy consumed when operating out of main memory is too high, cache is interposed between the processor and main memory to decrease the system's energy consumption.

The data in Table 7.1 should give some intuition about the design choice. If a cache can reduce the number of accesses made to the next level down in the hierarchy, then it potentially reduces both execution time and energy consumption for an application. The gain is only potential because these numbers are valid only for certain technology parameters. For example, many designs use large SRAM caches that consume much more energy than several DRAM chips combined, but because the caches can reduce execution time they are used in systems where performance is critical, even at the expense of energy consumption.

It is important to note at this point that, even though the term "cache" is usually interpreted to mean SRAM, a cache is merely a concept and as such imposes no expectations on its implementation. Caches are best thought of as compact databases, as shown in Figure 7.2. They contain data and, optionally, metadata such as the unique ID (address) of each data block in the array, whether it has been updated recently, etc. Caches can be built from SRAM, DRAM, disk, or virtually any storage technology. They can be managed completely in hardware and thus can be transparent to the running application and even to the memory system itself; and at the other extreme they can be explicitly managed by the running application. For instance, Figure 7.2 shows that there is an optional block of metadata, which if implemented in hardware would be called the cache's *tags*. In that instance, a key is passed to the tags array, which produces either the location of the corresponding item in the data array (a *cache hit*) or an indication that the item is not in the data array (a *cache miss*). Alternatively, software can be written to index the array explicitly, using direct cache-array addresses, in which case the key lookup (as well as its associated tags array) is unnecessary. The configuration chosen for the cache is called its *organization*. Cache organizations exist at all spots along the continuum between these two extremes. Clearly, the choice of organization will significantly impact the cache's performance and energy consumption.

Predictability of access time is another common figure of merit. It is a special aspect of performance that is very important when building real-time systems or systems with highly

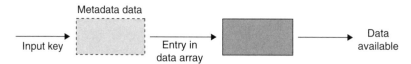

Figure 7.2: An idealized cache lookup. A cache is logically comprised of two elements: the data array and some management information that indicates what is in the data array (labeled "metadata"). Note that the key information may be virtual, i.e., data addresses can be embedded in the software using the cache, in which case there is no explicit key lookup, and only the data array is needed.

orchestrated data movement. DRAM is occasionally in a state where it needs to ignore external requests so that it can guarantee the integrity of its stored data (this is called *refresh*). Such hiccups in data movement can be disastrous for some applications. For this reason, many microprocessors, such as digital signal processors (DSPs) and processors used in embedded control applications (called *microcontrollers*), often have special caches that look like small main memories. These are *scratch-pad RAMs* whose implementation lies toward the end of the spectrum at which the running application manages the cache explicitly. DSPs typically have two of these scratch-pad SRAMs so that they can issue on every cycle a new *multiply-accumulate (MAC)* operation, an important DSP instruction whose repeated operation on a pair of data arrays produces its dot product. Performing a new MAC operation every cycle requires the memory system to load new elements from two different arrays simultaneously in the same cycle. This is most easily accomplished by having two separate data busses, each with its own independent data memory and each holding the elements of a different array.

Perhaps the most familiar example of a software-managed memory is the processor's *register file*, an array of storage locations that is indexed directly by bits within the instruction and whose contents are dictated entirely by software. Values are brought into the register file explicitly by software instructions, and old values are only overwritten if done so explicitly by software. Moreover, the register file is significantly smaller than most on-chip caches and typically consumes far less energy. Accordingly, software's best bet is often to optimize its use of the register file [Postiff & Mudge 1999].

7.1.2 Important Figures of Merit

The following issues have been touched on during the previous discussion, but at this point it would be valuable to formally present the various figures of merit that are important to a designer of memory systems. Depending on the environment in which the memory system will be used (supercomputer, departmental server, desktop, laptop, signal-processing system, embedded control system, etc.), each metric will carry more or less weight. Though most academic studies tend to focus on one axis at a time (e.g., performance), the design of a memory system is a multi-dimensional optimization problem, with all the adherent complexities of analysis. For instance, to analyze something in this design space or to consider one memory system over another, a designer should be familiar with concepts such as Pareto optimality (described later in this chapter). The various figures of merit, in no particular order other than performance being first due to its popularity, are performance, energy consumption and power dissipation, predictability of behavior (i.e., real time), manufacturing costs, and system reliability. This section describes them briefly, collectively. Later sections will treat them in more detail.

7.1.2.1 Performance

The term "performance" means many things to many people. The performance of a system is typically measured in the time it takes to execute a task (i.e., task *latency*), but it can also be measured in the number of tasks that can be handled in a unit time period (i.e., task *bandwidth*). Popular figures of merit for performance include the following:[4]

- Cycles per Instruction (CPI)

$$= \frac{\text{Total execution cycles}}{\text{Total user-level instructions committed}}$$

- Memory-system CPI overhead

$$= \text{Real CPI} \ - \ \text{CPI assuming perfect memo}$$

- Memory Cycles per Instruction (MCPI)

$$= \frac{\text{Total cycles spent in memory system}}{\text{Total user-level instructions committed}}$$

- Cache miss rate $= \dfrac{\text{Total cache misses}}{\text{Total cache accesses}}$

- Cash hit rate $= 1 - \text{Cashe miss rate}$

- Average access time

$$= (\text{hit rate} \cdot \text{average to service hit}) +$$
$$(\text{miss rate} \cdot \text{average to service miss})$$

- Million Instructions per Second (MIPS)

$$= \frac{\text{Instruction executed (seconds)}}{10^6 \cdot \text{Average required for execution}}$$

---

[4] Note that the MIPS metric is easily abused. For instance, it is inappropriate for comparing different instruction-set architectures, and marketing literature often takes the definition of "instructions executed" to mean any particular given window of time as opposed to the full execution of an application. In such cases, the metric can mean the highest possible issue rate of instructions that the machine can achieve (but not necessarily sustain for any realistic period of time).

A cautionary note: using a metric of performance for the memory system that is independent of a processing context can be very deceptive. For instance, the MCPI metric does not take into account how much of the memory system's activity can be overlapped with processor activity, and, as a result, memory system A which has a worse MCPI than memory system B might actually yield a computer system with better total performance. As Figure 7.5 in a later section shows, there can be significantly different amounts of overlapping activity between the memory system and CPU execution.

How to average a set of performance metrics correctly is still a poorly understood topic, and it is very sensitive to the weights chosen (either explicitly or implicitly) for the various benchmarks considered [John 2004]. Comparing performance is always the least ambiguous when it means the amount of time saved by using one design over another. When we ask the question *this machine is how much faster than that machine?* the implication is that we have been using *that* machine for some time and wish to know how much time we would save by using *this* machine instead. The true measure of performance is to compare the total execution time of one machine to another, with each machine running the benchmark programs that represent the user's typical workload as often as a user expects to run them. For instance, if a user compiles a large software application ten times per day and runs a series of regression tests once per day, then the total execution time should count the compiler's execution ten times more than the regression test.

7.1.2.2 Energy Consumption and Power Dissipation

Energy consumption is related to work accomplished (e.g., how much computing can be done with a given battery), whereas power dissipation is the rate of consumption. The instantaneous power dissipation of CMOS (complementary metal-oxide-semiconductor) devices, such as microprocessors, is measured in watts (W) and represents the sum of two components: *active power*, due to switching activity, and *static power*, due primarily to subthreshold leakage. To a first approximation, average power dissipation is equal to the following (we will present a more detailed model later):

$$P_{avg} = (P_{dynamic} + P_{static}) \equiv C_{tot} V^2_{dd} f + I_{leak} V_{dd} \qquad (7.1)$$

where C_{tot} is the total capacitance switched, V_{dd} is the power supply, f is the switching frequency, and I_{leak} is the leakage current, which includes such sources as subthreshold and gate leakage. With each generation in process technology, active power is decreasing on a device level and remaining roughly constant on a chip level. Leakage power, which used to be insignificant relative to switching power, increases as devices become smaller and has recently caught up to switching power in magnitude [Grove 2002]. In the future, leakage will be the primary concern.

Energy is related to power through time. The energy consumed by a computation that requires T seconds is measured in joules (J) and is equal to the integral of the instantaneous power over time T. If the power dissipation remains constant over T, the resultant energy consumption is simply the product of power and time.

$$E = (P_{avg} \cdot T) \equiv C_{tot} V_{dd}^2 N + I_{leak} V_{dd} T \tag{7.2}$$

where N is the number of switching events that occurs during the computation.

In general, if one is interested in extending battery life or reducing the electricity costs of an enterprise computing center, then *energy* is the appropriate metric to use in an analysis comparing approaches. If one is concerned with heat removal from a system or the thermal effects that a functional block can create, then *power* is the appropriate metric. In informal discussions (i.e., in common-parlance prose rather than in equations where units of measurement are inescapable), the two terms "power" and "energy" are frequently used interchangeably, though such use is technically incorrect. Beware, because this can lead to ambiguity and even misconception, which is usually unintentional, but not always so. For instance, microprocessor manufacturers will occasionally claim to have a "low-power" microprocessor that beats its predecessor by a factor of, say, two. This is easily accomplished by running the microprocessor at half the clock rate, which does reduce its power dissipation, but remember that power is the rate at which energy is consumed. However, to a first order, doing so doubles the time over which the processor dissipates that power. The net result is a processor that consumes the same amount of *energy* as before, though it is branded as having lower *power*, which is technically not a lie.

Popular figures of merit that incorporate both energy/power and performance include the following:

- Energy-Delay Product

$$= \left(\begin{matrix} \text{Energy required} \\ \text{to perform task} \end{matrix} \right) \cdot \left(\begin{matrix} \text{Time required} \\ \text{to perform task} \end{matrix} \right)$$

- Power-Delay Product

$$= \left(\begin{matrix} \text{Power required} \\ \text{to perform task} \end{matrix} \right)^m \cdot \left(\begin{matrix} \text{Time required} \\ \text{to perform task} \end{matrix} \right)^n$$

- MIPS per watt

$$= \frac{\text{Performance of benchmark in MIPS}}{\text{Average power dissipated by benchmark}}$$

The second equation was offered as a generalized form of the first (note that the two are equivalent when m = 1 and n = 2) so that designers could place more weight on the metric (time or energy/power) that is most important to their design goals [Gonzalez & Horowitz 1996, Brooks et al. 2000a].

7.1.2.3 Predictable (Real-Time) Behavior

Predictability of behavior is extremely important when analyzing real-time systems, because correctness of operation is often the primary design goal for these systems (consider, for example, medical equipment, navigation systems, anti-lock brakes, flight control systems, etc., in which failure to perform as predicted is not an option).

Popular figures of merit for expressing predictability of behavior include the following:

- Worst-Case Execution Time (WCET), taken to mean the longest amount of time a function could take to execute

- Response time, taken to mean the time between a stimulus to the system and the system's response (e.g., time to respond to an external interrupt)

- Jitter, the amount of deviation from an average timing value

These metrics are typically given as single numbers (average or worst case), but we have found that the probability density function makes a valuable aid in system analysis [Baynes et al. 2001, 2003].

7.1.2.4 Design (and Fabrication and Test) Costs

Cost is an obvious, but often unstated, design goal. Many consumer devices have cost as their primary consideration: if the cost to design and manufacture an item is not low enough, it is not worth the effort to build and sell it. Cost can be represented in many different ways (note that energy consumption is a measure of cost), but for the purposes of this book, by "cost" we mean the cost of producing an item: to wit, the cost of its design, the cost of testing the item, and/ or the cost of the item's manufacture. Popular figures of merit for cost include the following:

- Dollar cost (best, but often hard to even approximate)

- Design size, e.g., die area (cost of manufacturing a VLSI (very large scale integration) design is proportional to its area cubed or more)

- Packaging costs, e.g., pin count

- Design complexity (can be expressed in terms of number of logic gates, number of transistors, lines of code, time to compile or synthesize, time to verify or run DRC (design-rule check), and many others, including a design's impact on clock cycle time [Palacharla et al. 1996])

- Cost is often presented in a relative sense, allowing differing technologies or approaches to be placed on equal footing for a comparison.

- Cost per storage bit/byte/KB/MB/etc. (allows cost comparison between different storage technologies)

- Die area per storage bit (allows size-efficiency comparison within same process technology)

- In a similar vein, cost is especially informative when combined with performance metrics. The following are variations on the theme:

- Bandwidth per package pin (total sustainable bandwidth to/from part, divided by total number of pins in package)

- Execution-time-dollars (total execution time multiplied by total cost; note that cost can be expressed in other units, e.g., pins, die area, etc.)

An important note: cost should incorporate *all* sources of that cost. Focusing on just one source of cost blinds the analysis in two ways: first, the true cost of the system is not considered, and second, solutions can be unintentionally excluded from the analysis. If cost is expressed in pin count, then all pins should be considered by the analysis; the analysis should not focus solely on data pins, for example. Similarly, if cost is expressed in die area, then all sources of die area should be considered by the analysis; the analysis should not focus solely on the number of banks, for example, but should also consider the cost of building control logic (decoders, muxes, bus lines, etc.) to select among the various banks.

7.1.2.5 Reliability

Like the term "performance," the term "reliability" means many things to many different people. In this book, we mean reliability of the data stored within the memory system: how easily is our stored data corrupted or lost, and how can it be protected from corruption or loss? Data integrity is dependent upon physical devices, and physical devices can fail.

Approaches to guarantee the integrity of stored data typically operate by storing redundant information in the memory system so that in the case of device failure, some but not all of the data will be lost or corrupted. If enough redundant information is stored, then the missing data can be reconstructed. Popular figures of merit for measuring reliability characterize both device fragility and robustness of a proposed solution. They include the following:

- Mean Time Between Failures (MTBF):[5] given in time (seconds, hours, etc.) or number of uses

- Bit-error tolerance, e.g., how many bit errors in a data word or packet the mechanism can correct, and how many it can detect (but not necessarily correct)

- Error-rate tolerance, e.g., how many errors per second in a data stream the mechanism can correct

- Application-specific metrics, e.g., how much radiation a design can tolerate before failure, etc.

Note that values given for MTBF often seem astronomically high. This is because they are not meant to apply to individual devices, but to system-wide device use, as in a large installation. For instance, if the expected service lifetime of a device is several years, then that device is expected to fail in several years. If an administrator swaps out devices every few years (before the service lifetime is up), then the administrator should expect to see failure frequencies consistent with the MTBF rating.

7.1.3 The Goal of a Memory Hierarchy

As already mentioned, a well-implemented hierarchy allows a memory system to approach simultaneously the performance of the fastest component, the cost per bit of the cheapest component, and the energy consumption of the most energy-efficient component. A modern memory system typically has performance close to that of on-chip cache, the fastest component in the system. The rate at which microprocessors fetch and execute their instructions is measured in nanoseconds or fractions of a nanosecond. A modern low-end desktop machine has several hundred gigabytes of storage and sells for under $500, roughly half of which goes to the on-chip caches, off-chip caches, DRAM, and disk. This represents an average cost of several dollars per gigabyte—very close to that of disk, the cheapest

---

[5] A common variation is "Mean Time To Failure (MTTF)."

component. Modern desktop systems have an energy cost that is typically in the low tens of nanojoules per instruction executed—close to that of on-chip SRAM cache, the least energy-costly component in the system (on a per-access basis).

The goal for a memory-system designer is to create a system that behaves, on average and from the point of view of the processor, like a big cache that has the price tag of a disk. A successful memory hierarchy is much more than the sum of its parts; moreover, successful memory-system design is nontrivial.

How the system is built, how it is used (and what parts of it are used more heavily than others), and on which issues an engineer should focus most of his effort at design time—all these are highly dependent on the target application of the memory system. Two common categories of target applications are (a) general-purpose systems, which are characterized by their need for universal applicability for just about any type of computation, and (b) embedded systems, which are characterized by their tight design restrictions along multiple axes (e.g., cost, correctness of design, energy consumption, reliability) and the fact that each executes only a single, dedicated software application its entire lifespan, which opens up possibilities for optimization that are less appropriate for general-purpose systems.

7.1.3.1 General-Purpose Computer Systems

General-purpose systems are what people normally think of as "computers." These are the machines on your desktop, the machines in the refrigerated server room at work, and the laptop on the kitchen table. They are designed to handle any and all tasks thrown at them, and the software they run on a day-to-day basis is radically different from machine to machine.

General-purpose systems are typically overbuilt. By definition they are expected by the consumer to run all possible software applications with acceptable speed, and therefore, they are built to handle the average case very well and the worst case at least tolerably well. Were they optimized for any particular task, they could easily become less than optimal for all dissimilar tasks. Therefore, general-purpose systems are optimized for everything, which is another way of saying that they are actually optimized for nothing in particular. However, they make up for this in raw performance, pure number-crunching. The average notebook computer is capable of performing orders of magnitude more operations per second than that required by a word processor or email client, tasks to which the average notebook is frequently relegated, but because the general-purpose system may be expected to handle virtually anything at any time, it must have significant spare number-crunching ability, just in case.

It stands to reason that the memory system of this computer must also be designed in a Swiss-army-knife fashion. Figure 7.3 shows the organization of a typical personal computer, with

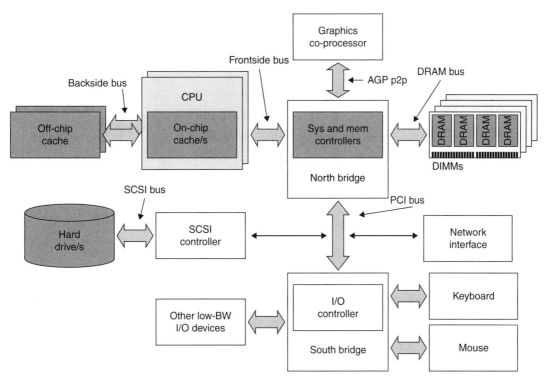

Figure 7.3: Typical PC organization. The memory subsystem is one part of a relatively complex whole. This figure illustrates a two-way multiprocessor, with each processor having its own dedicated off-chip cache. The parts most relevant to this text are shaded in grey: the CPU and its cache system, the system and memory controllers, the DIMMs and their component DRAMs, and the hard drive/s.

the components of the memory system highlighted in grey boxes. The cache levels are found both on-chip (i.e., integrated on the same die as the microprocessor core) and off-chip (i.e., on a separate die). The DRAM system is comprised of a memory controller and a number of DRAM chips organized into DIMMs (dual in-line memory modules, printed circuit boards that contain a handful of DRAMs each). The memory controller can be located on-chip or off-chip, but the DRAMs are always separate from the CPU to allow memory upgrades. The disks in the system are considered peripheral devices, and so their access is made through one or more levels of controllers, each representing a potential chip-to-chip crossing (e.g., here a disk request passes through the system controller to the PCI (peripheral component interconnect) bus controller, to the SCSI (small computer system interface) controller, and finally to the disk itself).

The software that runs on a general-purpose system typically executes in the context of a robust operating system, one that provides virtual memory. Virtual memory is a mechanism whereby the operating system can provide to all running user-level software (i.e., email clients, web browsers, spreadsheets, word-processing packages, graphics and video editing software, etc.) the illusion that the user-level software is in direct control of the computer, when in fact its use of the computer's resources is managed by the operating system. This is a very effective way for an operating system to provide simultaneous access by large numbers of software packages to small numbers of limited-use resources (e.g., physical memory, the hard disk, the network, etc.).

The virtual memory system is the primary constituent of the memory system, in that it is the primary determinant of the manner/s in which the memory system's components are used by software running on the computer. Permanent data is stored on the disk, and the operating store, DRAM, is used as a cache for this permanent data. This DRAM-based cache is explicitly managed by the operating system. The operating system decides what data from the disk should be kept, what should be discarded, what should be sent back to the disk, and, for data retained, where it should be placed in the DRAM system. The primary and secondary caches are usually transparent to software, which means that they are managed by hardware, not software (note, however, the use of the word "usually"—later sections will delve into this in more detail). In general, the primary and secondary caches hold *demand-fetched* data, i.e., running software demands data, the hardware fetches it from memory, and the caches retain as much of it as possible. The DRAM system contains data that the operating system deems worthy of keeping around, and because fetching data from the disk and writing it back to the disk are such time-consuming processes, the operating system can exploit that lag time (during which it would otherwise be stalled, doing nothing) to use sophisticated heuristics to decide what data to retain.

7.1.3.2 Embedded Computer Systems

Embedded systems differ from general-purpose systems in two main aspects. First and foremost, the two are designed to suit very different purposes. While general-purpose systems run a myriad of unrelated software packages, each having potentially very different performance requirements and dynamic behavior compared to the rest, embedded systems perform a single function their entire lifetime and thus execute the same code day in and day out until the system is discarded or a software upgrade is performed. Second, while performance is the primary (in many instances, the only) figure of merit by which a general-purpose system is judged, optimal embedded-system designs usually represent trade-offs between several goals, including manufacturing cost (e.g., die area), energy consumption, and performance.

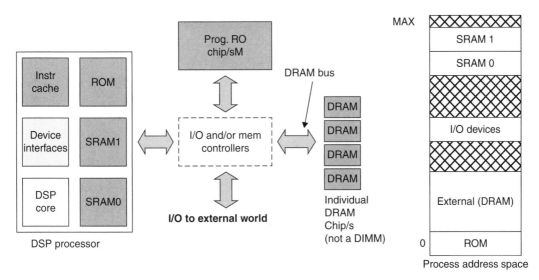

Figure 7.4: DSP-style memory system
(Example based on Texas Instruments' TMS320C3x DSP family)

As a result, we see two very different design strategies in the two camps. As mentioned, general-purpose systems are typically overbuilt; they are optimized for nothing in particular and must make up for this in raw performance. On the other hand, embedded systems are expected to handle only one task that is known at design time. Thus, it is not only possible, but highly beneficial to optimize an embedded design for its one suited task. If general-purpose systems are *overbuilt*, the goal for an embedded system is to be *appropriately* built. In addition, because effort spent at design time is amortized over the life of a product, and because many embedded systems have long lifetimes (tens of years), many embedded design houses will expend significant resources up front to optimize a design, using techniques not generally used in general-purpose systems (for instance, compiler optimizations that require many days or weeks to perform).

The memory system of a typical embedded system is less complex than that of a general-purpose system.[6] Figure 7.4 illustrates an average digital signal-processing system with dual tagless SRAMs on-chip, an off-chip programmable ROM (e.g., PROM, EPROM, flash ROM, etc.) that holds the executable image, and an off-chip DRAM that is used for computation and holding variable data. External memory and device controllers can be used, but many

---

[6] Note that "less complex" does not necessarily imply "small," e.g., consider a typical iPod (or similar MP3 player), whose primary function is to store gigabytes' worth of a user's music and/or image files.

embedded microprocessors already have such controllers integrated onto the CPU die. This cuts down on the system's die count and thus cost. Note that it would be possible for the entire hierarchy to lie on the CPU die, yielding a single-chip solution called a *system-on-chip*. This is relatively common for systems that have limited memory requirements. Many DSPs and microcontrollers have programmable ROM embedded within them. Larger systems that require megabytes of storage (e.g., in Cisco routers, the instruction code alone is more than a 12 MB) will have increasing numbers of memory chips in the system.

On the right side of Figure 7.4 is the software's view of the memory system. The primary distinction is that, unlike general-purpose systems, is that the SRAM caches are visible as separately addressable memories, whereas they are transparent to software in general-purpose systems.

Memory, whether SRAM or DRAM, usually represents one of the more costly components in an embedded system, especially if the memory is located on-CPU because once the CPU is fabricated, the memory size cannot be increased. In nearly all system-on-chip designs and many microcontrollers as well, memory accounts for the lion's share of available die area. Moreover, memory is one of the primary consumers of energy in a system, both on-CPU and off-CPU. As an example, it has been shown that, in many digital signal-processing applications, the memory system consumes more of both energy and die area than the processor datapath. Clearly, this is a resource on which significant time and energy is spent performing optimization.

7.2 Four Anecdotes on Modular Design

It is our observation that computer-system design in general, and memory-hierarchy design in particular, has reached a point at which it is no longer sufficient to design and optimize subsystems in isolation. Because memory systems and their subsystems are so complex, it is now the rule, and not the exception, that the subsystems we thought to be independent actually interact in unanticipated ways. Consequently, our traditional design methodologies no longer work because their underlying assumptions no longer hold. Modular design, one of the most widely adopted design methodologies, is an oft-praised engineering design principle in which clean functional interfaces separate subsystems (i.e., modules) so that subsystem design and optimization can be performed independently and in parallel by different designers. Applying the principles of modular design to produce a complex product can reduce the time and thus the cost for system-level design, integration, and test; optimization at the modular level guarantees optimization at the system level, provided that the system-level architecture and resulting module-to-module interfaces are optimal.

That last part is the sticking point: the principle of modular design assumes no interaction between module-level implementations and the choice of system-level architecture, but that is exactly the kind of interaction that we have observed in the design of modern, high-performance memory systems. Consequently, though modular design has been a staple of memory-systems design for decades, allowing cache designers to focus solely on caches, DRAM designers to focus solely on DRAMs, and disk designers to focus solely on disks, we find that, going forward, modular design is no longer an appropriate methodology.

Earlier we noted that, in the design of memory systems, many of the underlying implementation issues have begun to affect the higher level design process quite significantly: cache design is driven by interconnect physics; DRAM design is driven by circuit-level limitations that have dramatic system-level effects; and modern disk performance is dominated by the on-board caching and scheduling policies. As hierarchies and their components grow more complex, we find that the bulk of performance is lost not in the CPUs or caches or DRAM devices or disk assemblies themselves, but in the subtle interactions between these subsystems and in the manner in which these subsystems are connected. The bulk of lost performance is due to poor configuration of system-level parameters such as bus widths, granularity of access, scheduling policies, queue organizations, and so forth.

This is extremely important, so it bears repeating: the bulk of lost performance is not due to the number of CPU pipeline stages or functional units or choice of branch prediction algorithm or even CPU clock speed; the bulk of lost performance is due to poor configuration of system-level parameters such as bus widths, granularity of access, scheduling policies, queue organizations, etc. Today's computer-system performance is dominated by the manner in which data is moved between subsystems, i.e., the scheduling of transactions, and so it is not surprising that seemingly insignificant details can cause such a headache, as scheduling is known to be highly sensitive to such details.

Consequently, one can no longer attempt system-level optimization by designing/optimizing each of the parts in isolation (which, unfortunately, is often the approach taken in modern computer design). In subsystem design, nothing can be considered "outside the scope" and thus ignored. Memory-system design must become the purview of architects, and a subsystem designer must consider the system-level ramifications of even the slightest low-level design decision or modification. In addition, a designer must understand the low-level implications of system-level design choices. A simpler form of this maxim is as follows:

A designer must consider the system-level ramifications of circuit- and device-level decisions as well as the circuit- and device-level ramifications of system-level decisions.

To illustrate what we mean and to motivate our point, we present several anecdotes. Though they focus on the DRAM system, their message is global, and the relationships they uncover are certainly not restricted to the DRAM system alone.

7.2.1 Anecdote I: Systemic Behaviors Exist

In 1999–2001, we performed a study of DRAM systems in which we explicitly studied only system-level effects—those that had nothing to do with the CPU architecture, DRAM architecture, or even DRAM interface protocol. In this study, we held constant the CPU and DRAM architectures and considered only a handful of parameters that would affect how well the two communicate with each other. Figure 7.5 shows some of the results [Cuppu & Jacob 1999, 2001, Jacob 2003]. The varied parameters in Figure 7.5 are all seemingly innocuous parameters, certainly not the type that would account for up to 20% differences in system performance (execution time) if one parameter was increased or decreased by a small amount, which is indeed the case. Moreover, considering the top two graphs, all of the choices represent intuitively "good" configurations. None of the displayed values represent strawmen, machine configurations that one would avoid putting on one's own desktop. Nonetheless, the performance variability is significant. When the analysis considers a wider range of bus speeds and burst lengths, the problematic behavior increases. As shown in the bottom graph, the ratio of best to worst execution times can be a factor of three, and the local optima are both more frequent and more exaggerated. Systems with relatively low bandwidth (e.g., 100, 200, 400 MB/s) and relatively slow bus speeds (e.g., 100, 200 MHz), if configured well, can match or exceed the performance of system configurations with much faster hardware that is poorly configured.

Intuitively, one would expect the design space to be relatively smooth: as system bandwidth increases, so should system performance. Yet the design space is far from smooth. Performance variations of 20% or more can be found in design points that are immediately adjacent to one another. The variations from best-performing to worst-performing design exceed a factor of three across the full space studied, and local minima and maxima abound. Moreover, the behaviors are related. Increasing one parameter by a factor of two toward higher expected performance (e.g., increasing the channel width) can move the system off a local optimum, but local optimality can be restored by changing other related parameters to follow suit, such as increasing the burst length and cache block size to match the new channel width. This complex interaction between parameters previously thought to be independent arises because of the complexity of the system under study, and so we have named these "systemic" behaviors.[7] This study represents the moment we realized that systemic behaviors exist and that they are significant. Note that the behavior is not restricted to the DRAM

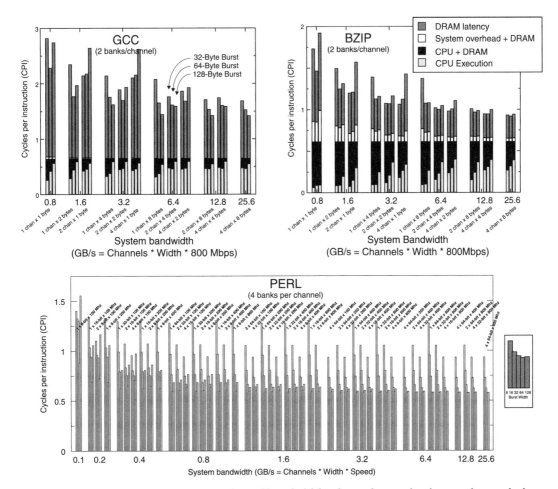

Figure 7.5: Execution time as a function of bandwidth, channel organization, and granularity of access

(Top two graphs from Cuppu & Jacob [2001] (© 2001 *IEEE*); bottom graph from Jacob [2003] (© 2003 *IEEE*).)

system. We have seen it in the disk system as well, where the variations in performance from one configuration to the next are even more pronounced.

---

[7] There is a distinction between this type of behavior and what in complex system theory is called "emergent system" behaviors or properties. Emergent system behaviors are those of individuals within a complex system, behaviors that an individual may perform in a group setting that the individual would never perform alone. In our environment, the behaviors are observations we have made of the design space, which is derived from the system as a whole.

Recall that this behavior comes from the varying of parameters that are seemingly unimportant in the grand scheme of things—at least they would certainly seem to be far less important than, say, the cache architecture or the number of functional units in the processor core. The bottom line, as we have observed, is that systemic behaviors—unanticipated interactions between seemingly innocuous parameters and mechanisms—cause significant losses in performance, requiring in-depth, detailed design-space exploration to achieve anything close to an optimal design given a set of technologies and limitations.

7.2.2 Anecdote II: The DLL in DDR SDRAM

Beginning with their first generation, DDR (double data rate) SDRAM devices have included a circuit-level mechanism that has generated significant controversy within JEDEC (Joint Electron Device Engineering Council), the industry consortium that created the DDR SDRAM standard. The mechanism is a delay-locked loop (DLL), whose purpose is to more precisely align the output of the DDR part with the clock on the system bus. The controversy stems from the cost of the technology versus its benefits.

The system's global clock signal, as it enters the chip, is delayed by the DLL so that the chip's internal clock signal, after amplification and distribution across the chip, is exactly in-phase with the original system clock signal. This more precisely aligns the DRAM part's output with the system clock. The trade-off is extra latency in the datapath as well as a higher power and heat dissipation because the DLL, a dynamic control mechanism, is continuously running. By aligning each DRAM part in a DIMM to the system clock, each DRAM part is effectively de-skewed with respect to the other parts, and the DLLs cancel out timing differences due to process variations and thermal gradients.

Figure 7.6 illustrates a small handful of alternative solutions considered by JEDEC, who ultimately chose Figure 7.6(b) for the standard. The interesting thing is that the data strobe is not used to capture data at the memory controller, bringing into question its purpose if the DLL is being used to help with data transfer to the memory controller. There is significant disagreement over the value of the chosen design; an anonymous JEDEC member, when asked "what is the DLL doing on the DDR chip?" answered with a grin, "burning power." In applications that require low latency and low power dissipation, designers turn off the DLL entirely and use only the data strobe for data capture, ignoring the system clock (as in Figure 7.6(a)) [Kellogg 2002, Lee 2002, Rhoden 2002].

The argument for the DLL is that it de-skews the DRAM devices on a DIMM and provides a path for system design that can use a global clocking scheme, one of the simplest system designs known. The argument against the DLL is that it would be unnecessary if a designer

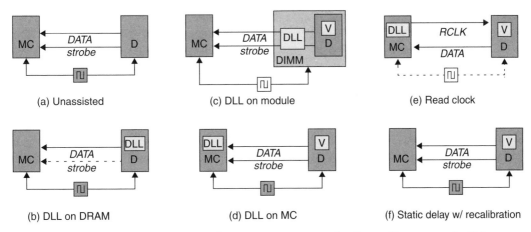

Figure 7.6: Several alternatives to the per-DRAM DLL. The figure illustrates a half dozen different timing conventions (a dotted line indicates a signal is unused for capturing data): (a) the scheme in single data rate SDRAM; (b) the scheme chosen for DDR SDRAM; (c) moving the DLL onto the module, with a per-DRAM static delay element (Vernier); (d) moving the DLL onto the memory controller, with a per-DRAM static delay; (e) using a separate read clock per DRAM or per DIMM; and (f) using only a static delay element and recalibrating periodically to address dynamic changes.

learned to use the data strobe—this would require a more sophisticated system design, but it would achieve better performance at a lower cost. At the very least, it is clear that a DLL is a circuit-oriented solution to the problem of system-level skew, which could explain the controversy.

7.2.3 Anecdote III: A Catch-22 in the Search for Bandwidth

With every DRAM generation, timing parameters are added. Several have been added to the DDR specification to address the issues of power dissipation and synchronization.

t_{FAW} (*Four-bank Activation Window*) and t_{RRD} (*Row-to-Row activation Delay*) put a ceiling on the maximum current draw of a single DRAM part. These are protocol-level limitations whose values are chosen to prevent a memory controller from exceeding circuit-related thresholds.

t_{DQS} is our own name for the DDR system-bus turnaround time; one can think of it as the DIMM-to-DIMM switching time that has implications only at the system level (i.e., it has no meaning or effect if considering read requests in a system with but a single DIMM). By obeying t_{DQS}, one can ensure that a second DIMM will not drive the data bus at the same time as a first when switching from one DIMM to another for data output.

These are per-device timing parameters that were chosen to improve the behavior (current draw, timing uncertainty) of individual devices. However, they do so at the expense of a significant loss in system-level performance. When reading large amounts of data from the DRAM system, an application will have to read, and thus will have to *activate*, numerous DRAM rows. At this point, the t_{FAW} and t_{RRD} timing parameters kick in and limit the available read bandwidth. The t_{RRD} parameter specifies the minimum time between two successive row activation commands to the same DRAM device (which implies the same DIMM, because all the DRAMs on a DIMM are slaved together[8]). The t_{FAW} parameter represents a sliding window of time during which no more than four row activation commands to the same device may appear.

The parameters are specified in nanoseconds and not bus cycles, so they become increasingly problematic at higher bus frequencies. Their net effect is to limit the bandwidth available from a DIMM by limiting how quickly one can get the data out of the DRAM's storage array, irrespective of how fast the DRAM's I/O circuitry can ship the data back to the memory controller. At around 1 GBps, sustainable bandwidth hits a ceiling and remains flat no matter how fast the bus runs because the memory controller is limited in how quickly it can activate a new row and start reading data from it.

The obvious solution is to interleave data from different DIMMs on the bus. If one DIMM is limited in how quickly it can read data from its arrays, then one should populate the bus with many DIMMs and move through them in a round-robin fashion. This should bring the system bandwidth up to maximum. However, the function of t_{DQS} is to prevent exactly that: t_{DQS} is the bus turnaround time, inserted to account for skew on the bus and to prevent different bus masters from driving the bus at the same time. To avoid such collisions, a second DIMM must wait at least t_{DQS} after a first DIMM has finished before driving the bus. So we have a catch:

- One set of parameters limits device-level bandwidth and expects a designer to go to the system level to reclaim performance.

- The other parameter limits system-level bandwidth and expects a designer to go to the device level to reclaim performance.

The good news is that the problem is solvable, but this is nonetheless a very good example of low-level design decisions that create headaches at the system level.

---

[8] This is a minor oversimplification. We would like to avoid having to explain details of DRAM-system organization, such as the concept of *rank,* at this point.

7.2.4 Anecdote IV: Proposals to Exploit Variability in Cell Leakage

The last anecdote is an example of a system-level design decision that ignores circuit- and device-level implications. Ever since DRAM was invented, it has been observed that different DRAM cells exhibit different data-retention time characteristics, typically ranging between hundreds of milliseconds to tens of seconds. DRAM manufacturers typically set the refresh requirement conservatively and require that every row in a DRAM device be refreshed at least once every 64 or 32 ms to avoid losing data. Though refresh might not seem to be a significant concern, in mobile devices researchers have observed that refresh can account for one-third of the power in otherwise idle systems, prompting action to address the issue. Several recent papers propose moving the refresh function into the memory controller and refreshing each row only when needed. During an initialization phase, the controller would characterize each row in the memory system, measuring DRAM data-retention time on a row-by-row basis, discarding leaky rows entirely, limiting its DRAM use to only those rows deemed nonleaky, and refreshing once every tens of seconds instead of once every tens of milliseconds.

The problem is that these proposals ignore another, less well-known phenomenon of DRAM cell variability, namely that a cell with a long retention time can suddenly (in the time frame of seconds) exhibit a short retention time [Yaney et al. 1987, Restle et al. 1992, Ueno et al. 1998, Kim 2004]. Such an effect would render these power-efficient proposals functionally erroneous. The phenomenon is called *variable retention time* (VRT), and though its occurrence is infrequent, it is nonzero. The occurrence rate is low enough that a system using one of these reduced-refresh proposals could protect itself against VRT by using error correcting codes, but none of the proposals so far discuss VRT or ECC.

7.2.5 Perspective

To summarize so far:

Anecdote I: Systemic behaviors exist and are significant (they can be responsible for factors of two to three in execution time).

Anecdote II: The DLL in DDR SDRAM is a circuit-level solution chosen to address system-level skew.

Anecdote III: t_{DQS} represents a circuit-level solution chosen to address system-level skew in DDR SDRAM; t_{FAW} and t_{RRD} are circuit-level limitations that significantly limit system-level performance.

Anecdote IV: Several research groups have recently proposed system-level solutions to the DRAM-refresh problem, but fail to account for circuit-level details that might compromise the correctness of the resulting system.

Anecdotes II and III show that a common practice in industry is to focus at the level of devices and circuits, in some cases ignoring their system-level ramifications. Anecdote IV shows that a common practice in research is to design systems that have device- and circuit-level ramifications while abstracting away the details of the devices and circuits involved. Anecdote I illustrates that both approaches are doomed to failure in future memory-systems design.

It is clear that in the future we will have to move away from modular design; one can no longer safely abstract away details that were previously considered "out of scope." To produce a credible analysis, a designer must consider many different subsystems of a design and many different levels of abstraction—one must consider the forest when designing trees and consider the trees when designing the forest.

7.3 Cross-Cutting Issues

Though their implementation details might apply at a local level, most design decisions must be considered in terms of their system-level effects and side-effects before they become part of the system/hierarchy. For instance, power is a cross-cutting, system-level phenomenon, even though most power optimizations are specific to certain technologies and are applied locally; reliability is a system-level issue, even though each level of the hierarchy implements its own techniques for improving it; and, as we have shown, performance optimizations such as widening a bus or increasing support for concurrency rarely result in system performance that is globally optimal. Moreover, design decisions that locally optimize along one axis (e.g., power) can have even larger effects on the system level when all axes are considered. Not only can the global power dissipation be thrown off optimality by blindly making a local decision, it is even easier to throw the system off a global optimum when more than one axis is considered (e.g., power/ performance).

Designing the best system given a set of constraints requires an approach that considers multiple axes simultaneously and measures the system-level effects of all design choices. Such a holistic approach requires an understanding of many issues, including cost and performance models, power, reliability, and software structure. The following sections provide overviews of these cross-cutting issues.

7.3.1 Cost/Performance Analysis

To perform a cost/performance analysis correctly, the designer must define the problem correctly, use the appropriate tools for analysis, and apply those tools in the manner for which they were designed. This section provides a brief, intuitive look at the problem. Herein, we will use *cost* as an example of problem definition, *Pareto optimality* as an example of an appropriate tool, and *sampled averages* as an example to illustrate correct tool usage.

7.3.1.1 Problem Definition: Cost

A designer must think in an all-inclusive manner when accounting for cost. For example, consider a cost-performance analysis of a DRAM system wherein performance is measured in sustainable bandwidth and cost is measured in pin count.

To represent the cost correctly, the analysis should consider *all* pins, including those for control, power, ground, address, and data. Otherwise, the resulting analysis can incorrectly portray the design space, and workable solutions can get left out of the analysis. For example, a designer can reduce latency in some cases by increasing the number of address and command pins, but if the cost analysis only considers data pins, then these optimizations would be cost-free. Consider DRAM addressing, which is done half of an address at a time. A 32-bit physical address is sent to the DRAM system 16 bits at a time in two different commands; one could potentially decrease DRAM latency by using an SRAM-like wide address bus and sending the entire 32 bits at once. This represents a *real* cost in design and manufacturing that would be higher, but an analysis that accounts only for data pins would not consider it as such.

Power and ground pins must also be counted in a cost analysis for similar reasons. High-speed chip-to-chip interfaces typically require more power and ground pins than slower interfaces. The extra power and ground signals help to isolate the I/O drivers from each other and the signal lines from each other, both improving signal integrity by reducing crosstalk, ground bounce, and related effects. I/O systems with higher switching speeds would have an unfair advantage over those with lower switching speeds (and thus fewer power/ ground pins) in a cost-performance analysis if power and ground pins were to be excluded from the analysis. The inclusion of these pins would provide for an effective and easily quantified trade-off between cost and bandwidth.

Failure to include address, control, power, and ground pins in an analysis, meaning failure to be all-inclusive at the conceptual stages of design, would tend to blind a designer to possibilities. For example, an architecturally related family of solutions that at first glance

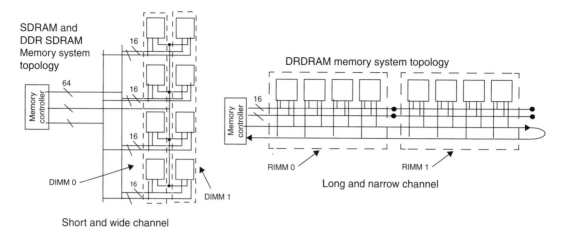

Figure 7.7: Difference in topology between SDRAM and Rambus memory systems

gives up total system bandwidth so as to be more cost-effective might be thrown out at the conceptual stages for its intuitively lower performance. However, considering all sources of cost in the analysis would allow a designer to look more closely at this family and possibly to recover lost bandwidth through the addition of pins.

Comparing SDRAM and Rambus system architectures provides an excellent example of considering cost as the total number of pins leading to a continuum of designs. The Rambus memory system is a narrow-channel architecture, compared to SDRAM's wide-channel architecture, pictured in Figure 7.7. Rambus uses fewer address and command pins than SDRAM and thus incurs an additional latency at the command level. Rambus also uses fewer data pins and occurs an additional latency when transmitting data as well. The trade-off is the ability to run the bus at a much higher bus frequency, or *pin-bandwidth* in bits per second per pin, than SDRAM. The longer channel of the DRDRAM (direct Rambus DRAM) memory system contributes directly to longer read-command latencies and longer bus turnaround times. However, the longer channel also allows for more devices to be connected to the memory system and reduces the likelihood that consecutive commands access the same device. The width and depth of the memory channels impact the bandwidth, latency, pin count, and various cost components of the respective memory systems. The effect that these organizational differences have on the DRAM access protocol is shown in Figure 7.8 which illustrates a row activation and column read command for both DDR SDRAM and Direct Rambus DRAM.

Contemporary SDRAM and DDR SDRAM memory chips operating at a frequency of 200 MHz can activate a row in 3 clock cycles. Once the row is activated, memory controllers

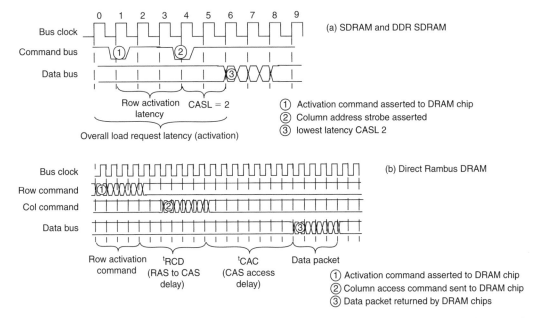

Figure 7.8: Memory access latency in SDRAM and DDR SDRAM memory systems (top) and DRDRAM (bottom)

in SDRAM or DDR SDRAM memory systems can retrieve data using a simple column address strobe command with a latency of 2 or 3 clock cycles. In Figure 7.8(a), Step 1 shows the assertion of a row activation command, and Step 2 shows the assertion of the column address strobe signal. Step 3 shows the relative timing of a high-performance DDR SDRAM memory module with a CASL (CAS latency) of 2 cycles. For a fair comparison against the DRDRAM memory system, we include the bus cycle that the memory controller uses to assert the load command to the memory chips. With this additional cycle included, a DDR SDRAM memory system has a read latency of 6 clock cycles (to critical data). In a SDRAM or DDR SDRAM memory system that operates at 200 MHz, 6 clock cycles translate to 30 ns of latency for a memory load command with row activation latency inclusive. These latency values are the same for high-performance SDRAM and DDR SDRAM memory systems.

The DRDRAM memory system behaves very differently from SDRAM and DDR SDRAM memory systems. Figure 7.8(b) shows a row activation command in Step 1, followed by a column access command in Step 2. The requested data is then returned by the memory chip to the memory controller in Step 3. The row activation command in Step 1 is transmitted by the memory controller to the memory chip in a packet format that spans 4 clock cycles.

The minimum delay between the row activation and column access is 7 clock cycles, and, after an additional (also minimum) CAS (column address strobe) latency of 8 clock cycles, the DRDRAM chip begins to transmit the data to the memory controller. One caveat to the computation of the access latency in the DRDRAM memory system is that CAS delay in the DRDRAM memory system is a function of the number of devices on a single DRDRAM memory channel. On a DRDRAM memory system with a full load of 32 devices on the data bus, the CAS-latency delay may be as large as 12 clock cycles. Finally, it takes 4 clock cycles for the DRDRAM memory system to transport the data packet. Note that we add half the transmission time of the data packet in the computation of the latency of a memory request in a DRDRAM memory system due to the fact that the DRDRAM memory system does not support critical word forwarding, and the critically requested data may exist in the latter parts of the data packet; on average, it will be somewhere in the middle. This yields a total latency of 21 cycles, which, in a DRDRAM memory system operating at 600 MHz, translates to a latency of 35 ns.

The Rambus memory system trades off a longer latency for fewer pins and higher pin bandwidth (in this example, three times higher bandwidth). How do the systems compare in performance?

Peak bandwidth of any interface depends solely on the channel width and the operating frequency of the channel. In Table 7.2, we summarize the statistics of the interconnects

Table 7.2: Peak bandwidth statistics of SDRAM, DDR SDRAM, and DRDRAM memory systems

	Operating Frequency (Data)	Data Channel Pin Count	Data Channel Bandwidth	Control Channel Pin Count	Command Channel Bandwidth	Address Channel Pin Count	Address Channel Bandwidth
SDRAM controller	133	64	1064 MB/s	28	465 MB/s	30	500 MB/s
DDR SDRAM controller	2 * 200	64	3200 MB/s	42	1050 MB/s	30	750 MB/s
DRDRAM controller	2 * 600	16	2400 MB/s	9	1350 MB/s	8	1200 MB/s
x16 SDRAM chip	133	16	256 MB/s	9	150 MB/s	15	250 MB/s
x16 DDR SDRAM chip	2 *200	16	800 MB/s	11	275 MB/s	15	375 MB/s

and compute the peak bandwidths of the memory systems at the interface of the memory controller and at the interface of the memory chips as well.

Table 7.3 compares a 133-MHz SDRAM, a 200-MHz DDR SDRAM system, and a 600-MHz DRDRAM system. The 133-MHz SDRAM system, as represented by a PC-133 compliant SDRAM memory system on an AMD Athlon-based computer system, has a theoretical peak bandwidth of 1064 MB/s. The maximum sustained bandwidth for the single channel of SDRAM, as measured by the use of the add kernel in the STREAM benchmark, reaches 540 MB/s. The maximum sustained bandwidth for DDR SDRAM and DRDRAM was also measured on STREAM, yielding 1496 and 1499 MB/s, respectively. The pin cost of each system is factored in, yielding bandwidth per pin on both a per-cycle basis and a per-nanosecond basis.

7.3.1.2 Appropriate Tools: Pareto Optimality

It is convenient to represent the "goodness" of a design solution, a particular system configuration, as a single number so that one can readily compare the number with the goodness ratings of other candidate design solutions and thereby quickly find the "best" system configuration. However, in the design of memory systems, we are inherently dealing with a multi-dimensional design space (e.g., one that encompasses performance, energy consumption, cost, etc.), and so using a single number to represent a solution's worth is not really appropriate, unless we can assign exact weights to the various figures of merit (which is dangerous and will be discussed in more detail later) or we care about one aspect to the exclusion of all others (e.g., performance at any cost).

Assuming that we do not have exact weights for the figures of merit and that we do care about more than one aspect of the system, a very powerful tool to aid in system analysis is the concept of *Pareto optimality* or *Pareto efficiency,* named after the Italian economist Vilfredo Pareto, who invented it in the early 1900s.

Table 7.3: Cross-comparison of SDRAM, DDR SDRAM, and DRDRAM memory systems

DRAM Technology	Operating Frequency (Data Bus)	Pin Count per Channel	Peak Bandwidth	Sustained BW on StreamAdd	Bits per Pin per Cycle (Peak)	Bits per Pin per Cycle (Sustained)
SDRAM	133	152	1064 MB/s	540 MB/s	0.4211	0.2139
DDR SDRAM	2 * 200	171	3200 MB/s	1496 MB/s	0.3743	0.1750
DRDRAM	2 * 600	117	2400 MB/s	1499 MB/s	0.1368	0.0854

Pareto optimality asserts that one candidate solution to a problem is better than another candidate solution only if the first *dominates* the second, i.e., if the first is better than or equal to the second in *all* figures of merit. If one solution has a better value in one dimension but a worse value in another, then the two candidates are Pareto equivalent. The best solution is actually a set of candidate solutions: the set of Pareto-equivalent solutions that is not dominated by any solution.

Figure 7.9(a) shows a set of candidate solutions in a two-dimensional space that represent a cost/ performance metric. The *x*-axis represents system performance in execution time

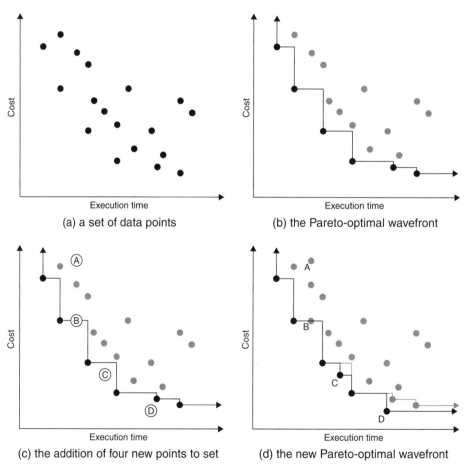

(a) a set of data points

(b) the Pareto-optimal wavefront

(c) the addition of four new points to set

(d) the new Pareto-optimal wavefront

Figure 7.9: Pareto optimality

(Members of the Pareto-optimal set are shown in solid black; nonoptimal points are grey)

(smaller numbers are better), and the *y*-axis represents system cost in dollars (smaller numbers are better). Figure 7.9(b) shows the Pareto-optimal set in solid black and connected by a line; nonoptimal data points are shown in grey. The Pareto-optimal set forms a wavefront that approaches both axes simultaneously. Figures 7.9(c) and (d) show the effect of adding four new candidate solutions to the space: one lies inside the wavefront, one lies on the wavefront, and two lie outside the wavefront. The first two new additions, A and B, are both dominated by at least one member of the Pareto-optimal set, and so neither is considered Pareto optimal. Even though B lies on the wavefront, it is not considered Pareto optimal. The point to the left of B has better performance than B at equal cost. Thus, it dominates B.

Point C is not dominated by any member of the Pareto-optimal set, nor does it dominate any member of the Pareto-optimal set. Thus, candidate-solution C is added to the optimal set, and its addition changes the shape of the wavefront slightly. The last of the additional points, D, is dominated by no members of the optimal set, but it *does* dominate several members of the optimal set, so D's inclusion in the optimal set excludes those dominated members from the set. As a result, candidate-solution D changes the shape of the wave front more significantly than candidate-solution C.

7.3.1.3 Tool Use: Taking Sampled Averages Correctly

In many fields, including the field of computer engineering, it is quite popular to find a *sampled average,* i.e., the average of a sampled set of numbers, rather than the average of the entire set. This is useful when the entire set is unavailable, difficult to obtain, or expensive to obtain. For example, one might want to use this technique to keep a running performance average for a real microprocessor, or one might want to sample several windows of execution in a terabyte-size trace file. Provided that the sampled subset is representative of the set as a whole, and provided that the technique used to collect the samples is correct, this mechanism provides a low-cost alternative that can be very accurate.

The discussion will use as an example a mechanism that samples the miles-per-gallon performance of an automobile under way. The trip we will study is an out and back trip with a brief pit stop, as shown in Figure 7.10. The automobile will follow a simple course that is easily analyzed:

1. The auto will travel over even ground for 60 miles at 60 mph, and it will achieve 30 mpg during this window of time.

2. The auto will travel uphill for 20 miles at 60 mph, and it will achieve 10 mpg during this window of time.

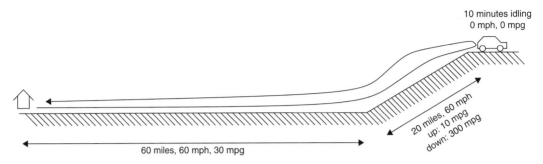

Figure 7.10: Course taken by the automobile in the example

3. The auto will travel downhill for 20 miles at 60 mph, and it will achieve 300 mpg during this window of time.

4. The auto will travel back home over even ground for 60 miles at 60 mph, and it will achieve 30 mpg during this window of time.

5. In addition, before returning home, the driver will sit at the top of the hill for 10 minutes, enjoying the view, with the auto idling, consuming gasoline at the rate of 1 gallon every 5 hours. This is equivalent to 1/300 gallon per minute or 1/30 of a gallon during the 10-minute respite. Note that the auto will achieve 0 mpg during this window of time.

Our car's algorithm samples evenly in time, so for our analysis we need to break down the segments of the trip by the amount of time that they take:

- Outbound: 60 minutes

- Uphill: 20 minutes

- Idling: 10 minutes

- Downhill: 20 minutes

- Return: 60 minutes

This is displayed graphically in Figure 7.11, in which the time for each segment is shown to scale. Assume, for the sake of simplicity, that the sampling algorithm samples the car's miles-per-gallon every minute and adds that sampled value to the running average (it could just as easily sample every second or millisecond). Then the algorithm will sample the value 30 mpg 60 times during the first segment of the trip, the value 10 mpg 20 times during the second segment of the trip, the value 0 mpg 10 times during the third segment of the trip, and so on.

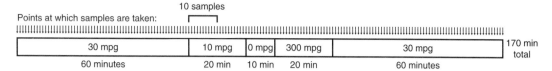

Figure 7.11: Sampling miles-per-gallon (mpg) over time. The figure shows the trip in time, with each segment of time labeled with the average miles-per-gallon for the car during that segment of the trip. Thus, whenever the sampling algorithm samples miles-per-gallon during a window of time, it will add that value to the running average.

Over the trip, the car is operating for a total of 170 minutes. Thus, we can derive the sampling algorithm's results as follows:

$$\frac{60}{170}30 + \frac{20}{170}10 + \frac{10}{170}0 + \frac{20}{170}300 + \frac{60}{170}30 = 57.5 \text{ mpg} \qquad (7.3)$$

The sampling algorithm tells us that the auto achieved 57.5 mpg during our trip. However, a quick reality check will demonstrate that this cannot be correct; somewhere in our analysis we have made an invalid assumption. What is the correct answer, the correct approach? The reader is encouraged to figure the answer out for him- or herself.

7.3.2 Power and Energy

Power has become a "first-class" design goal in recent years within the computer architecture and design community. Previously, low-power circuit, chip, and system design was considered the purview of specialized communities, but this is no longer the case, as even high-performance chip manufacturers can be blindsided by power dissipation problems.

7.3.2.1 Power Dissipation in Computer Systems

Power dissipation in CMOS circuits arises from two different mechanisms: *static power*, which is primarily *leakage power* and is caused by the transistor not completely turning off, and *dynamic power*, which is largely the result of switching capacitive loads between two different voltage states. Dynamic power is dependent on frequency of circuit activity, since no power is dissipated if the node values do not change, while static power is independent of the frequency of activity and exists whenever the chip is powered on. When CMOS circuits were first used, one of their main advantages was the negligible leakage current flowing with the gate at DC or steady state. Practically all of the power consumed by CMOS gates was due to dynamic power consumed during the transition of the gate. But as transistors become increasingly smaller, the CMOS leakage current starts to become significant and is projected to be larger than the dynamic power, as shown in Figure 7.12.

In charging a load capacitor C up ΔV volts and discharging it to its original voltage, a gate pulls an amount of current equal to $C \cdot \Delta V$ from the V_{dd} supply to charge up the capacitor and then sinks this charge to ground discharging the node. At the end of a charge/discharge cycle, the gate/capacitor combination has moved $C \cdot \Delta V$ of charge from V_{dd} to ground, which uses an amount of energy equal to $C \cdot \Delta V \cdot V_{dd}$ that is independent of the cycle time. The average dynamic power of this node, the average rate of its energy consumption, is given by the following equation [Chandrakasan & Brodersen 1995]:

$$P_{dynamic} = C \cdot \Delta V \cdot V_{dd} \cdot \alpha \cdot f \qquad (7.4)$$

Dividing by the charge/discharge period (i.e., multiplying by the clock frequency f) produces the rate of energy consumption over that period. Multiplying by the expected *activity ratio* α, the probability that the node will switch (in which case it dissipates dynamic power; otherwise, it does not), yields an average power dissipation over a larger window of time for which the activity ratio holds (e.g., this can yield average power for an entire hour of computation, not just a nanosecond). The dynamic power for the whole chip is the sum of this equation over all nodes in the circuit.

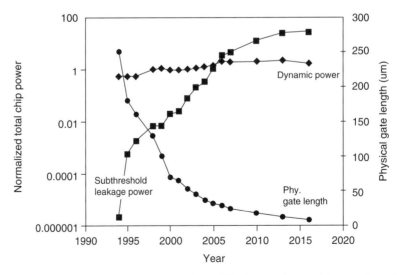

Figure 7.12: Projections for dynamic and leakage, along with gate length

(Figure taken from Kim et al. [2004a])

It is clear from equation 7.4 what can be done to reduce the dynamic power dissipation of a system. We can either reduce the capacitance being switched, the voltage swing, the power supply voltage, the activity ratio, or the operating frequency. Most of these options are available to a designer at the architecture level.

Note that, for a specific chip, the voltage swing ΔV is usually proportional to V_{dd}, so equation 7.4 is often simplified to the following:

$$P_{dynamic} = C \cdot V^2_{dd} \cdot \alpha \cdot f \qquad (7.5)$$

Moreover, the activity ratio α is often approximated as 1/2, giving the following form:

$$P_{dynamic} = \frac{1}{2} \cdot C \cdot V^2_{dd} \cdot f \qquad (7.6)$$

Static leakage power is due to our inability to completely turn off the transistor, which leaks current in the subthreshold operating region [Taur & Ning 1998]. The gate couples to the active channel mainly through the gate oxide capacitance, but there are other capacitances in a transistor that couple the gate to a "fixed charge" (charge which cannot move) present in the bulk and not associated with current flow [Peckerar et al. 1979, 1982]. If these extra capacitances are large (note that they increase with each process generation as physical dimensions shrink), then changing the gate bias merely alters the densities of the fixed charge and will not turn the channel off. In this situation, the transistor becomes a leaky faucet; it does not turn off no matter how hard you turn it.

Leakage power is proportional to V_{dd}. It is a linear, not a quadratic, relationship. For a particular process technology, the per-device leakage power is given as follows [Butts & Sohi 2000]:

$$P_{static} = I_{leakage} \cdot V^2_{dd} \qquad (7.7)$$

Leakage energy is the product of leakage power times the duration of operation.

It is clear from EQ 7.7 what can be done to reduce the leakage power dissipation of a system: reduce leakage current and/or reduce the power supply voltage. Both options are available to a designer at the architecture level.

Heat in VLSI circuits is becoming a significant and related problem. The rate at which physical dimensions such as gate length and gate oxide thickness have been reduced is faster than for other parameters, especially voltage, resulting in higher power densities on the chip surface. To lower leakage power and maintain device operation, voltage levels are set according to the silicon bandgap and intrinsic built-in potentials, in spite of the conventional scaling algorithm. Thus, power densities are increasing exponentially for next-generation chips. For instance, the power density of Intel's Pentium chip line has already surpassed that of a hot plate with the introduction of the Pentium Pro [Gelsinger 2001]. The problem of power and heat dissipation now extends to the DRAM system, which traditionally has represented low power densities and low costs. Today, higher end DRAMs are dynamically throttled when, due to repeated high-speed access to the same devices, their operating temperatures surpass design thresholds. The next-generation memory system embraced by the DRAM community, the Fully Buffered DIMM architecture, specifies a per-module controller that, in many implementations, requires a heatsink. This is a cost previously unthinkable in DRAM-system design.

Disks have many components that dissipate power, including the spindle motor driving the platters, the actuator that positions the disk heads, the bus interface circuitry, and the microcontroller/s and memory chips. The spindle motor dissipates the bulk of the power, with the entire disk assembly typically dissipating power in the tens of watts.

7.3.2.2 Schemes for Reducing Power and Energy

There are numerous mechanisms in the literature that attack the power dissipation and/or energy consumption problem. Here, we will briefly describe three: dynamic voltage scaling, the powering down of unused blocks, and circuit-level approaches for reducing leakage power.

Dynamic Voltage Scaling. Recall that total energy is the sum of switching energy and leakage energy, which, to a first approximation, is equal to the following:

$$E_{\text{tot}} = \left[(C_{\text{tot}} \cdot V^2_{\text{dd}} \cdot \alpha \cdot f) + (N_{\text{tot}} \cdot I_{\text{leakage}} \cdot V_{\text{dd}}) \right] \cdot T \qquad (7.8)$$

T is the time required for the computation, and N_{tot} is the total number of devices leaking current. Variations in processor utilization affect the amount of switching activity (the activity ratio α). However, a light workload produces an idle processor that wastes clock cycles and energy because the clock signal continues propagating and the operating voltage remains the same. Gating the clock during idle cycles reduces the switched capacitance C_{tot} during idle cycles. Reducing the frequency f during periods of low workload eliminates most idle cycles altogether.

None of the approaches, however, affects $C_{tot}V^2_{dd}$ for the actual computation or substantially reduces the energy lost to leakage current. Instead, reducing the supply voltage V_{dd} in conjunction with the frequency f achieves savings in switching energy and reduces leakage energy. For high-speed digital CMOS, a reduction in supply voltage increases the circuit delay as shown by the following equation [Baker et al. 1998, Baker 2005]:

$$T_d = \frac{C_L V_{dd}}{\mu C_{ox}(W/L)(V_{dd} - V_t)^2} \tag{7.9}$$

where

- T_d is the delay or the reciprocal of the frequency f

- V_{dd} is the supply voltage

- C_L is the total node capacitance

- μ is the carrier mobility

- C_{ox} is the oxide capacitance

- V_t is the threshold voltage

- W/L is the width-to-length ratio of the transistors in the circuit

This can be simplified to the following form, which gives the maximum operating frequency as a function of supply and threshold voltages:

$$f_{MAX} \sim \frac{(V_{dd} - V_t)^2}{V_{dd}} \tag{7.10}$$

As mentioned earlier, the threshold voltage is closely tied to the problem of leakage power, so it cannot be arbitrarily lowered. Thus, the right-hand side of the relation ends up being a constant proportion of the operating voltage for a given process technology. Microprocessors typically operate at the maximum speed at which their operating voltage level will allow, so there is not much headroom to arbitrarily lower V_{dd} by itself. However, V_{dd} can be lowered if the clock frequency is also lowered in the same proportion. This mechanism is called *dynamic voltage scaling (DVS)* [Pering & Broderson 1998] and is appearing in nearly every modern microprocessor. The technique sets the microprocessor's frequency to the most appropriate level for performing each task at hand, thus avoiding hurry-up-and-wait scenarios that consume more energy than is required for the computation (see Figure 7.13). As Weiser points out, idle time represents wasted energy, even if the CPU is stopped [Weiser et al. 1994].

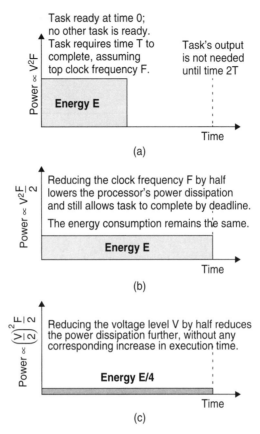

Figure 7.13: Dynamic voltage scaling. Not every task needs the CPU's full computational power. In many cases, for example, the processing of video and audio streams, the only performance requirement is that the task meet a deadline, see (a). Such cases create opportunities to run the CPU at a lower performance level and achieve the same perceived performance while consuming less energy. As (b) shows, reducing the clock frequency of a processor reduces power dissipation but simply spreads a computation out over time, thereby consuming the same total energy as before. As (c) shows, reducing the voltage level as well as the clock frequency achieves the desired goal of reduced energy consumption and appropriate performance level. Figure and caption from Varma et al. [2003].

Note that it is not sufficient to merely have a chip that *supports* voltage scaling. There must be a heuristic, either implemented in hardware or software, that decides when to scale the voltage and by how much to scale it. This decision is essentially a prediction of the near-future computational needs of the system and is generally made on the basis of the recent computing requirements of all tasks and threads running at the time. The development of good heuristics is a tricky problem (pointed out by Weiser et al. [1994]). Heuristics that closely track performance requirements save little energy, while those that save the most energy tend to do so at the expense of performance, resulting in poor response time, for example.

Most research quantifies the effect that DVS has on reducing dynamic power dissipation because dynamic power follows V_{dd} in a quadratic relationship: reducing V_{dd} can significantly reduce dynamic power. However, lowering V_{dd} also reduces leakage power, which is becoming just as significant as dynamic power. Though the reduction is only linear, it is nonetheless a reduction.

Note also that even though DVS is commonly applied to microprocessors, it is perfectly well suited to the memory system as well. As a processor's speed is decreased through application of DVS, it requires less speed out of its associated SRAM caches, whose power supply can be scaled to keep pace. This will reduce both the dynamic and the static power dissipation of the memory circuits.

Powering-Down Unused Blocks. A popular mechanism for reducing power is simply to turn off functional blocks that are not needed. This is done at both the circuit level and the chip or I/O-device level.

At the circuit level, the technique is called *clock gating*. The clock signal to a functional block (e.g., an adder, multiplier, or predictor) passes through a gate, and whenever a control circuit determines that the functional block will be unused for several cycles, the gate halts the clock signal and sends a nonoscillating voltage level to the functional block instead. The latches in the functional block retain their information; do not change their outputs; and, because the data is held constant to the combinational logic in the circuit, do not switch. Therefore, it does not draw current or consume energy.

Note that, in the naïve implementation, the circuits in this instance are still powered up, so they still dissipate static power; clock gating is a technique that only reduces dynamic power. Other gating techniques can reduce leakage as well. For example, in caches, unused blocks can be powered down using Gated-V_{dd} [Powell et al. 2000] or Gated-ground [Powell et al. 2000] techniques. Gated-V_{dd} puts the power supply of the SRAM in a series with a transistor as shown in Figure 7.14. With the stacking effect introduced by this transistor, the leakage current is reduced drastically. This technique benefits from having both low-leakage current and a simpler fabrication process requirement since only a single threshold voltage is conceptually required (although, as shown in Figure 7.14, the gating transistor can also have a high threshold to decrease the leakage even further at the expense of process complexity).

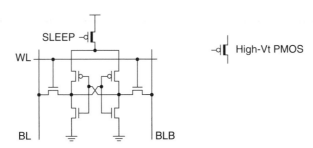

Figure 7.14: Gated-V_{dd} technique using a high-V_t transistor to gate V_{dd}

At the device level, for instance in DRAM chips or disk assemblies, the mechanism puts the device into a low-activity, low-voltage, and/or low-frequency mode such as *sleep* or *doze* or, in the case of disks, *spin-down*. For example, microprocessors can dissipate anywhere from a fraction of a watt to over 100 W of power; when not in use, they can be put into a low-power sleep or doze mode that consumes milli-watts. The processor typically expects an interrupt to cause it to resume normal operation, for instance, a clock interrupt, the interrupt output of a watchdog timer, or an external device interrupt. DRAM chips typically consume on the order of 1 W each; they have a low-power mode that will reduce this by more than an order of magnitude. Disks typically dissipate power in the tens of watts, the bulk of which is in the spindle motor. When the disk is placed in the "spin-down" mode (i.e., it is not rotating, but it is still responding to the disk controller), the disk assembly consumes a total of a handful of watts [Gurumurthi et al. 2003].

Leakage Power in SRAMs. Low-power SRAM techniques provide good examples of approaches for lowering leakage power. SRAM designs targeted for low power have begun to account for the increasingly larger amount of power consumed by leakage currents.

One conceptually simple solution is the use of multi-threshold CMOS circuits. This involves using process-level techniques to increase the threshold voltage of transistors to reduce the leakage current. Increasing this threshold serves to reduce the gate overdrive and reduces the gate's drive strength, resulting in increased delay. Because of this, the technique is mostly used on the noncritical paths of the logic, and fast, low-V_t transistors are used for the critical paths. In this way the delay penalty involved in using higher V_t transistors can be hidden in the noncritical paths, while reducing the leakage currents drastically. For example, multi-V_t transistors are selectively used for memory cells since they represent a majority of the circuit, reaping the most benefit in leakage power consumption with a minor penalty in the access time. Different multi-V_t configurations are shown in Figure 7.15, along with the leakage current path that each configuration is designed to minimize.

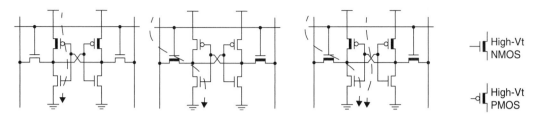

Figure 7.15: Different multi-V_t configurations for the 6T memory cell showing which leakage currents are reduced for each configuration

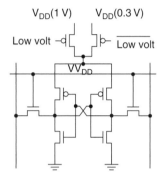

Figure 7.16: A drowsy SRAM cell containing the transistors that gate the desired power supply

Another technique that reduces leakage power in SRAMs is the Drowsy technique [Kim et al. 2004a]. This is similar to gated-V_{dd} and gated-ground techniques in that it uses a transistor to conditionally enable the power supply to a given part of the SRAM. The difference is that this technique puts infrequently accessed parts of the SRAM into a *state-preserving*, low-power mode. A second power supply with a lower voltage than the regular supply provides power to memory cells in the "drowsy" mode. Leakage power is effectively reduced because of its dependence on the value of the power supply. An SRAM cell of a drowsy cache is shown in Figure 7.16.

7.3.3 Reliability

Like performance, reliability means many things to many people. For example, embedded systems are computer systems, typically small, that run dedicated software and are embedded within the context of a larger system. They are increasingly appearing in the place of traditional electromechanical systems, whose function they are replacing because one can now find chip-level computer systems which can be programmed to perform virtually any function at a price of pennies per system. The reliability problem stems from the fact that the embedded system is a state machine (piece of software) executing within the context of a relatively complex state machine (real-time operating system) executing within the context of an extremely complex state machine (microprocessor and its memory system). We are replacing simple electromechanical systems with ultra-complex systems whose correct function cannot be guaranteed. This presents an enormous problem for the future, in which systems will only get more complex and will be used increasingly in safety-critical situations, where incorrect functioning can cause great harm.

This is a very deep problem, and one that is not likely to be solved soon. A smaller problem that we *can* solve right now—one that engineers currently do—is to increase the reliability of data within the memory system. If a datum is stored in the memory system, whether in a cache, in a DRAM, or on disk, it is reasonable to expect that the next time a processor reads that datum, the processor will get the value that was written.

How could the datum's value change? Solid-state memory devices (e.g., SRAMs and DRAMs) are susceptible to both hard failures and soft errors in the same manner that other semiconductor-based electronic devices are susceptible to both hard failures and soft failures. Hard failures can be caused by electromigration, corrosion, thermal cycling, or electrostatic shock. In contrast to hard failures, soft errors are failures where the physical device remains functional, but random and transient electronic noises corrupt the value of the stored information in the memory system. Transient noise and upset comes from a multitude of sources, including circuit noise (e.g., crosstalk, ground bounce, etc.), ambient radiation (e.g., even from sources within the computer chassis), clock jitter, or substrate interactions with high-energy particles. Which of these is the most common is obviously very dependent on the operating environment.

Figure 7.17 illustrates the last of these examples. It pictures the interactions between high-energy alpha particles and neutrons with the silicon lattice. The figure shows that when high-energy alpha particles pass through silicon, the alpha particle leaves an ionized trail, and the length of that ionized trail depends on the energy of the alpha particle. The figure also illustrates that when high-energy neutrons pass through silicon, some neutrons pass through without affecting operations of the semiconductor device, but some neutrons collide with nuclei in the silicon lattice. The atomic collision can result in the creation of multiple ionized trails as the secondary particles generated in the collision scatter in the silicon lattice. In the presence of an electric field, the ionized trails of electron-hole pairs behave as temporary surges in current or as charges that can change the data values in storage cells. In addition, charge from the ionized trails of electron-hole pairs can impact the voltage level of bit lines

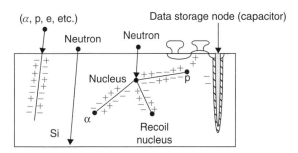

Figure 7.17: Generation of electron-hole pairs in silicon by alpha particles and high-energy neutrons

as the value of the stored data is resolved by the sense amplifiers. The result is that the *soft error rate (SER)* of a memory-storage device depends on a combination of factors including the type, number, and energy distribution of the incident particles as well as the process technology design of the storage cells, design of the bit lines and sense amplifiers, voltage level of the device, as well as the design of the logic circuits that control the movement of data in the DRAM device.

Table 7.4 compares the failure rates for SRAM, DRAM, and disk. SRAM device error rates have historically tracked DRAM devices and did so up until the 180-nm process generation. The combination of reduced supply voltage and reduced critical cell charge means that SRAM SERs have climbed dramatically for the 180-nm and 130-nm process generations. In a recent publication, Monolithic System Technology, Inc. (MoSys) claimed that for the 250-nm process generation, SRAM SERs were reported to be in the range of 100 failures per million device-hours per megabit, while SERs were reported to be in the range of 100,000 failures per megabit for the 130-nm process generation. The generalized trend is expected to continue to increase as the demand for low power dissipation forces a continued reduction in supply voltage and reduced critical charge per cell.

Solid-state memory devices (SRAMs and DRAMs) are typically protected by error detection codes and/or ECC. These are mechanisms wherein data redundancy is used to detect and/or recover from single- and even multi-bit errors. For instance, parity is a simple scheme that adds a bit to a protected word, indicating the number of even or odd bits in the word. If the read value of the word does not match the parity value, then the processor knows that the read value does not equal the value that was initially written, and an error has occurred. Error correction is achieved by encoding a word such that a bit error moves the resulting word some distance away from the original word (in the Hamming-distance sense) into an invalid encoding. The encoding space is chosen such that the new, invalid word is closest in the space to the original, valid word. Thus, the original word can always be derived from an invalid code-word, assuming a maximum number of bit errors.

Due to SRAM's extreme sensitivity to soft errors, modern processors now ship with parity and single-bit error correction for the SRAM caches. Typically, the tag arrays are protected by parity, whereas the data arrays are protected by single-bit error correction. More sophisticated multi-bit ECC algorithms are typically not deployed for on-chip SRAM caches in modern processors since the addition of sophisticated computation circuitry can add to the die size and cause significant delay relative to the timing demands of the on-chip caches. Moreover, caches store frequently accessed data, and in case an uncorrectable error is detected, a processor simply has to re-fetch the data from memory. In this sense, it can be considered unnecessary to detect and correct multi-bit errors, but sufficient to simply detect multi-bit errors. However, in the physical design of modern SRAMs, often designers will intentionally place capacitors above the SRAM cell to improve SER.

Disk reliability is a more-researched area than data reliability in disks, because data stored in magnetic disks tends to be more resistant to transient errors than data stored in solid-state memories. In other words, whereas reliability in solid-state memories is largely concerned with correcting soft errors, reliability in hard disks is concerned with the fact that disks occasionally die, taking most or all of their data with them. Given that the disk drive performs the function of permanent store, its reliability is paramount, and, as Table 7.4 shows, disks tend to last several years. This data is corroborated by a recent study from researchers at Google [Pinheiro et al. 2007]. The study tracks the behavior and environmental parameters of a fleet of over 100,000 disks for five years.

Table 7.4: Cross-comparison of failure rates for SRAM, DRAM, and disk

Technology	Failure Rate[a] (SRAM & DRAM: at 0.13μm)	Frequency of Multi-bit Errors (Relative to Single-bit Errors)	Expected Service Life
SRAM	100 per million device-hours		Several years
DRAM	1 per million device-hours	10–20%	Several years
Disk	1 per million device-hours		Several years

[a] Note that failure rate, i.e., a variation of mean-time-between-failures, says nothing about the expected performance of a single device. However, taken with the expected service life of a device, it can give a designer or administrator an idea of expected performance. If the service life of a device is 5 years, then the part will last about 5 years. A very large installation of those devices (e.g., in the case of disks or DRAMs, hundreds or more) will collectively see the expected failure rate: i.e., several hundred disks will collectively see several million device hours of operation before a single disk fails.

Reliability in the disk system is improved in much the same manner as ECC: data stored in the disk system is done so in a redundant fashion. RAID (redundant array of inexpensive disks) is a technique wherein encoded data is striped across multiple disks, so that even in the case of a disk's total failure the data will always be available.

7.3.4 Virtual Memory

Virtual memory is the mechanism by which the operating system provides executing software access to the memory system. In this regard, it is the primary consumer of the memory system: its procedures, data structures, and protocols dictate how the components of the memory system are used by all software that runs on the computer. It therefore behooves the reader to know what the virtual memory system does and how it does it. This section provides a brief overview of the mechanics of virtual memory. More detailed treatments of the topic can also be found on-line in articles by the author [Jacob & Mudge 1998a–c].

In general, programs today are written to run on no particular hardware configuration. They have no knowledge of the underlying memory system. Processes execute in imaginary address spaces that are mapped onto the memory system (including the DRAM system and disk system) by the operating system. Processes generate instruction fetches and loads and stores using imaginary or "virtual" names for their instructions and data. The ultimate home for the process's address space is nonvolatile *permanent store*, usually a disk drive; this is where the process's instructions and data come from and where all of its permanent changes go to. Every hardware memory structure between the CPU and the permanent store is a cache for the instructions and data in the process's address space. This includes main memory—main memory is really nothing more than a cache for a process's virtual address space. A cache operates on the principle that a small, fast storage device can hold the most important data found on a larger, slower storage device, effectively making the slower device look fast. The large storage area in this case is the process address space, which can range from kilobytes to gigabytes or more in size. Everything in the address space initially comes from the program file stored on disk or is created on demand and defined to be zero. This is illustrated in Figure 7.18.

7.3.4.1 Address Translation

Translating addresses from virtual space to physical space is depicted in Figure 7.19. Addresses are mapped at the granularity of *pages*. Virtual memory is essentially a mapping of *virtual page numbers* (*VPNs*) to *page frame numbers* (*PFNs*). The mapping is a function, and any virtual page can have only one location. However, the inverse map is not necessarily

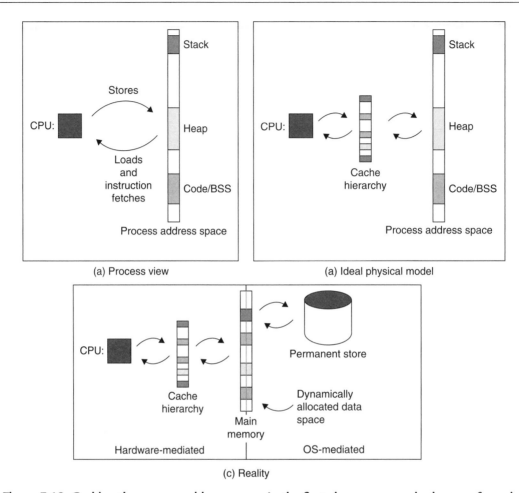

Figure 7.18: Caching the process address space. In the first view, a process is shown referencing locations in its address space. Note that all loads, stores, and fetches use virtual names for objects. The second view illustrates that a process references locations in its address space indirectly through a hierarchy of caches. The third view shows that the address space is not a linear object stored on some device, but is instead scattered across hard drives and dynamically allocated when necessary.

a function. It is possible and sometimes advantageous to have several virtual pages mapped to the same page frame (to share memory between processes or threads or to allow different views of data with different protections, for example). This is depicted in Figure 7.19 by mapping two virtual pages (0x00002 and 0xFFFFC) to PFN 12.

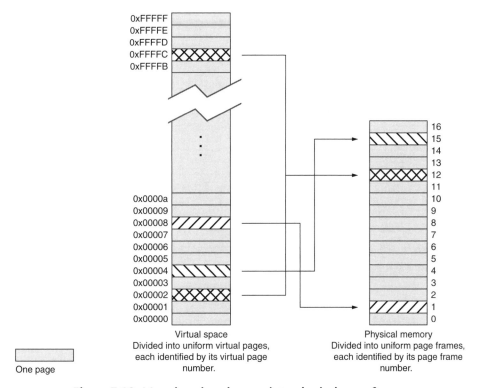

Figure 7.19: Mapping virtual pages into physical page frames

If DRAM is a cache, what is its organization? For example, an idealized *fully associative* cache (one in which any datum can reside at any location within the cache's data array) is pictured in Figure 7.20. A data tag is fed into the cache. The first stage compares the input tag to the tag of every piece of data in the cache. The matching tag points to the data's location in the cache. However, DRAM is not physically built like a cache. For example, it has no inherent concept of a tags array: one merely tells memory what data location one wishes to read or write, and the datum at that location is read out or overwritten. There is no attempt to match the address against a tag to verify the contents of the data location. However, if main memory is to be an effective cache for the virtual address space, the tags mechanism must be implemented *somewhere*. There is clearly a myriad of possibilities, from special DRAM designs that include a hardware tag feature to software algorithms that make several memory references to look up one datum. Traditional virtual memory has the tags array implemented in software, and this software structure often holds more entries than there are entries in the data array (i.e., pages in main memory). The software structure is called a *page table;* it is a database of mapping information.

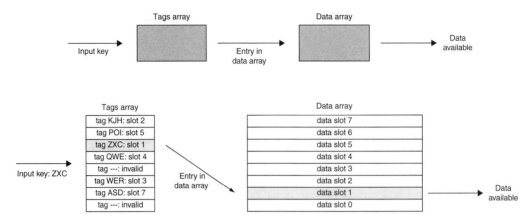

Figure 7.20: An idealized cache lookup. A cache is comprised of two parts: the tag's array and the data array. In the example organization, the tags act as a database. They accept as input a key (an address) and output either the location of the item in the data array or an indication that the item is not in the data array.

The page table performs the function of the tags array depicted in Figure 7.20. For any given memory reference, it indicates where in main memory (corresponding to "data array" in the figure) that page can be found. There are many different possible organizations for page tables, most of which require only a few memory references to find the appropriate tag entry. However, requiring more than one memory reference for a page table lookup can be very costly, and so access to the page table is sped up by caching its entries in a special cache called the *translation lookaside buffer* (*TLB*) [Lee 1960], a hardware structure that typically has far fewer entries than there are pages in main memory. The TLB is a hardware cache which is usually implemented as a content addressable memory (CAM), also called a fully associative cache.

The TLB takes as input a VPN, possibly extended by an address-space identifier, and returns the corresponding PFN and protection information. This is illustrated in Figure 7.21. The address-space identifier, if used, extends the virtual address to distinguish it from similar virtual addresses produced by other processes. For a load or store to complete successfully, the TLB must contain the mapping information for that virtual location. If it does not, a *TLB miss* occurs, and the system[9] must search the page table for the appropriate entry and place

---

[9] In the discussions, we will use the generic term "system" when the acting agent is implementation-dependent and can refer to either a hardware state machine or the operating system. For example, in some implementations, the page table search immediately following a TLB miss is performed by the operating system (MIPS, Alpha); in other implementations, it is performed by the hardware (PowerPC, x86).

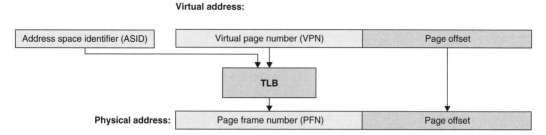

Figure 7.21: Virtual-to-physical address translation using a TLB

it into the TLB. If the system fails to find the mapping information in the page table, or if it finds the mapping but it indicates that the desired page is on disk, a *page fault* occurs. A page fault interrupts the OS, which must then retrieve the page from disk and place it into memory, create a new page if the page does not yet exist (as when a process allocates a new stack frame in virgin territory), or send the process an error signal if the access is to illegal space.

7.3.4.2 Shared Memory

Shared memory is a feature supported by virtual memory that causes many problems and gives rise to cache-management issues. It is a mechanism whereby two address spaces that are normally protected from each other are allowed to intersect at points, still retaining protection over the nonintersecting regions. Several processes sharing portions of their address spaces are pictured in Figure 7.22. The shared memory mechanism only opens up a pre-defined portion of a process's address space; the rest of the address space is still protected, and even the shared portion is only unprotected for those processes sharing the memory. For instance, in Figure 7.22, the region of A's address space that is shared with process B is unprotected from whatever actions B might want to take, but it is safe from the actions of any other processes. It is therefore useful as a simple, secure means for inter-process communication. Shared memory also reduces requirements for physical memory, as when the text regions of processes are shared whenever multiple instances of a single program are run or when multiple instances of a common library are used in different programs.

The mechanism works by ensuring that shared pages map to the same physical page. This can be done by simply placing the same PFN in the page tables of two processes sharing a page. An example is shown in Figure 7.23. Here, two very small address spaces are shown overlapping at several places, and one address space overlaps with itself; two of its virtual pages map to the same physical page. This is not just a contrived example. Many operating systems allow this, and it is useful, for example, in the implementation of user-level threads.

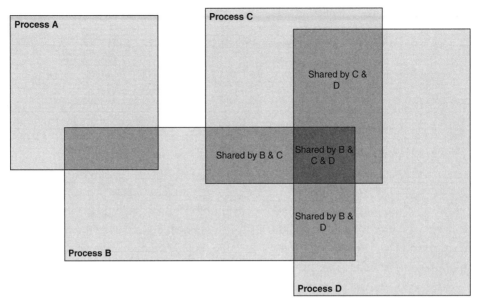

Figure 7.22: Shared memory. Shared memory allows processes to overlap portions of their address space while retaining protection for the nonintersecting regions. This is a simple and effective method for inter-process communication. Pictured are four process address spaces that have overlapped. The darker regions are shared by more than one process, while the lightest regions are still protected from other processes.

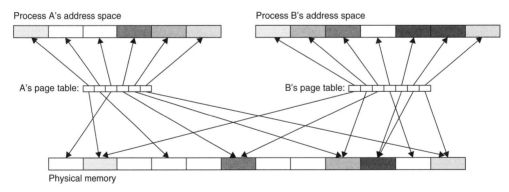

Figure 7.23: An example of shared memory. Two process address spaces—one comprised of six virtual pages and the other of seven virtual pages—are shown sharing several pages. Their page tables maintain information on where virtual pages are located in physical memory. The darkened pages are mapped to several locations; note that the darkest page is mapped at two locations in the same address space.

7.3.4.3 Some Commercial Examples

A few examples of what has been done in industry can help to illustrate some of the issues involved.

MIPS Page Table Design. MIPS [Heinrich 1995, Kane & Heinrich 1992] eliminated the page table-walking hardware found in traditional memory management units and, in doing so, demonstrated that software can table-walk with reasonable efficiency. It also presented a simple hierarchical page table design, shown in Figure 7.24. On a TLB miss, the VPN of the address that missed the TLB is used as an index into the user page table, which is accessed using a virtual address. The architecture provides hardware support for this activity, storing the virtual address of the base of the user-level page table in a hardware register and forming the concatenation of the base address with the VPN. This is illustrated in Figure 7.25. On a TLB miss, the hardware creates a virtual address for the mapping PTE in the user page table, which must be aligned on a 2-MB virtual boundary for the hardware's lookup address to work. The base pointer, called *PTEBase*, is stored in a hardware register and is usually changed on context switch.

PowerPC Segmented Translation. The IBM 801 introduced a segmented design that persisted through the POWER and PowerPC architectures (Chang & Mergen 1988, IBM & Motorola 1993, May et al. 1994, Weiss & Smith 1994). It is illustrated in Figure 7.26. Applications generate 32-bit "effective" addresses that are mapped onto a larger "virtual" address space at the granularity of *segments*, 256-MB virtual regions. Sixteen segments

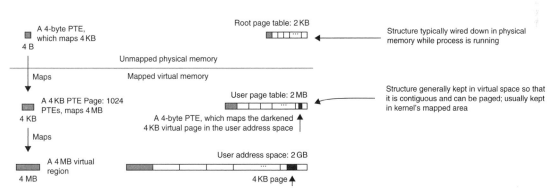

Figure 7.24: The MIPS 32-bit hierarchical page table. MIPS hardware provides support for a 2-MB linear virtual page table that maps the 2-GB user address space by constructing a virtual address from a faulting virtual address that indexes the mapping PTE (page-table entry) in the user page table. This 2-MB page table can easily be mapped by a 2-KB user root page table.

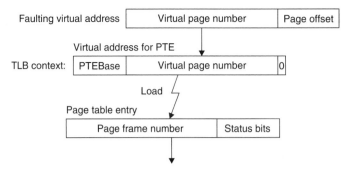

Figure 7.25: The use of the MIPS TLB context register. The VPN of the faulting virtual address is placed into the context register, creating the virtual address of the mapping PTE. This PTE goes directly into the TLB.

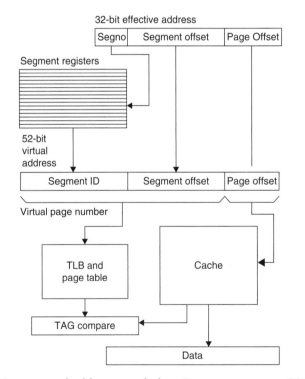

Figure 7.26: PowerPC segmented address translation. Processes generate 32-bit effective addresses that are mapped onto a 52-bit address space via 16 segment registers, using the top 4 bits of the effective address as an index. It is this extended virtual address that is mapped by the TLB and page table. The segments provide address space protection and can be used for shared memory.

comprise an application's address space. The top four bits of the effective address select a segment identifier from a set of 16 registers. This segment ID is concatenated with the bottom 28 bits of the effective address to form an extended virtual address. This extended address is used in the TLB and page table. The operating system performs data movement and relocation at the granularity of pages, not segments.

The architecture does not use explicit address-space identifiers; the segment registers ensure address space protection. If two processes duplicate an identifier in their segment registers, they share that virtual segment by definition. Similarly, protection is guaranteed if identifiers are *not* duplicated. If memory is shared through global addresses, the TLB and cache need not be flushed on context switch[10] because the system behaves like a single address space operating system.

7.4 An Example Holistic Analysis

Disk I/O accounts for a substantial fraction of an application's execution time and power dissipation. A new DRAM technology called *Fully Buffered DIMM* (*FB-DIMM*) has been in development in the industry [Vogt 2004a, b, Haas & Vogt 2005], and, though it provides storage scalability significantly beyond the current DDRx architecture, FB-DIMM has met with some resistance due to its high power dissipation. Our modeling results show that the energy consumed in a moderate-size FB-DIMM system is indeed quite large, and it can easily approach the energy consumed by a disk.

This analysis looks at a trade-off between storage in the DRAM system and in the disk system, focusing on the disk-side write buffer; if configured and managed correctly, the write buffer enables a system to approach the performance of a large DRAM installation at half the energy. Disk-side caches and write buffers have been proposed and studied, but their effect upon total system behavior has not been studied. We present the impact on total system execution time, CPI, and memory-system power, including the effects of the operating system. Using a full-system, execution-based simulator that combines Bochs, Wattch, CACTI, DRAMsim, and DiskSim and boots the RedHat Linux 6.0 kernel, we have investigated the memory-system behavior of the SPEC CPU2000 applications. We study the

---

[10] Flushing is avoided until the system runs out of identifiers and must reuse them. For example, the address-space identifiers on the MIPS R3000 and Alpha 21064 are six bits wide, with a maximum of 64 active processes [Digital 1994, Kane & Heinrich 1992]. If more processes are desired, identifiers must be constantly reassigned, requiring TLB and virtual-cache flushes.

disk-side cache in both single-disk and RAID-5 organizations. Cache parameters include size, organization, whether the cache supports write caching or not, and whether it prefetches read blocks or not. Our results are given in terms of L1/L2 cache accesses, power dissipation, and energy consumption; DRAM-system accesses, power dissipation, and energy consumption; disk-system accesses, power dissipation, and energy consumption; and execution time of the application plus operating system, in seconds. The results are not from sampling, but rather from a simulator that calculates these values on a cycle-by-cycle basis over the entire execution of the application.

7.4.1 Fully-Buffered DIMM vs. the Disk Cache

It is common knowledge that disk I/O is expensive in both power dissipated and time spent waiting on it. What is less well known is the system-wide breakdown of disk power versus cache power versus DRAM power, especially in light of the newest DRAM architecture adopted by industry, the FB-DIMM. This new DRAM standard replaces the conventional memory bus with a narrow, high-speed interface between the memory controller and the DIMMs. It has been shown to provide performance similar to that of DDRx systems, and thus, it represents a relatively low-overhead mechanism (in terms of execution time) for scaling DRAM-system capacity. FB-DIMM's latency degradation is not severe. It provides a noticeable bandwidth improvement, and it is relatively insensitive to scheduling policies [Ganesh et al. 2007].

FB-DIMM was designed to solve the problem of storage scalability in the DRAM system, and it provides scalability well beyond the current JEDEC-style DDRx architecture, which supports at most two to four DIMMs in a fully populated dual-channel system (DDR2 supports up to two DIMMs per channel; proposals for DDR3 include limiting a channel to a single DIMM). The daisy-chained architecture of FB-DIMM supports up to eight DIMMs per channel, and its narrow bus requires roughly one-third the pins of a DDRx SDRAM system. Thus, an FB-DIMM system supports an order of magnitude more DIMMs than DDRx. This scalability comes at a cost, however. The DIMM itself dissipates almost an order of magnitude more power than a traditional DDRx DIMM. Couple this with an order-of-magnitude increase in DIMMs per system, and one faces a serious problem.

To give an idea of the problem, Figure 7.27 shows the simulation results of an entire execution of the *gzip* benchmark from SPEC CPU2000 on a complete-system simulator. The memory system is only moderate in size: one channel and four DIMMs, totalling a half-gigabyte. The graphs demonstrate numerous important issues, but in this book we are concerned with two items in particular:

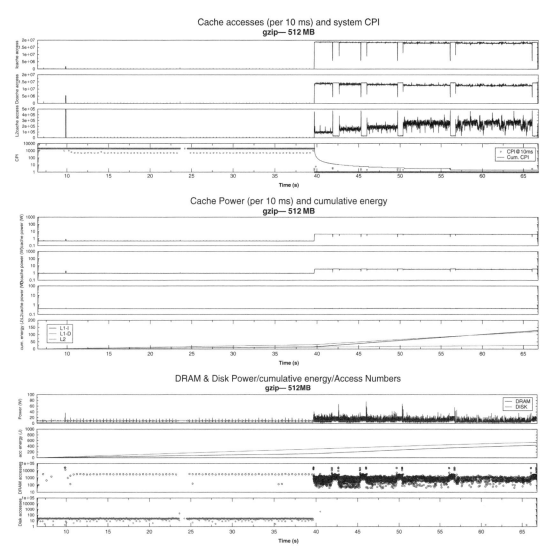

Figure 7.27: Full execution of Gzip. The figure shows the entire run of gzip. System configuration is a 2-GHz Pentium processor with 512 MB of DDR2-533 FB-DIMM main memory and a 12k-RPM disk drive with built-in disk cache. The figure shows the interaction between all components of the memory system, including the L1 instruction and data caches, the unified L2 cache, the DRAM system, and the disk drive. All graphs use the same x-axis, which represents execution time in seconds. The x-axis does not start at zero; the measurements exclude system boot time, invocation of the shell, etc. Each data point represents aggregated (not sampled) activity within a 10-ms epoch. The CPI graph shows two system CPI values: one is the average CPI for each 10-ms epoch, and the other is the cumulative average CPI. A duration with no CPI data point indicates that no instructions were executed due to I/O latency. During such a window the CPI is essentially infinite, and thus, it is possible for the cumulative average to range higher than the displayed instantaneous CPI. Note that the CPI, the DRAM accesses, and the disk accesses are plotted on log scales.

Program initialization is lengthy and represents a significant portion of an application's run time. As the CPI graph shows, the first two-thirds of execution time are spent dealing with the disk, and the corresponding CPI (both average and instantaneous) ranges from the 100s to the 1000s. After this initialization phase, the application settles into a more compute-intensive phase in which the CPI asymptotes down to the theoretical sustainable performance, the single-digit values that architecture research typically reports.

By the end of execution, the total energy consumed in the FB-DIMM DRAM system (a half a kilojoule) almost equals that of the energy consumed by the disk, and it is twice that of the L1 data cache, L1 instruction cache, and unified L2 cache combined.

Currently, there is substantial work happening in both industry and academia to address the latter issue, with much of the work focusing on access scheduling, architecture improvements, and data migration. To complement this work, we look at a wide range of organizational approaches, i.e., attacking the problem from a parameter point of view rather than a system-redesign, component-redesign, or new-proposed-mechanism point of view, and find significant synergy between the disk cache and the memory system. Choices in the disk-side cache affect both system-level performance and system-level (in particular, DRAM-subsystem-level) energy consumption. Though disk-side caches have been proposed and studied, their effect upon the total system behavior, namely execution time or CPI or total memory-system power including the effects of the operating system, is as yet unreported. For example, Zhu and Hu [2002] evaluate disk built-in cache using both real and synthetic workloads and report the results in terms of average response time. Smith [1985a and b] evaluates a disk cache mechanism with real traces collected in real IBM mainframes on a disk cache simulator and reports the results in terms of miss rate. Huh and Chang [2003] evaluate their RAID controller cache organization with a synthetic trace. Varma and Jacobson [1998] and Solworth and Orji [1990] evaluate destaging algorithms and write caches, respectively, with synthetic workloads. This study represents the first time that the effects of the disk-side cache can be viewed at a system level (considering both application and operating-system effects) and compared directly to all the other components of the memory system.

We use a full-system, execution-based simulator combining Bochs [Bochs 2006], Wattch [Brooks et al. 2000], CACTI [Wilton & Jouppi 1994], DRAMsim [Wang et al. 2005, September], and DiskSim [Ganger et al. 2006]. It boots the RedHat Linux 6.0 kernel and therefore can capture all application behavior, and all operating-system behavior, including I/O activity, disk-block buffering, system-call overhead, and virtual memory overhead such as translation, table walking, and page swapping. We investigate the disk-side cache in both single-disk and RAID-5 organizations. Cache parameters include size, organization, whether the cache supports write caching or not, and whether it prefetches read blocks or not. Additional parameters include disk rotational speed and DRAM-system capacity.

We find a complex trade-off between the disk cache, the DRAM system, and disk parameters like rotational speed. The disk cache, particularly its write-buffering feature, represents a very powerful tool enabling significant savings in both energy and execution time. This is important because, though the cache's support for write buffering is often enabled in desktop operating systems (e.g., Windows and some but not all flavors of Unix/Linux [Ng 2006]), it is typically disabled in enterprise computing applications [Ng 2006], and these are the applications most likely to use FB-DIMMs [Haas & Vogt 2005]. We find substantial improvement between existing implementations and an ideal write buffer (i.e., this is a limit study). In particular, the disk cache's write-buffering ability can offset the total energy consumption of the memory system (including caches, DRAMs, and disks) by nearly a factor of two, while sacrificing a small amount of performance.

7.4.2 Fully Buffered DIMM: Basics

The relation between a traditional organization and a FB-DIMM organization is shown in Figure 7.28, which motivates the design in terms of a graphics-card organization. The first two drawings show a multi-drop DRAM bus next to a DRAM bus organization typical of graphics cards, which use point-to-point soldered connections between the DRAM and memory controller to achieve higher speeds. This arrangement is used in FB-DIMM. A slave memory controller has been added onto each DIMM, and all connections in the system are point to point. A narrow, high-speed channel connects the master memory controller to the DIMM-level memory controllers (called *Advanced Memory Buffers* or AMBs). Since each DIMM-to-DIMM connection is a point-to-point connection, a channel becomes a *de facto* multi-hop store and forward network. The FB-DIMM architecture limits the channel length to eight DIMMs, and the narrower inter-module bus requires roughly one-third as many pins as a traditional organization. As a result, an FB-DIMM organization can handle roughly 24 times the storage capacity of a single-DIMM DDR3-based system, without sacrificing any bandwidth and even leaving headroom for increased intra-module bandwidth.

The AMB acts like a pass-through switch, directly forwarding the requests it receives from the controller to successive DIMMs and forwarding frames from southerly DIMMs to northerly DIMMs or the memory controller. All frames are processed to determine whether the data and commands are for the local DIMM. The FB-DIMM system uses a serial packet-based protocol to communicate between the memory controller and the DIMMs. Frames may contain data and/or commands. Commands include DRAM commands such as row

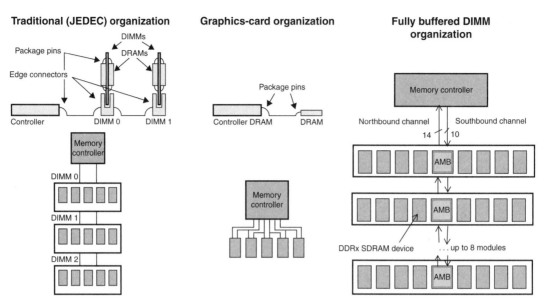

Figure 7.28: FB-DIMM and its motivation. The first two pictures compare the memory organizations of a JEDEC SDRAM system and a graphics card. Above each design is its side-profile, indicating potential impedance mismatches (sources of reflections). The organization on the far right shows how the FB-DIMM takes the graphics-card organization as its *de facto* DIMM. In the FB-DIMM organization, there are no multi-drop busses; DIMM-to-DIMM connections are point to point. The memory controller is connected to the nearest AMB via two unidirectional links. The AMB is, in turn, connected to its southern neighbor via the same two links.

activate (RAS), column read (CAS), refresh (REF) and so on, as well as channel commands such as write to configuration registers, synchronization commands, etc. Frame scheduling is performed exclusively by the memory controller. The AMB only converts the serial protocol to DDRx-based commands without implementing any scheduling functionality.

The AMB is connected to the memory controller and/or adjacent DIMMs via unidirectional links: the southbound channel which transmits both data and commands and the northbound channel which transmits data and status information. The southbound and northbound datapaths are 10 bits and 14 bits wide, respectively. The FB-DIMM channel clock operates at six times the speed of the DIMM clock; i.e., the link speed is 4 Gbps for a 667-Mbps DDRx system. Frames on the north- and southbound channel require 12 transfers (6 FB-DIMM channel clock cycles) for transmission. This 6:1 ratio ensures that the FB-DIMM frame rate matches the DRAM command clock rate.

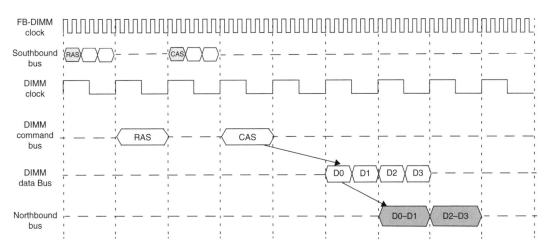

Figure 7.29: Read transaction in an FB-DIMM system. The figure shows how a read transaction is performed in an FB-DIMM system. The FB-DIMM serial busses are clocked at six times the DIMM busses. Each FB-DIMM frame on the southbound bus takes six FB-DIMM clock periods to transmit. On the northbound bus a frame comprises two DDRx data bursts.

Southbound frames comprise both data and commands and are 120 bits long; northbound frames are data only and are 168 bits long. In addition to the data and command information, the frames also carry header information and a frame CRC (cyclic redundancy check) checksum that is used to check for transmission errors. A northbound read-data frame transports 18 bytes of data in 6 FB-DIMM clocks or 1 DIMM clock. A DDRx system can burst back the same amount of data to the memory controller in two successive beats lasting an entire DRAM clock cycle. Thus, the read band-width of an FB-DIMM system is the same as that of a single channel of a DDRx system. Due to the narrower southbound channel, the write bandwidth in FB-DIMM systems is one-half that available in a DDRx system. However, this makes the *total* bandwidth available in an FB-DIMM system 1.5 times that of a DDRx system.

Figure 7.29 shows the processing of a read transaction in an FB-DIMM system. Initially, a command frame is used to transmit a command that will perform row activation. The AMB translates the request and relays it to the DIMM. The memory controller schedules the CAS command in a following frame. The AMB relays the CAS command to the DRAM devices which burst the data back to the AMB. The AMB bundles two consecutive bursts of data into a single northbound frame and transmits it to the memory controller. In this example, we assume a burst length of four corresponding to two FB-DIMM data frames. Note that although the figures do not identify parameters like t_CAS, t_RCD, and t_CWD, the memory controller must ensure that these constraints are met.

The primary dissipater of power in an FB-DIMM channel is the AMB, and its power depends on its position within the channel. The AMB nearest to the memory controller must handle its own traffic and repeat all packets to and from all downstream AMBs, and this dissipates the most power. The AMB in DDR2-533 FB-DIMM dissipates 6 W, and it is currently 10 W for 800 Mbps DDR2 [Staktek 2006]. Even if one averages out the activity on the AMB in a long channel, the eight AMBs in a single 800-Mbps channel can easily dissipate 50 W. Note that this number is for the AMBs only; it does not include power dissipated by the DRAM devices.

7.4.3 Disk Caches: Basics

Today's disk drives all come with a built-in cache as part of the drive controller electronics, ranging in size from 512 KB for the micro-drive to 16 MB for the largest server drives. Figure 7.30 shows the cache and its place within a system. The earliest drives had no cache memory, as they had little control electronics. As the control of data transfer migrated from the host-side control logic to the drive's own controller, a small amount of memory was needed to act as a speed-matching buffer, because the disk's media data rate is different from that of the interface. Buffering is also needed because when the head is at a position ready to do data transfer, the host or the interface may be busy and not ready to receive read data. DRAM is usually used as this buffer memory.

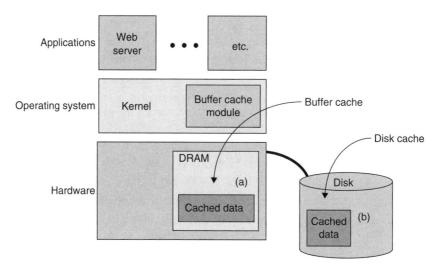

Figure 7.30: Buffer caches and disk caches. Disk blocks are cached in several places, including (a) the operating system's *buffer cache* in main memory and (b), on the disk, in another DRAM buffer, called a *disk cache*.

In a system, the host typically has some memory dedicated for caching disk data, and if a drive is attached to the host via some external controller, that controller also typically has a cache. Both the system cache and the external cache are much larger than the disk drive's internal cache. Hence, for most workloads, the drive's cache is not likely to see too many reuse cache hits. However, the disk-side cache is very effective in opportunistically prefetching data, as only the controller inside the drive knows the state the drive is in and when and how it can prefetch without adding any cost in time. Finally, the drive needs cache memory if it is to support write caching/buffering.

With write caching, the drive controller services a write request by transferring the write data from the host to the drive's cache memory and then reports back to the host that the write is "done," even though the data has not yet been written to the disk media (data not yet written out to disk is referred to as *dirty*). Thus, the service time for a cached write is about the same as that for a read cache hit, involving only some drive controller overhead and electronic data transfer time but no mechanical time. Clearly, write caching does not need to depend on having the right content in the cache memory for it to work, unlike read caching. Write caching will always work, i.e., a write command will always be a cache hit, as long as there is available space in the cache memory. When the cache becomes full, some or all of the dirty data are written out to the disk media to free up space. This process is commonly referred to as *destage*.

Ideally, destage should be done while the drive is idle so that it does not affect the servicing of read requests. However, this may not be always possible. The drive may be operating in a high-usage system with little idle time ever, or the writes often arrive in bursts which quickly fill up the limited memory space of the cache. When destage must take place while the drive is busy, such activity adds to the load of drive at that time, and a user will notice a longer response time for his requests. Instead of providing the full benefit of cache hits, write caching in this case merely delays the disk writes.

Zhu and Hu (2002) have suggested that large disk built-in caches will not significantly benefit the overall system performance because all modern operating systems already use large file system caches to cache reads and writes. As suggested by Przybylski (1990), the reference stream missing a first-level cache and being handled by a second-level cache tends to exhibit relatively low locality. In a real system, the reference stream to the disk system has missed the operating system's buffer cache, and the locality in the stream tends to be low. Thus, our simulation captures all of this activity. In our experiments, we investigate the disk cache, including the full effects of the operating system's file-system caching.

7.4.4 Experimental Results

Figure 7.27 showed the execution of the GZIP benchmark with a moderate-sized FB-DIMM DRAM system: half a gigabyte of storage. At 512 MB, there is no page swapping for this application. When the storage size is cut in half to 256 MB, page swapping begins but does not affect the execution time significatly. When the storage size is cut to one-quarter of its original size (128 MB), the page swapping is significant enough to slow the application down by an order of magnitude. This represents the hard type of decision that a memory-systems designer would have to face: if one can reduce power dissipation by cutting the amount of storage and feel negligible impact on performance, then one has too much storage to begin with.

Figure 7.31 shows the behavior of the system when storage is cut to 128 MB. Note that all aspects of system behavior have degraded; execution time is longer, *and* the system consumes more energy. Though the DRAM system's energy has decreased from 440 J to just under 410 J, the execution time has increased from 67 to 170 seconds, the total cache energy has increased from 275 to 450 J, the disk energy has increased from 540 to 1635 J, and the total energy has doubled from 1260 to 2515 J. This is the result of swapping activity—not enough to bring the system to its knees, but enough to be relatively painful.

We noticed that there exists in the disk subsystem the same sort of activity observed in a microprocessor's load/store queue: reads are often stalled waiting for writes to finish, despite the fact that the disk has a 4-MB read/write cache on board. The disk's cache is typically organized to prioritize prefetch activity over write activity because this tends to give the best performance results and because the write buffering is often disabled by the operating system. The solution to the write-stall problem in microprocessors has been to use write buffers; we therefore modified DiskSim to implement an ideal write buffer on the disk side that would not interfere with the disk cache. Figure 7.32 indicates that the size of the cache seems to make little difference to the behavior of the system. The important thing is that a cache is present. Thus, we should not expect read performance to suddenly increase as a result of moving writes into a separate write buffer.

Figure 7.33 shows the behavior of the system with 128 MB and an ideal write buffer. As mentioned, the performance increase and energy decrease is due to the writes being buffered, allowing read requests to progress. Execution time is 75 seconds (compared to 67 seconds for a 512 MB system); and total energy is 1100 J (compared to 1260 J for a 512-MB system). For comparison, to show the effect of faster read and write throughput, Figure 7.34 shows the behavior of the system with 128 MB and an 8-disk RAID-5 system. Execution time is 115 seconds, and energy consumption is 8.5 KJ. This achieves part of the performance effect as

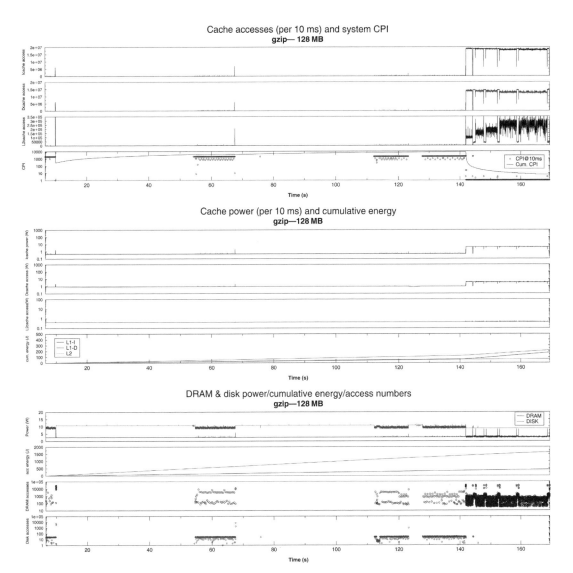

Figure 7.31: Full execution of GZIP, 128 MB DRAM. The figure shows the entire run of GZIP. System configuration is a 2 GHz Pentium processor with 128 MB of FB-DIMM main memory and a 12 K-RPM disk drive with built-in disk cache. The figure shows the interaction between all components of the memory system, including the L1 instruction cache, the L1 data cache, the unified L2 cache, the DRAM system, and the disk drive. All graphs use the same x-axis, which represents the execution time in seconds. The x-axis does not start at zero; the measurements exclude system boot time, invocation of the shell, etc. Each data point represents aggregated (not sampled) activity within a 10-ms epoch. The CPI graph shows 2 system CPI values: one is the average CPI for each 10-ms epoch, the other is the cumulative average CPI. A duration with no CPI data point indicates that no instructions were executed due to I/O latency. The application is run in single-user mode, as is common for SPEC measurements; therefore, disk delay shows up as stall time. Note that the CPI, the DRAM accesses, and the Disk accesses are plotted on log scales.

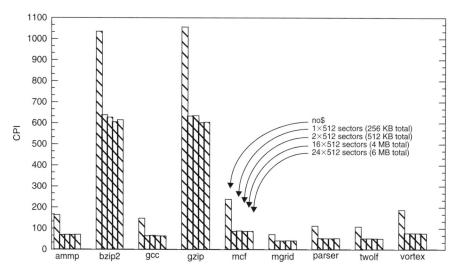

Figure 7.32: The effects of disk cache size by varying the number of segments. The figure shows the effects of a different number of segments with the same segment size in the disk cache. The system configuration is 128 MB of DDR SDRAM with a 12 k-RPM disk. There are five bars for each benchmark, which are (1) no cache, (2) 1 segment of 512 sectors each, (3) 2 segments of 512 sectors each, (4) 16 segment of 512 sectors each, and (5) 24 segment of 512 sectors each. Note that the CPI values are for the disk-intensive portion of application execution, not the CPU-intensive portion of application execution (which could otherwise blur distinctions).

write buffering by improving write time, thereby freeing up read bandwidth sooner. However, the benefit comes at a significant cost in energy.

Table 7.5 gives breakdowns for **gzip** in tabular form, and the graphs beneath the table give the breakdowns for **gzip**, **bzip2**, and **ammp** in graphical form and for a wider range of parameters (different disk RPMs). The applications all demonstrate the same trends: to cut down the energy of a 512-MB system by reducing the memory to 128 MB which causes both the performance and the energy to get worse. Performance degrades by a factor of 5–10; energy increases by 1.5× to 10×. Ideal write buffering can give the best of both worlds (performance of a large memory system and energy consumption of a small memory system), and its benefit is independent of the disk's RPM. Using a RAID system does not gain significant performance improvement, but it consumes energy proportionally to the number of disks. Note, however, that this is a uniprocessor model running in single-user mode, so RAID is not expected to shine.

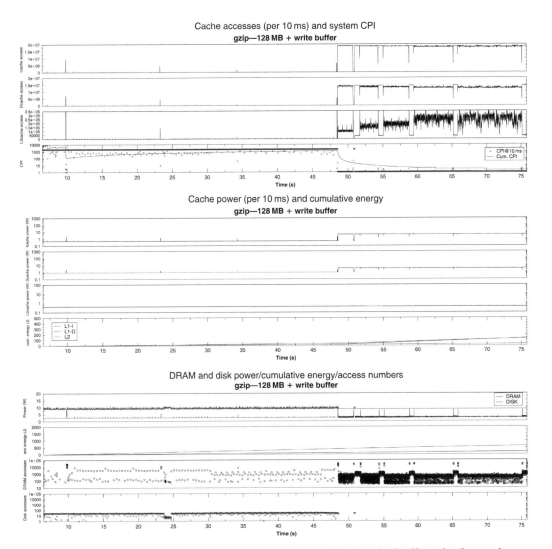

Figure 7.33: Full execution of GZIP, 128 MB DRAM and ideal write buffer. The figure shows the entire run of GZIP. System configuration is a 2 GHz Pentium processor with 128 MB of FB-DIMM main memory and a 12 K-RPM disk drive with built-in disk cache. The figure shows the interaction between all components of the memory system, including the L1 instruction cache, the L1 data cache, the unified L2 cache, the DRAM system, and the disk drive. All graphs use the same *x*-axis, which represents the execution time in seconds. The *x*-axis does not start at zero; the measurements exclude system boot time, invocation of the shell, etc. Each data point represents aggregated (not sampled) activity within a 10-ms epoch. The CPI graph shows two system CPI values: one is the average CPI for each 10-ms epoch, the other is the cumulative average CPI. A duration with no CPI data point indicates that no instructions were executed due to I/O latency. The application is run in single-user mode, as is common for SPEC measurements; therefore, disk delay shows up as stall time. Note that the CPI, the DRAM accesses, and the Disk accesses are plotted on log scales.

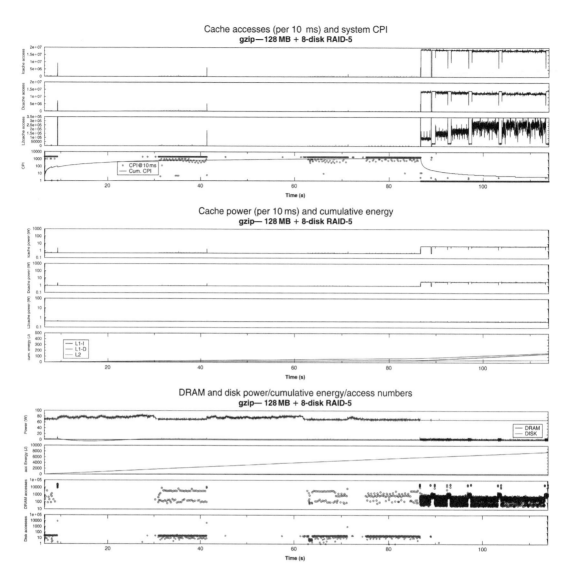

Figure 7.34: Full execution of GZIP, 128 MB DRAM and RAID-5 disk system. The figure shows the entire run of GZIP. System configuration is a 2 GHz Pentium processor with 128 MB of FB-DIMM main memory and a RAID-5 system of eight 12-K-RPM disk drives with built-in disk cache. The figure shows the interaction between all components of the memory system, including the L1 instruction cache, the L1 data cache, the unified L2 cache, the DRAM system, and the disk drive. All graphs use the same x-axis, which represents the execution time in seconds. The x-axis does not start at zero; the measurements exclude system boot time, invocation of the shell, etc. Each data point represents aggregated (not sampled) activity within a 10-ms epoch. The CPI graph shows two system CPI values: one is the average CPI for each 10-ms epoch, the other is the cumulative average CPI. A duration with no CPI data point indicates that no instructions were executed due to I/O latency. The application is run in single-user mode, as is common for SPEC measurements; therefore, disk delay shows up as stall time. Note that the CPI, the DRAM accesses, and the Disk accesses are plotted on log scales.

Table 7.5: Execution time and energy breakdowns for GZIP and BZIP2

System Configuration (DRAM Size - Disk RPM - Option)	Ex. Time (sec)	L1-I Energy (J)	L1-D Energy (J)	L2 Energy (J)	DRAM Energy (J)	Disk Energy (J)	Total Energy (J)
GZIP							
512 MB–12 K	66.8	129.4	122.1	25.4	440.8	544.1	1261.8
128 MB–12 K	169.3	176.5	216.4	67.7	419.6	1635.4	2515.6
128 MB–12 K–WB	75.8	133.4	130.2	28.7	179.9	622.5	1094.7
128 MB–12 K–RAID	113.9	151	165.5	44.8	277.8	7830	8469.1

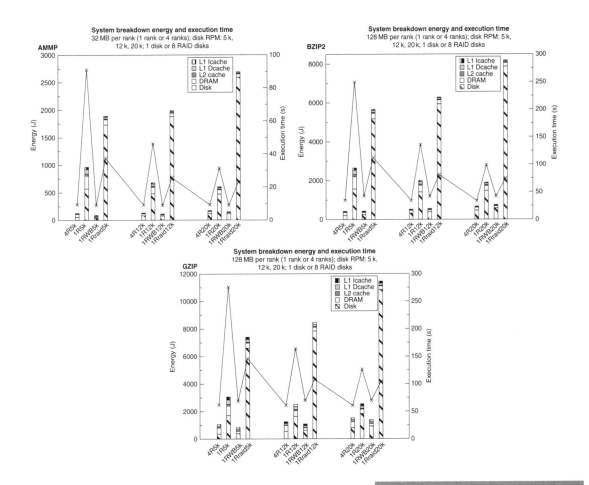

Figure 7.35 shows the effects of disk caching and prefetching on both single-disk and RAID systems. In RAID systems, disk caching has only marginal effects to both the CPI and the disk average response time. However, disk caching with prefetching has significant benefits. In a slow disk system (i.e., 5400 RPM), RAID has more tangible benefits over a non-RAID system. Nevertheless, the combination of using RAID, disk cache, and fast disks can improve the overall performance up to a factor of 10. For the average response time, even though the write time, response time in a RAID system is much higher than the write response time in a single-disk system, this trend does not translate directly into the overall performance. The write response time in a RAID system is higher due to parity calculations, especially the benchmarks with small writes. Despite the improvement in performance, care must be taken in applying RAID because RAID increases the energy proportion ally to the number of the disks.

Perhaps the most interesting result in Figure 7.35 is that the CPI values (top graph) track the disk's *average **read** response time* (bottom graph) and not the disk's *average response time* (which includes both reads and writes, also bottom graph). This observation holds true for both read-dominated applications and applications with significant write activity (as are **gzip** and **bzip2**). The reason this is interesting is that the disk community tends to report performance numbers in terms of average response time and not average *read* response time, presumably believing the former to be a better indicator of system-level performance than the latter. Our results suggest that the disk community would be better served by continuing to model the effects of write traffic (as it affects read latency) by reporting performance as the average *read* response time.

7.4.5 Conclusions

We find that the disk cache can be an effective tool for improving performance at the system level. There is a significant interplay between the DRAM system and the disk's ability to buffer writes and prefetch reads. An ideal write buffer homed within the disk has the potential to move write traffic out of the way and begin working on read requests far sooner, with the result that a system can be made to perform nearly as well as one with four times the amount of main memory, but with roughly half the energy consumption of the configuration with more main memory.

This is extremely important, because FB-DIMM systems are likely to have significant power-dissipation problems, and because of this they will run at the cutting edge of the storage-performance trade-off. Administrators will configure these systems to use the least amount of storage available to achieve the desired performance, and thus a simple reduction in FB-DIMM storage will result in an unacceptable hit to performance. We have shown that an ideal write buffer in the disk system will solve this problem, transparently to the operating system.

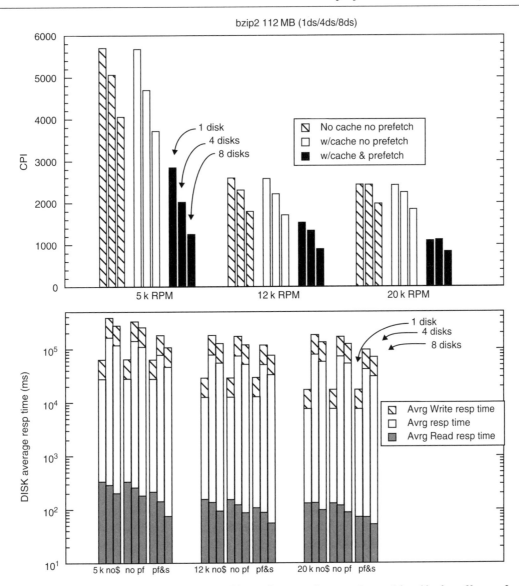

Figure 7.35: The effects of disk prefetching. The experiment tries to identify the effects of prefetching and caching in the disk cache. The configuration is 112 MB of DDR SDRAM running bzip2. The three bars in each group represent a single-disk system, 4-disk RAID-5 system, and 8-disk RAID-5 system. The figure above shows the CPI of each configuration, and the figure below shows the average response time of the disk requests. Note that the CPI axis is in linear scale, but the disk average response time axis is in log scale. The height of the each bar in the average response time graph is the absolute value.

7.5 What to Expect

What are the more important architecture-level issues in store for these technologies? On what problems should a designer concentrate?

For caches and SRAMs in particular, power dissipation and reliability are primary issues. A rule of thumb is that SRAMs typically account for at least one-third of the power dissipated by microprocessors, and the reliability for SRAM is the worst of the three technologies.

For DRAMs, power dissipation is becoming an issue with the high I/O speeds expected of future systems. The FB-DIMM, the only proposed architecture seriously being considered for adoption that would solve the capacity-scaling problem facing DRAM systems, dissipates roughly two orders of magnitude more power than a traditional organization (due to an order of magnitude higher per DIMM power dissipation and the ability to put an order of magnitude more DIMMs into a system).

For disks, miniaturization and development of heuristics for control are the primary considerations, but a related issue is the reduction of power dissipation in the drive's electronics and mechanisms. Another point is that some time this year, the industry will be seeing the first generation of hybrid disk drives: those with flash memory to do write caching. Initially, hybrid drives will be available only for mobile applications. One reason for a hybrid drive is to be able to have a disk drive in spin-down mode longer (no need to spin up to do a write). This will save more power and make the battery of a laptop last longer.

For memory systems as a whole, a primary issue is optimization in the face of subsytems that have unanticipated interactions in their design parameters.

From this book, a reader should expect to learn the details of operation and tools of analysis that are necessary for understanding the intricacies and optimizing the behavior of modern memory systems. The designer should expect of the future a memory-system design space that will become increasingly difficult to analyze simply and in which alternative figures of merit (e.g., energy consumption, cost, reliability) will become increasingly important. Future designers of memory systems will have to perform design-space explorations that consider the effects of design parameters in all subsystems of the memory hierarchy, and they will have to consider multiple dimensions of design criteria (e.g., performance, energy consumption, cost, reliability, and real-time behavior).

In short, a holistic approach to design that considers the whole hierarchy is warranted, but this is very hard to do. Among other things, it requires in-depth understanding at all the levels of the hierarchy. It is our goal that this book will enable just such an approach.

Bibliography

Baker, R.J., Li, H.W., & Boyce, D.E. (1998). *CMOS: Circuit Design, Layout, and Simulation.* New York: IEEE Press.

Baker, R.J. (2005). *CMOS: Circuit Design, Layout, and Simulation,* (2nd Ed.). New York: IEEE Press and Wiley-Interscience.

Baynes, K., Collins, C., Fiterman, E., Ganesh, B., Kohout, P., Smit, C., Zhang, T., & Jacob, B. (2001). The performance and energy consumption of three embedded real-time operating systems. In *Proc. Int. Conf. Compilers, Architecture, and Synthesis for Embedded Systems (CASES 2001)* (pp. 203–210), Atlanta, GA November.

Baynes, K., Collins, C., Fiterman, E., Smit, C., Zhang, T., & Jacob, B. (2003). The performance and energy consumption of embedded real-time operating systems. *IEEE Trans. Computers, 52*(11), 1454–1469.

Belady, L.A. (1966). A study of replacement algorithms for virtual storage. *IBM Syst. J., 5*(2), 78–101.

Bochs. (2006). *The Bochs IA-32 Emulator Project.* http://bochs.sourceforge.net.

Brooks, D., & Martonosi, M. (2000). Value-based clock gating and operation packing: Dynamic strategies for improving processor power and performance. *ACM Trans. Comput. Syst., 18*(2), 89–126.

Brooks, D.M., Bose, P., Schuster, S.E., Jacobson, H., Kudva, P.N., Buyuktosunoglu, A., Wellman, J.-D., Zyuban, V., Gupta, M., & Cook, P.W. (November/December 2000a). Power-aware microarchitecture: Design and modeling challenges for next-generation microprocessors. *IEEE Micro, 20*(6), 26–44.

Butts, J.A., & Sohi, G.S. (December 2000). A static power model for architects. In *Proc. 33rd Ann. Int. Symp. Microarchitecture (MICRO-33)* (pp. 191–201), Monterey, CA.

Cuppu, V., Jacob, B., Davis, B., & Mudge, T. (May 1999). A performance comparison of contemporary DRAM architectures. In *Proc. 26th Ann. Int. Symp. Computer Architecture (ISCA'99)* (pp. 222–233), Atlanta, GA.

Cuppu, V., & Jacob, B. (June 2001). Concurrency, latency, or system overhead: Which has the largest impact on uniprocessor dram-system performance? In *Proc. 28th Ann. Int. Symp. Computer Architecture (ISCA'01)* (pp. 62–71), Göteborg, Sweden.

Denning, P.J. (1970). Virtual Memory. *Computing Surveys, 2*(3), 153–189.

Ganesh, B., Jaleel, A., Wang, D., & Jacob, B. (February 2007). Fully-Buffered DIMM memory architectures: Understanding mechanisms, overheads and scaling. In *Proc. 13th Int. Symp. on High Performance Computer Architecture (HPCA 2007)*, Phoenix, AZ.

Ganger, G., Worthington, B., & Patt, Y. (2006). *The DiskSim Simulation Environment Version 2.0 Reference Manual.* http://www.pdl.cmu.edu/DiskSim/.

Gelsinger, P. (2001). Microprocessors for the new millennium: Challenges, opportunities, and new frontiers. *ISSCC*, 22–25.

Gonzalez, R., & Horowitz, M. (Sept. 1996). Energy dissipation in general purpose microprocessors. *IEEE J. Solid-State Circuits, 31*(9), 1277–1284.

Grove, A. (2002). Changing Vectors of Moore's Law. *Presented at International Electron Devices Meeting (IEDM)*, December.

Haas, J., & Vogt, P. (March 2005). Fully-buffered DIMM technology moves enterprise platforms to the next level. *Technlogy @Intel Magazine*.

Heinrich, J. (Ed.), (1995). *MIPS R10000 Microprocessor User's Manual, version 1.0.* Mountain View, CA: MIPS Technologies. Inc.

Huh, J., & Chang, T. (Dec. 2003). Hierarchical disk cache management in RAID 5 controller. *J. Computing Sciences in Colleges Archive, 19*(2), 47–59.

Jacob, B., & Mudge, T. (1998a). A look at several memory-management units, TLB-refill mechanisms, and page table organizations. In *Proc. Eighth Int. Conf. on Architectural Support for Programming Languages and Operating Systems (ASPLOs'98)* (pp. 295–306), San Jose, CA.

Jacob, B., & Mudge, T. (1998b). Virtual memory in contemporary microprocessors. *IEEE Micro, 18*(4), 60–75.

Jacob, B., & Mudge, T. (1998c). Virtual memory: Issues of implementation. *IEEE Computer, 31*(6), 33–43.

Jacob, B. (July/Aug. 2003). A case for studying DRAM issues at the system level. *IEEE Micro, 23*(4), 44–56.

John, L.K. (March 2004). More on finding a single number to indicate overall performance of a benchmark suite. *SIGARCH Computer Architecture News, 32*(1), 3–8.

Kane, G., & Heinrich, J. (1992). *MIPS RISC Architecture.* Englewood Cliffs, NJ: Prentice Hall.

Kellogg, M. (2002). Personal communication.

Kim, Y.I., Yang, K.H., & Lee, W.S. (2004). Thermal degradation of DRAM retention time: Characterization and improving techniques. In *Proc. 42nd Ann. Int. Reliability Physics Symposium* (pp. 667–668).

Lee, F.F. (Nov. 1960). Study of look aside memory. *IEEE Trans. Computers, 18*(11), 1062–1064.

Lee, T. (2002). Personal communication.

Ng, S. (2006). Personal communication.

Palacharla, S., Jouppi, N.P., & Smith, J.E. (November 19, 1996). *Quantifying the complexity of superscalar processors, Technical Report CS-TR-96-1328.* Madison: University of Wisconsin.

Peckerar, M., Fulton, R., Blaise, P., & Brown, D. (November–December 1979). Radiation effects in MOS devices caused by X-ray and e-beam lithography. *J. Vacuum Science Technology, 16*(6), 1658–1661.

Peckerar, M.C., Dozier, C.M., Brown, D.B., & Patterson, D. (Dec. 1982). Radiation effects introduced by X-ray lithography in MOS devices. *IEEE Trans. Nuclear Science, 29*(6), 1697–1701.

Pering, T., & Broderson, R. (June 1998). The simulation and evaluation of dynamic voltage scaling algorithms. In *Proc. Int. Symp. on Low-Power Electronics and Design (ISPLED'98).*

Pinheiro, E., Weber, W.-D., & Barroso, L.A. (February 2007). Failure trends in a large disk-drive population. In *Proc. Fifth USENIX Conf. on File and Storage Technoligies (FAST 2007),* San Jose, CA.

Postiff, M.A., & Mudge, T. (1999). Smart Register Files for High-Performance Microporcessors, Technial Report no. CSE-TR-403-99. University of Michigan.

Powell, M., et al. (2000). Gated-Vdd: A circuit technique to reduce leakage in deep-submicron cache memories. In *Proc. IEEE/ACM Int. Symp. on Low Power Electronics and Design (ISLPED)* (pp. 90–05).

Przybylski, S. (1990). *Cache and Memory Hierarchy Design: A Performance-Directed Approach.* San Mateo, CA: Morgan Kaufmann.

Restle, P.J., Park, J.W., & Lloyd, B.F. (1992). DRAM variable retention time. In *International Electron Devices Meeting Technical Digest,* pp. 807–810.

Rhoden, D. (2002). Personal communication.

Staktek. (2006). Personal communication.

Solworth, J.A., & Orji, C.U. (1990). Write-only disk caches. In *Proc. 1990 ACM SIGMOD Int. Conf. on Management of Data, (SIGMOD '90, pp. 123–132, Atlantic City, New Jersey, May 23–26, 1990, SIGMOD '90.* New York: ACM Press.

Taur, Y., & Ning, T. (1998). *Fundamentals of Modern VLSI Devices.* Cambridge, England: Cambridge University Press.

Ueno, S., Yamashita, T., Oda, H., Komori, S., Inoue, Y., & Nishimura, T. (1998). Leakage current observation on irregular local Pn junction forming the tail distribution of DRAM retention time characteristics. In *International Electron Devices Meeting Technical Digest (IEDM)* (pp. 153–156).

Varma, A., & Jacobson, Q. (Feb. 1998). Destage algorithms for disk arrays with nonvolatile caches. *IEEE Trans. Comput, 47*(2), 228–235.

Vogt, P. (February 2004a). Fully buffered DIMM (FB-DIMM) server memory architecture: Capacity, performance, reliability, and longevity. *Intel Developer Forum, Session OSAS008.*

Vogt, P. (May 2004b). Fully buffered DIMM (FB-DIMM) Architecture. Denali MemCon. Westford, MA: Denali MemCon.

Wang, D., Ganesh, B., Tuaycharoen, N., Baynes, K., Jaleel, A., & Jacob, B. (Sept. 2005). DRAMsim: A memory-system simulator. *SIGARCH Computer Architecture News, 33*(4), 100–107.

Weiser, M., Welch, B., Demers, A., & Shenker, S. (November 1994). Scheduling for reduced CPU energy. In *Proc. First USENIX Symp. on Operating Systems Design and Implementation (OSDI'94)* (pp. 13–23), Monterey, CA.

Wilton, S., & Jouppi, N. (1994). An Enhanced Access and Cycle Time Model for On-chip Caches. WRL Research Report 93/5, DEC Western Research Laboratory.

Yaney, D.S., Lu, C.Y., Kohler, R.A., Kelly, M.J., & Nelson, J.T. (1987). A Meta-Stable Leakage Phenonmenon in DRAM Charge Storage – Variable Hold Time. In *Int. Electron Devices Meeting Technical Digest (IEDM)* (pp. 336–338).

Zhu, Y., & Hu, Y. (June 15–19, 2002). Can large disk built-in caches really improve system performance? *In Proc. ACM SIGMETRICS 2002 (extended abstract)* (pp. 284–285), Marina Del Rey, CA.

CHAPTER 8

Storage in Mobile Consumer Electronics Devices

Thomas Coughlin

Remember when cell phones were all about making calls? Now they're all about data.

As consumers have decided they want to carry their audio, video and photo libraries with them in their handheld devices, the amount of memory these devices require has exploded. The number of ways to store that data has also proliferated. You've still got SRAM for cache, DRAM for working storage, and flash and hard disks (HDDs) for more permanent storage. You can use NOR flash for execute-in-place code and NAND flash for data storage. Or you can use NOR with a NAND interface or NAND with a NOR interface. Do you want to use that alone or as a buffer to a hard disk? Portable designers have more and more storage choices, but the choices can be confusing.

Tom Coughlin doesn't get down to quite that nitty-gritty a level (for that turn to the previous chapter). He takes a top-down view of storage requirements, which are dictated by the architecture of the device you're designing. He examines the different memory requirements and trade-offs involved in automotive electronics, portable media players (PMPs), digital cameras, camcorders, cell phones and portable gaming and navigation devices. Coughlin's use of case studies, teardowns and roadmaps for the development of each category makes the material useful both now and into the future.

Cars use a wide range of media for storage. Control, diagnostic and safety systems generally use embedded solid-state memory, which is quite robust. Entertainment and navigation systems to date use optical media, CDs and DVDs to store songs, movies and maps. But as real-time traffic and road information becomes more available—as well as ads for local restaurants and other attractions, with high-resolution images, of course—navigation systems will migrate to large HDDs.

Pretty much all portable consumer electronics devices are designed for high-quality music, video and photos. Storing them can be quite a trick on devices with tiny batteries. For example, 20,000 4-megapixel photos, 10,000 MP3 songs and 100 VGA movies require about 130 GB of storage. Personally, I'm not sure I've ever heard 10,000 songs, but there seems to be no lack of teenage cell phone owners who wouldn't consider walking around with less. Up the ante a bit as technology advances—say to 20,000 8-megapixel photos, 10,000 lossless compressed songs (CD quality) and 100 DVD (MPEG-2) movies—and now you need about 597 GB of storage.

This is clearly a nontrivial problem, and addressing it starts with an understanding of the applications you're trying to address. The trade-offs and solutions vary between device types, which Coughlin addresses clearly in this chapter.

—**John Donovan**

8.1 Introduction

This chapter looks at the design and digital storage requirements of consumer devices in the home to mobile consumer devices. Applications such as automobiles, mobile music and video players, and digital still and video cameras have significant challenges in environmental reliability, battery life in normal use, and resolution requirements. These characteristics must be taken into account to design the right type of digital storage for these applications.

This chapter explores the unique drivers in mobile consumer product digital storage, both at the present time and for likely developments in the future. It gives some insights into what sort of storage capacity will be available and how much this will cost for the available storage options. It ends by briefly considering new uses of digital storage in mobile devices that could increase the required storage considerably both in these devices and in the home.

8.2 Automobile Consumer Electronics Storage

Automobiles represent one of the most challenging environments in consumer electronics. I was raised in the upper Midwestern United States and I experienced temperatures that ranged from as low as $-30°F$ ($-34°C$), not including wind chill, to as high as $125°F$ ($46°C$). In a car, temperatures can get even higher with the windows rolled up. Furthermore, electronic products designed for this environment must also deal with the regular vibrations, accelerations, and shocks that are common to automobiles.

Storage devices designed for these applications must be very robust to survive in such an environment. Because of the long life expected for embedded automobile components in such a harsh environment, qualification times tend to be rather long, often better measured in years rather than months.

**Table 8.1: Automobile environmental conditions of interest for
most consumer electronics applications**

Environmental Factor	Range
Temperature	−40°F to 120°F (Internal Automobile Temperature)
Relative Humidity	10% to 100%
Vibration	Can Be Very Severe, Usually Damped Around Electronics
Shock	Can Be Very Severe, Usually Absorbed Around Electronics

This section will explore the requirements for automobile applications, as well as storage devices designed to be used in automobiles. Common storage devices used in automotive consumer applications are optical discs, hard disk drives, and flash memory.

8.2.1 Digital Storage for the Automobile

Table 8.1 shows some characteristics that could be expected for storage devices in an automobile environment. These are not comprehensive, and many of these environmental conditions such as shock and vibration are reduced at the location of the electronic device to increase the reliability.

Automobiles are incorporating more and more electronics. With the rising price of gasoline, it is likely that within the next 10 years most automobiles will be run at least partially off of electricity rather than combustion engines. Electronics is making its way into automobile control systems, safety and diagnostic systems, as well as navigation and entertainment systems.

Most digital storage in automobile control, diagnostic, and safety systems is embedded solid state memory since these storage requirements are generally not too great and the robustness required is high. Automobile navigation and entertainment systems are the exception, because they could require very large amounts of storage. This storage could be removable or fixed, depending upon the purpose it is used for.

Navigation systems use digital storage devices for two basic purposes:

1. To supply data to the navigation system of basic routes, locations, and maps.

2. To acquire and store real-time traffic updates and suggest alternative routes when appropriate.

The first approach uses relatively fixed data that changes slowly with time. Data for these applications is often supplied on removable media such as an optical disc (a CD or DVD).

Table 8.2: Comparison of digital storage requirements for various resolution maps

Description of Map	Storage Capacity Needed (MB)	Source
California Highway Map (PDF)	7.73	http://www.metrotown.info/
Santa Cruz County, California Topographic Base Map	9.8	http://www.usgs.gov/
Salinas, California Area Landsat Image	~55	http://www.nasa.gov/

Alternately, it could be downloaded into a mass storage device in the automobile that may use a hard disk drive or a flash memory. Depending upon the resolution of the maps, route, and destination information, an entire country could be represented in basic maps of different locations on one or two DVD discs (or one blue laser DVD). If greater resolution maps are required, then the storage requirements can increase even further.

If audio or video clips about possible destinations are included, then the storage capacity for even this relatively static content can become rather large. It is easy to see that a simple static navigation system could grow to be many gigabytes in size. As the storage capacity of optical storage (or a mass storage download location) grows, such navigation could become much richer, becoming to some extent part of the entertainment system. Table 8.2 gives some idea of the amount of storage required to store various resolution maps of a geographic region.

Real-time traffic update systems for automobile navigation have become very popular in Japan due to the narrowness of the streets in Tokyo and the propensity for traffic to become congested. For this reason, **Vehicle Information and Communications Systems** (VICSs) have become very popular. Several million of these systems have been installed in automobiles in Japan and other Asian countries. These in-car navigation systems are combined with a sensor and radio network located on the streets of the town that give real-time traffic condition information and with the help of the navigation system provide the user with alternative routes to avoid congestion and delays. The real-time route updates are stored on an onboard storage device.

Usually these storage devices are ruggedized hard disk drives that also contain map, route, and perhaps destination information. These hard drives allow for rapid updating of the route information as well as ongoing changes in the general map and destination information. It is likely that hard disk drives could be augmented by a flash memory write cache (a hybrid drive) or even substituted by a solid state drive in future navigation systems depending upon the storage requirements and price point of the application.

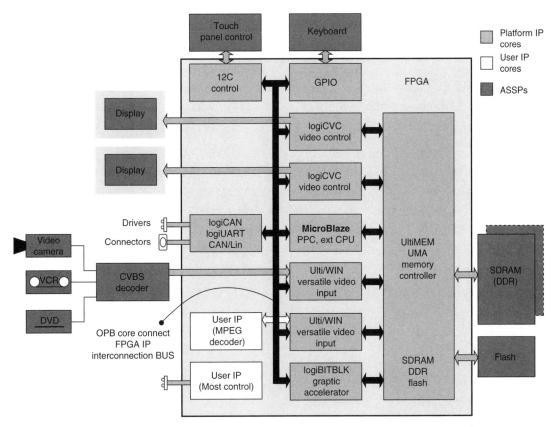

Figure 8.1: Backseat car FPGA-based entertainment system schematic[1]

8.2.2 Basic Layout of Automobile Navigation and Entertainment System

Let's look at some examples of schematics for an automobile entertainment and navigation system. Figure 8.1 shows a Floating Point Gate Array (FPGA) implementation of a multimedia backseat entertainment system. This system is capable of multiple content displays simultaneously. The Composite Video Blank and Sync (CVBS) analog signal is converted to an ITU656 digital signal by a composite video signal decoder, an ASSP component. The MOST MAC (Media Access Control) and MPEG decoders are customer-added IP cores. The I²C (Inter-IC bus) and GPIO (General Purpose Input Output) IP cores are used for the keyboard and touch controller interface. The UltiMEM IP core is configured for unified memory architecture and six access ports. The 32-bit DDRAM (Double Data Random Access Memory) ensures

---

[1]After Xilinx block diagram.

800 MBps bandwidth for display refresh, video streaming, graphics acceleration, and CPU program execution. The logiCAN and logiUART-local interconnect network (LIN) IP cores are used for network connections with in-car body electronics.

A GPS navigation system reference design schematic is shown in Figure 8.2. This system uses CD/DVDs for the map information, but other storage devices could be substituted for the optical media.

The ***Global Positioning System*** (GPS) works on the principle that if you know your distance from several locations, then you can calculate your location. The known locations are the 24 satellites located in six orbital planes at an altitude of 20,200 km. These satellites circle the Earth every 12 hours and broadcast a data stream at the primary frequency L1 of 1.575 GHz. This data stream carries the ***Coarse Acquisition*** (C/A) encoded signal to the ground. The GPS receiver measures the time of arrival of the C/A code to a fraction of a millisecond, and thus determines the distance to the satellite.

The major components of the GPS receiver are:

- *Front End*: The GPS L1 signals (Maximum=24 signals) at 1.575 GHz are received at the antenna and amplified by the ***Low-Noise-Amplifier*** (LNA). The RF front end then filters, mixes, and amplifies ***Analog Gain Control*** (AGC) to bring the signal down to the IF frequency where it is digitally sampled by an ***Analog to Digital Converter*** (ADC).

- *Digital Signal Processor/CPU*: The ADC samples of GPS C/A code signals are correlated by the ***Digital Signal Processor*** (DSP) and then formulated to make range measurements to the GPS satellites. The DSP is interfaced with a general-purpose CPU that handles tracking channels and controls user interfaces.

- *Memory*: The processor runs applications stored in memory. The operating system is stored in nonvolatile memory such as EEPROM/flash/ROM. Applications may be loaded in flash or DRAM.

- *User Interface*: The user interface allows the user to input/output data from the receiver using input commands via microphone and touch screen, and output MP3 to the earplug.

- *Connectivity*: Connectivity allows the receiver to connect to the USB port.

- *Power Conversion*: Converts input power (battery or wall plug) to run various functional blocks.

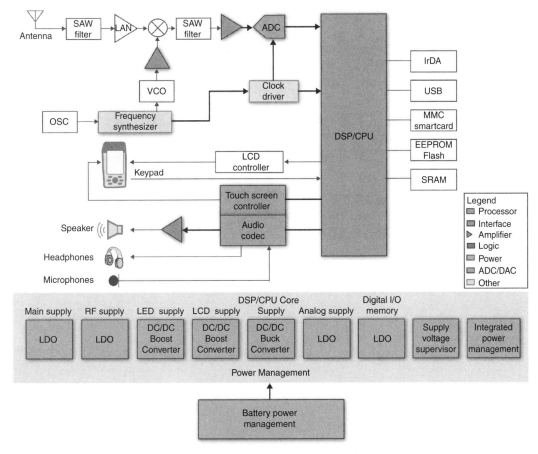

Figure 8.2: GPS receiver block diagram[2]

8.2.3 Storage Device Trade-offs and Options for the Automobile

Automobile entertainment systems typically use optical disc formats such as CDs and DVDs[3] today to store music and movies, but this could change in the future if consumers demand large amounts of onboard mass storage to store movie libraries and other information. Also, as mentioned earlier, destination audio and video information could be available through a navigation system to learn more about a destination before reaching it.

---

[2] Courtesy of Texas Instruments.

[3] The use of consumer DVDs in automobiles required the addition of the Automotive DVD Playback System to the DVD Copy Control Association's Content Scramble System.

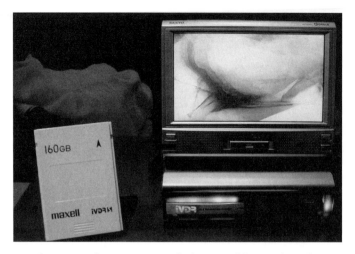

Figure 8.3: iVDR storage device cartridge and reader

A real-time navigation system requires digital storage media that can be written and read from many times. These are popular in Japan where sensors in the streets detect local traffic and road conditions and these signals are used to tell subscribing automobiles what route should best be taken to reach a destination. Real-time navigation systems, as well as systems that require lots of content for entertainment or navigation, often use a storage device such as a hard disk drive.

Hard disk drives used in automobiles live in a very harsh environment, very different from those in computers and even most consumer applications. In order to operate under the temperature extremes, vibration, and shock conditions found in automobiles, the track density and linear density of hard disk drives are reduced to give lower storage capacity but a more rugged storage product with performance margin to spare. For the very extreme temperatures in which hard disk drives are expected to operate, they must be heated or cooled to stay within a specified environmental tolerance.

A shock absorbing case can be used around a hard disk drive to decrease shock and vibration sensitivity further. There have been several designs of such cases. One that has been promoted by several Japanese companies is the *Information Versatile Disk for Removable Usage* (iVDR) standard.[4] Figure 8.3 shows a product using this design. The iVDR provides a modest amount of additional shock protection beyond what the initial disk drive would have, and allows for a removable storage option using a hard disk drive.

---

[4]See http://www.ivdr.org.

Although the most popular storage media for automobile applications is currently optical discs, hard disk drives and flash memory are starting to show up in automobiles as well. Although the flash memory device is more robust at extremes of temperature and shock, the price per gigabyte is three to four times higher, and likely to remain so for several years to come. Although the various power specifications for the hard disk drives appear to be higher, in actual use, the hard drive stays on less than five percent of the time (low duty cycle), filling a semiconductor buffer memory that is used for actual playback in an entertainment system. As a consequence, the total power usage is not very different for many flash based devices than for a hard disk drive based equivalent device.

8.2.4 Road Map for Automobile Digital Storage Requirements

With the increasing price of gasoline, it is likely that electronics will become more important in automobile design. By the next decade, it is likely that there will be a substantial population of automobiles that may use electric power sources for all or part of their motive power. With electronic controls and power sources increasing in their capability and numbers, the integration of other functions into the electronics for navigation, entertainment, accident and liability records (automotive black box), and so forth will be natural additional capabilities.

Over the next couple of decades it is even likely that automated driving (automotive autopilot) could move to the consumer sector. There are many successful university autonomous driving programs that may find implementation in military and dangerous environments in the next few years. Once such autonomous driving systems go into consumer transportation, riders will have time to do more than just watch the road while they drive. This opens up even more need for communication and entertainment technology to use this time productively or for entertainment. These changes and the increasing amount of time that many people spend in their cars will drive new requirements for digital storage.

Real-time traffic and road updates using a wireless network or radio system and a rewritable storage device such as a hard disk drive will become more widespread, particularly for autonomous vehicles or autopilots. As an aid in navigation and as a source of information to the passengers, it is likely that information on locations, commercial information in the area, and other information automatically recorded in an automobile as it approaches a destination, will become more common. Base maps of roads will probably consume considerably less than 1 GB and cover a very large area and a fair amount of detail. As indicated in Table 8.2, topographical information and satellite images would require much larger amounts of digital storage.

Satellite and other high resolution images and even video taken from a location may also be uploaded on automobiles. Combining these needs with traditional entertainment, especially

HD video, could require fairly significant storage either in the automobile itself or available to it through wireless networking. It is not difficult to see a need for several hundred gigabytes or even a terabyte in an automobile navigation and entertainment system in the next few years.

When automobiles are parked near to their base (your home), they will need to connect to your home network, probably using a wireless network interface. This will allow updating and synchronization of content between your home and your automobile. The automobile will likely only contain copies of a subset of your entire data, since the larger bulk of your content is available relatively conveniently.

If DRM issues can be resolved, it would be very convenient for consumers to be able to copy their CDs and DVDs to their home network to make them available in their automobiles without having to handle physical disks that can become damaged and lost. A much larger library of content can be used and organized for easier use when they can be put on a network storage device. Thus in the long run I expect that optical media will not be the preferred way to use content in automobiles.

Hard disk drives and flash memory are more likely to be the primary storage in automobiles in the future. Flash is attractive because it has a greater operating temperature range and it is more shock and vibration insensitive. However, it appears likely that flash memory will remain about two to four times more expensive than an equivalent storage capacity available on a small form factor hard disk drive. For large content storage, even in an automobile environment, a role will probably remain for hard disk drives.

Flash memory may be a way to transfer content in and out of a car (like a USB drive) or be used for certain applications within the automobile that need a more robust storage product. I also consider it likely that hybrid storage devices that contain a flash memory cache built into a hard drive or on the car electronics motherboard could provide some of the faster response under extreme environments that flash can provide, while adding only a little to the cost of the hard disk drive.

8.3 Mobile Media Players

MP3 and video portable players are some of the most popular consumer devices. These products store local copies of music, photographs, or video content and so digital storage is an important element in their design.

In this section we will describe the trade-offs in design and performance that must be made in battery powered consumer devices. Many of the observations made here will also be applicable to other mobile devices.

8.3.1 Mobile Media Player Designs

We will investigate two types of portable media player designs and the digital storage and other design requirements for these devices. These two player devices are (1) mobile music players and (2) the more powerful multimedia player devices that also play video content. Both of these devices can store personal as well as commercial content.

8.3.1.1 Basic Layout of Mobile Media Players

A media player is a simple device in concept. It consists of a sound system that is often an amplifier with a headphone jack, connected to a codec that decodes data read back from a storage device. Typically, a battery provides the power for the device. Since the power from a battery is limited, power consumption is a critical constraint in the amount of time that a customer can use a device between charges. Thus power saving modes and various other techniques are used to reduce power consumption and increase the time that the device can be used on a single charge. In addition to the power requirements, customers often prefer to have devices that are more compact, have a lighter weight, are more durable under likely usage conditions, have an easy way for users to obtain content in the device, look good, and of course are within the budget of the user.

Most mobile portable media players are temporary storage devices that contain a copy of the content that is copied from some other device owned by the consumer, such as a computer. Since the player can be lost or damaged, it is wise to retain the original copy on a static device that stays in the home of the user, where it is hopefully less subject to damage and loss.

Figure 8.4 shows a block diagram of a mobile music player (with voice recorder capability) and Figure 8.5 is an image of a teardown of a portable music player.

The MP3/Voice Recorder-Player performs noise reduction, speech compression, and MP3 stereo recording/playback at all rates and options required by the MPEG/speech compression standards. Following is a description of the components for such a device.

- The DSP performs the audio encode functions, executes post-processing algorithms like equalization and bass management, and system related tasks like file management and the user interface control.

- The *memory* stores executing code plus data and parameters such as content metadata.

- The *peripheral interface* allows the user to control input/output (I/Os) and the display.

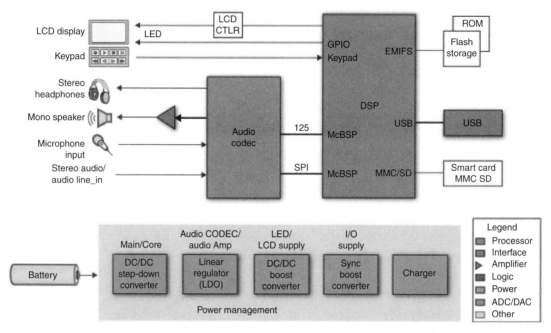

Figure 8.4: MP3 player and voice recorder block diagram[5]

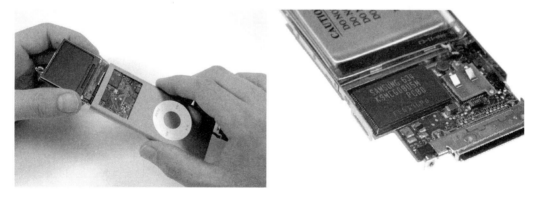

Figure 8.5: 8 GB iPod Nano teardown showing Flash memory storage[6]

- The audio *encoder/decoder* or *codec* interfaces with microphone, radio signals (if the device included a radio receiver) and other audio input with the headphone and possible speaker for audio output.

---

[5]Courtesy of Texas Instruments.
[6]Image courtesy of iFixit.

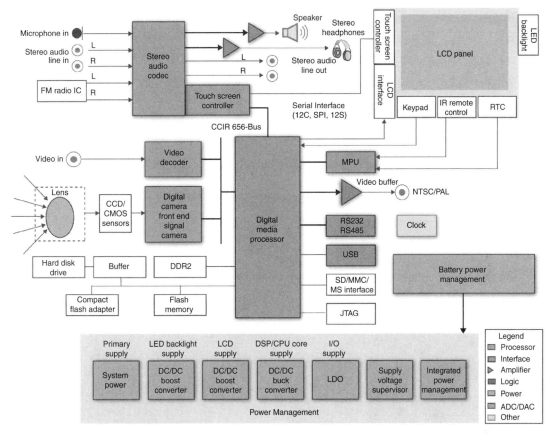

Figure 8.6: Portable personal media player block diagram[7]

Power conversion changes the provided power to run various functions in the device.

8.3.1.2 Basic Layout of a Personal Multimedia Player

Music players have been the most popular of the mobile players so far, but with larger storage capacities available at low cost video players are becoming increasingly powerful and affordable. Multimedia players that play commercial video as well as audio content and can also be used to show user generated content such as digital pictures or home videos are becoming more popular and will probably eventually displace many of the music only players. Figure 8.6 shows a block diagram of a portable media player that handles video content. Figure 8.7 shows images of a teardown of a personal multimedia player.

---

[7]Courtesy of Texas Instruments.

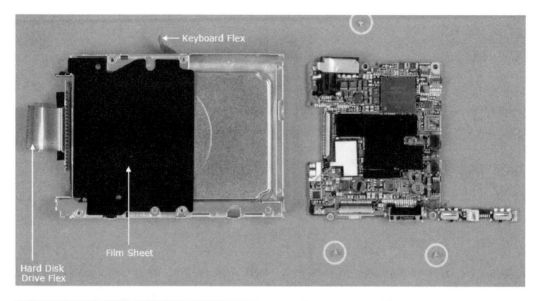

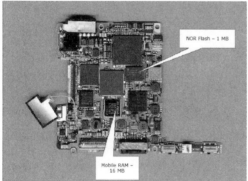
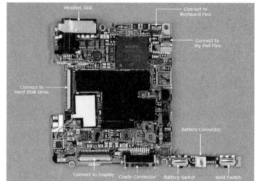

Figure 8.7: Teardown images of Sony Walkman (hard-disk-drive based) personal media player identifying the various components[8]

The *Personal Media Player* (PMP) is a handheld audio/video system that can record and play back *Audio/Video* (A/V) from TV, DVD player, camera, or a media file downloaded from internet. A description of the various components in this block diagram is given below:

• The DSPs and CPU (the digital media processor) performs digital audio and video processing for all standard media formats. The CPU executes operating system software and controls data transfers and user interface.

---

[8]Courtesy of Portelligent, Inc.—teardown.com.

- The digital video input comes from the analog video decoder or the CMOS/CCD camera. The analog video decoder converts the NTSC/PAL/S-Video analog input to raw digital video.

- The stereo audio codec converts the audio input from microphones, FM receiver, and stereo audio line sources into digital audio.

- The CCIR656 Bus moves raw digital video content to the DSP.

- The serial interface enables file transfers with the host computer via standard interface formats such as USB, RS232, and JTAG interfaces (this could eventually include serial ATA).

- The memory interface handles direct memory interfaces with various portable memory media including compact flash cards, SD cards, MMC cards, memory sticks, and so on.

- The IDE interface handles data transfers with a hard disk drive.

- The user interfaces allow users to control the personal media player using a keypad, IR remote control, voice, or touch screen.

- The power management and conversion system converts input power from a battery or wall plug to the various functional blocks of the PMP.

8.3.2 Display Technologies in Mobile Devices

Creating inexpensive and easy to use personal high resolution display technologies has implications in power use, physio-psychological effects, as well as weight, convenience, and price. Nevertheless, we expect that there will be considerable effort expended to make higher resolution content available to mobile users.

Figure 8.8 shows a display technology road map (including mobile or wearable devices) through 2015 (after a table in an Optoelectronic Industry and Technology Development Association Report). Based on these road maps we might expect initial products with a useful mobile high resolution display device to be available sometime toward the end of the decade and this will drive requirements for higher resolution digital content and thus higher capacity storage devices.

A very interesting option for getting a large format display from a mobile battery powered device is the pocket projector technology from Novalux.[9] This technology uses surface

---

[9]See http://www.novalux.com.

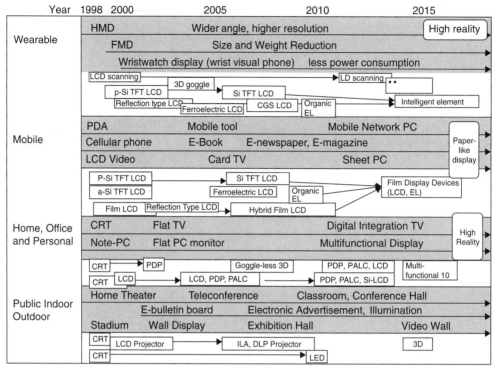

Figure 8.8: Road map of electronic display technologies[10]

emitting infrared lasers and frequency doubling to achieve 1.5 W scanning RGB lasers that can provide 100–300 lumen projection of images from battery powered devices. This display technology automatically projects and focuses on any surface allowing very interesting potential applications. Figure 8.9 illustrates how such a device can be used. The company says that this technology should be introduced in cell phones, PDAs, and other mobile devices by 2009.

From a storage point of view, the pocket projector is interesting since it allows watching larger format content on a mobile device than is possible using the small LCD screens typical on PMPS, PDAs, or cell phones. This translates into a capability of watching higher resolution content from a mobile device. Higher resolution content translates into larger files and that means larger capacity storage needs. The storage required could be embedded in the product or else available on a local network, perhaps in personal network storage architecture such as Seagate's DAVE or LSI's Blue Onyx.

---

[10]After Optoelectronic Industry and Technology Development Association Chart.

Figure 8.9: Demonstration of uses of Novalux Pocket Projection technology[11]

8.3.3 Power Requirements for Mobile Devices

Several factors need to be taken into account in comparing the power used for a flash based and a hard disk based media player.

- The required data rate for continuous streamed content.

- The data rate out of the storage device.

- The power available from the battery of mobile power source.

- The power budget for the mobile device including storage, electronics, audio player, and display technologies.

In mobile devices power is a serious design parameter. All the components in a mobile player require power. The power requirements for different system components (after evaluating and optimizing performance and usage life trade-offs for the expected products applications) can be used to create a power budget that then defines the mobile power supply needs of the device.

---

[11]Image courtesy of Novalux Corporation.

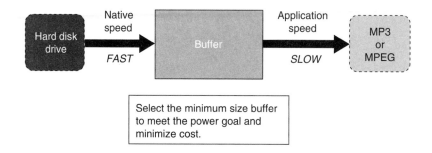

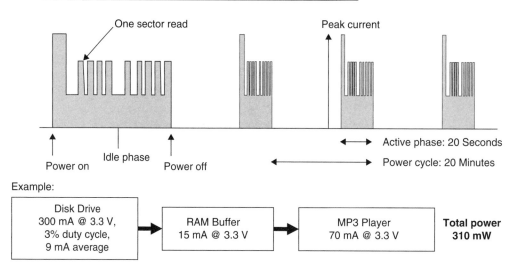

Figure 8.10: A buffer can be used to match the content delivery speed of a storage device such as a hard disk drive and the application[12]

Figure 8.10 gives us some insight into the process of evaluating the actual power usage of a personal multimedia player. The data rate of a storage device such as a hard disk drive (or a flash memory device) is much greater than required by an MP3 or MPEG player application. A buffer is used as a host speed matching device. The size of the host buffer will determine the power dissipation; more buffer means less power-hungry spin up/down drive cycles, resulting in lower average power dissipation.

Trade-offs in buffer size, power, and cost help host manufacturers select a buffer that meets the power dissipation goals of the application while minimizing cost. Typical music player

[12]After Duncan Furness, Agere, "Architecting Minature Hard Disk Drives for Battery Savings," Planet Analog, 2005.

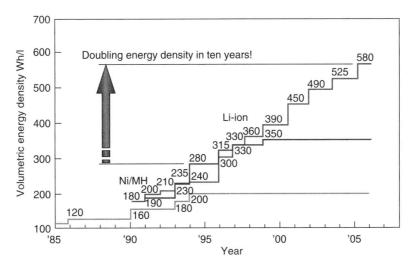

Figure 8.11: Battery energy volumetric density development

buffers are in the range of 8 to 16 MB. For instance, an MPEG-2 video file streams at less than 3 MBps while a 1.8-inch hard disk drive has a sustained data rate of 16 MBps (128 MBps).

Since the hard disk drive consumes a fair amount of power while reading (1.1 W) we would probably design a battery based hard disk drive media player so that the disk drive would not remain on all the time. Instead, such a media player would turn the disk drive on only a fraction of the total player power-on time (less than three percent). As a result, as shown in Figure 8.10, the total power used in this player would probably be close to 300 mW rather than the 1.1 W if the hard disk drive were constantly running (about a 70 percent power reduction). Note that flash memory designed devices may also use RAM buffers for play out since their data rate can exceed that required for the streaming content. By the way, it is interesting to note that almost half of the total power consumed by a hard disk drive is used in the electronics rather than the electromechanical components.

Power is one of the major constraints on mobile consumer electronic products. Conventional battery technology lasts no more than a few hours and promising technologies such as fuel cells have not yet shrunk to the point where they can be used in many mobile devices and consumer electronics companies and consumers have yet to be convinced that fuel cell powered devices will be safe or reliable. Battery technology has been developing over time as shown in Figure 8.11 with battery power density roughly doubling every 10 years.

Fuel cells will likely find some application in mobile devices by 2009–2010, perhaps as mobile rechargers if not the actual mobile power source. Many of these devices use methanol as the source of hydrogen, but there are some technologies that directly use hydrogen. Some issues needing resolution for the use of fuel cells include heat management and exhaust of by-products (such as water). Fuel cells could increase the energy density of mobile energy sources considerably.

Another interesting possibility is wireless power to recharge mobile devices. At the 2007 CES Conference, Herman Miller (a designer of office furniture) showed the eCoupled technology developed by Fulton Innovations.[13] This technology has a resonant circuit connected to the power supply for the mobile device that can resonate with a similar circuit in the furniture with the capability of transferring small amounts of power to recharge the battery of mobile devices. The technology only operates at a distance of about an inch, so the mobile device would probably have to sit in some sort of cradle. If such technologies became common in commercial furniture, it could help extend the life of mobile device batteries and help encourage more feature rich applications. MIT demonstrated at the 2007 CES their "Witricity" initiative which has been shown capable of transferring 60 W of electrical power 7 feet at 40 percent efficiency.

With more power available for mobile devices, features such as higher definition content projection (i.e., with Novalus personal projectors) may be possible, allowing viewing of such content and hence a requirement for greater storage capacity and read bandwidth to support playing such content.

8.3.4 How to Choose the Right Digital Storage for a Mobile Media Player

In addition to power there are other characteristics that are important to the use of one digital storage device versus another. These include:

- Price for a useful capacity (always very useful).

- Size and shape.

- Weight.

- Ruggedness (shock, vibration, and other environmental sensitivities).

- Heat generation.

We shall look at how choices can be made for each of these characteristics.

Figure 8.12 compares the price and storage capacity of flash memory and various form factor hard disk drives as of mid-2007. Flash memory devices have a minimum base price

---

[13]See http://www.engadget.com/2007/01/19/herman-miller-planning-desk-of-the-future/.

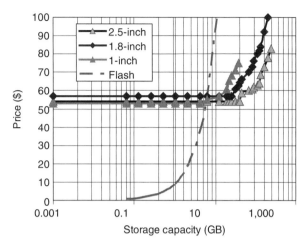

Figure 8.12: Comparison of price and storage capacity of Flash memory and hard disk drives in mid-2007

of $7 to $10 while the hard disk drive minimum base price is in the $40 to $50 range. Flash memory can be extended to larger storage capacities by adding more NAND flash chips with an increase in total price. Flash capacity increments are smaller than hard disk drive increments since each disk surface adds so much more capacity. We can see that for lower storage capacity requirements flash memory will be cheaper, while if the storage capacity requirements are large, hard disk drives can be cheaper. Of all the other factors, the price to buy needed sufficient storage for an application is paramount for most consumers. The figure also shows that there is a range of storage capacities that hard disk drives cannot address due to the incremental nature of hard disk drive capacity increases.

Hard disk drives have a definite length and width dimension defined by the size of the disk form factor used. The height is a function of the dimensions needed for the magnetic head and drive electronics. Flash memory is composed of collections of semiconductor chips and so the options for packing in storage capacity are greater and devices using flash memory can be more flexible in their dimensions. This can provide definite advantages for many applications if all other factors are equal.

The weight of flash memory is dependent upon the number of flash memory chips required as well as the packaging. Higher flash memory capacities will weigh more, but the weight generally is less than for a hard disk drive device unless the flash memory is very large.

Likewise, flash memory is generally more rugged than hard disk drives in terms of operating shock. However, as indicated earlier, the drive may be in a low power nonoperating state most of the time (more than 95 percent of the time) where the device plays out of a RAM buffer. The nonoperating shock specifications for small form factor hard disk drives are about those of a NAND flash memory device. If an accelerometer is included in the hard disk drive or the application device then even the operating shock specification can be improved further, since the drive could detect when it is in a free-fall state and retract the heads to their protective nonoperating state.

Vibration and temperature sensitivity can be lower for flash memory than for hard disk drives. Flash memory may also work better in many moist or even wet environments than hard disk drives.

In general, since the power usage for a hard disk drive consumer device can be similar to that of a flash memory device (with the proper power management features) the heat generated may be about the same and the battery life may also be similar.

In addition to these characteristics other criteria may be important for various consumer applications in the choice of a storage device. For instance, flash memory cells will wear out with multiple writes and the intrinsic read speed of flash memory is less than the write speed (this is also true of hard disk drives but with flash the difference is greater). These characteristics of flash memory may cause a designer to not use flash memory for write intensive applications such as DVR functions or real-time traffic navigation.

Let us look at the particular case of why Apple switched from using 1-inch hard disk drives to NAND flash memory in their music iPod product line. At the time the decision was made, 1-inch hard disk drives had larger storage capacity for the price compared to flash memory, but the storage capacity in a flash memory device that could be purchased for an affordable price by many consumers (less than $300) was approaching a point that many MP3 songs could be stored on it.

Since a flash memory player would be lighter weight and could be made thinner as well as smaller in other dimensions than a hard disk drive based device, the "portability" of the device would be better. Note that in general it is advisable to make a mobile consumer device with a specific gravity close to that of water (which is also close to our specific gravity) so they would not seem to weigh too much when on our bodies. As shown in Figure 8.13 the flash based iPod Nano was more "portable" than the iPod Mini (with a 1-inch hard disk drive) even though it couldn't carry as many songs. In addition, with the declining price of flash memory, a player with 10,000 or more tracks would be possible with flash memory in an attractive price range within a few years from the introduction of the iPod Nano in 2005.

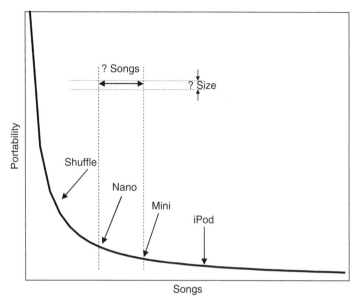

Figure 8.13: iPod portability versus song capacity[14]

The combination of greater *portability* and adequate storage capacities for the application probably played a significant role in the decision to switch the Apple music players to flash memory. It should also be mentioned that Apple was able to make contracts with Samsung and other large flash manufacturers to get its flash at considerably lower prices than they would be available to the rest of the market, thus allowing them to either gain higher profit margins on the product or to lower the price to remain competitive.

8.3.5 Road Map for Digital Storage in Mobile Content Players

8.3.5.1 Content Formats: Lossy and Lossless Formats

The MP3 music format is a lossy compressed music format created to economize on scarce and relatively expensive digital storage. MP3 and similar compression technologies generally compress music files to about 10 percent of the capacity size of the original content, removing most of the music "information" in the compressed file. Once this information is removed, it cannot be recovered—hence the term lossy. MP3 lossy compression takes advantage of the fact that the human auditory system doesn't notice certain types of signal degradation, but lossy compression can introduce artifacts that can be noticed by a keen ear, particularly in a

---

[14]Courtesy of Semico Research Corp.

quiet background. This format has become very popular for use in relatively unsophisticated sound systems and in noisy mobile applications. A 10,000 song MP3 player requires less than 40 GB of total storage capacity.

With the continuing decreases in storage prices it will become more common for users to move from MP3 to less compressed and even non-lossy compressed music in their portable devices. Lossless compression reduces the storage capacity of audio content without losing the original audio signal's integrity. Thus an audio track compressed with lossless compression can be decoded to its original uncompressed form without artifacts.

A lossless compressed music file is at most reduced to 50 percent of the size of the original music file. A 10,000 song personal stereo player with 50 percent lossless compression requires about 140 GB of storage. Lossless compressed music download services have become available and there are many programs to support lossless ripping of CDs.

Table 8.3 compares digital storage and streaming bandwidth requirements for various formats of music and video content. The numbers are approximations since compression rates and resolutions may vary somewhat. It is clear that richer digital content requires much greater storage capacity and bandwidth. Ultra-HD is assumed to be compressed as much as HD.

Table 8.3: Storage, streaming bandwidth and estimated power requirements for various formats of music and video content

Format	Bandwidth (MBps)	Storage Capacity/Hour (GB)
Music Formats		
MP3	~0.128	~0.576
Lossless Compressed CD	~0.700 min.	~0.315
CD Quality	1.400	0.630
DVD Audio	9.600 max.	4.320
Video Formats		
Format for iPod (MPEG-4)	~0.750	~0.337
DVD MPEG 2	11.080	2.700
MPEG 4	~1.400	~0.630
SDTV	~8.000	~2.000
Blu-ray/HD DVD	36.550	3.750
HDTV	~19.300	~8.890
Ultra-HDTV	~295.000	~133.000

A physical distribution media for mobile devices such as a small form factor optical disc or a flash memory device in a removable reader could be used for loading lossless content. We expect that with higher capacity storage the need to economize storage space using MP3 files will be less compelling, thus driving the higher end of the music player market to use higher capacity hard disk drives.

Figure 8.14 shows projections for the storage capacity that could be bought by an OEM for about $60 out to 2010 for flash memory as well as various form factor hard disk drives. Also on this chart are shown the storage capacity requirements for a 10,000 track MP3 and lossless compressed music formats. It is clear from this chart that the MP3 player market will probably be almost entirely flash memory dominated by 2010, but if higher resolution content is required, then a hard disk drive based device may make more sense from a price and capacity point of view.

The very large digital storage capacity that is available with small form factor hard disk drives (or larger capacity flash memory) will develop a new class of personal media player. This player will provide significantly higher resolution audio and video content on an appropriately sized screen or some external viewing device such as the personal projector technology described earlier in this section. Just as greater memory on personal computers led to new features and higher performance, so too these new consumer electronics products will provide higher resolution and lossless content storage, providing a more refined user experience. These sophisticated media player products will provide ready access to photographs and music files, as well as video files.

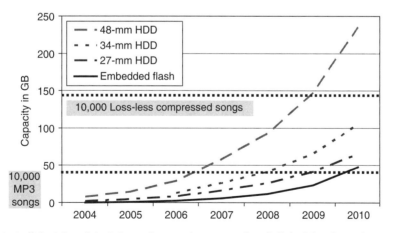

Figure 8.14: 1 disk/2 head (minimum) capacity versus hard disk drive form factor and flash memory projections for ~$60 OEM

In addition to playing prerecorded content, these devices will support DVR technology to record video and other content from cable, satellite, broadcast, or the internet. They could also be portable repositories for a collection of family photographs and video, increasing the capacity requirements considerably. The only limit to the resolution requirements of these devices is the size of the display available. If these mobile devices are used to play on higher resolution displays or projected images as discussed earlier, then there is almost no limit to the storage requirements that such devices could have. The following examples show the sort of storage capacity required:

- A combination 20,000 4-megapixel photo, 10,000 MP3 song, 100 VGA movie player would need about 130 GB.

- A combination 20,000 8-megapixel photo, 10,000 lossless compressed song (CD quality), 100 DVD (MPEG-2) movie player would need about 597 GB.

8.4 Cameras and Camcorders

Digital still cameras mostly use flash cards of various formats. Very high pixel count cameras used by professional photographers sometimes use removable 1-inch hard disk drives in a compact flash form factor, especially if photos are being stored in an uncompressed format. It is likely that photograph resolution for professional use will continue to increase (it still doesn't match the resolution available with traditional silver halide film) and thus I believe there will continue to be a niche market for removable hard disk drives for still cameras.

The camcorder market is relatively small today, averaging between 10–15 million units a year. This small market penetration is probably due to the fact that the $500 to $2,000 price tag that many of these units command is too high for a large number of users. These high prices are driven in part by the recorder's very intricate recording mechanism.

Hard disk drive based camcorders are starting to make a dent in the market. One such unit, the $1,300 JVC Everio, boasts a 30 GB hard disk drive capable of recording up to 7 hours of MPEG-2 DVD-quality video. The attraction of these devices is that they can store a lot of images, and data transfers of video files from the camera to a computer are as fast as other external disk drive file transfers.

Flash card based camcorders make lower cost camcorders possible. As the price of NAND in the near future supports the storage of reasonable amounts of video on an affordable quantity of NAND this market segment will take off. Once the expensive tape transport or other mechanical media handlers are designed out of camcorders, the subsequent cost reductions are quite likely to reduce average camcorder prices below $100, opening up the market to

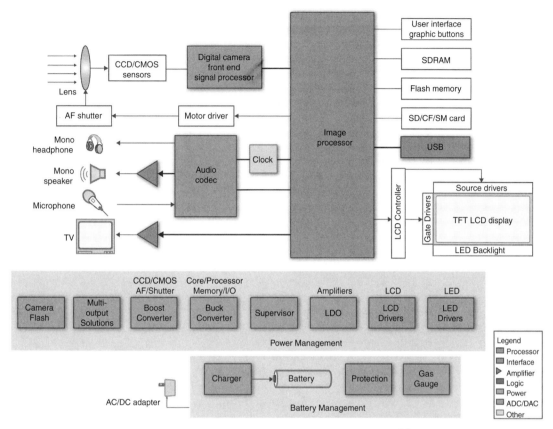

Figure 8.15: Digital still camera block diagram[15]

many more users than can currently justify such a purchase. It is quite likely that the market for $100 camcorders is 10 or more times the size of the market for $500–$2,000 models.

8.4.1 Layout of a Digital Still Camera

Figure 8.15 shows a block diagram for a digital still camera. Digital cameras serve many different market segments all the way from very inexpensive low resolution models to expensive professional units with resolutions of greater than 20 megapixels. Over time, the resolution of cameras in the popular price point of $200 to $400 has increased, with a doubling in average megapixels in digital still cameras roughly every four years. Figure 8.16 shows a teardown of a digital still camera showing various labeled components.

---

[15]Courtesy of Texas Instruments.

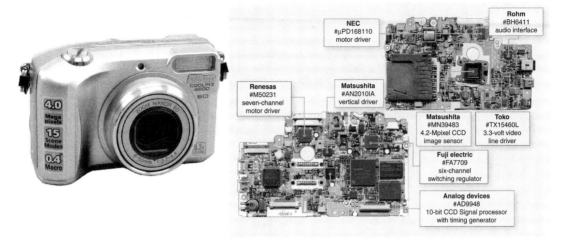

Figure 8.16: Teardown image of digital still camera (Nikon 4800 Digital Still Camera)[16]

The digital camera uses an image sensor (CCD or CMOS) to convert light directly into a series of pixel values that represent the image. The major components of a digital still camera are:

- The CCD or CMOS image sensor converts photons of light into electrons at the photosensitive sites in the CCD or CMOS image sensors.

- The front end processor filters, amplifies and digitizes the analog signal from the image sensor using high-speed analog to digital conversion.

- The digital image processor handles industry-standard computationally intensive imaging, audio and video algorithms. It also controls the timing relationship of the vertical/horizontal reference signals and the pixel clock.

- The various memory stores executing code and image data files.

- The audio codec performs digital audio recording/playback under the control of the DSP.

- The LCD controller receives digital images from the camera and images them on the LCD.

- The power conversion electronics converts input power from AC or USB to charge the camera battery and control the power management functional blocks.

---

[16]Courtesy of Portelligent, Inc.—teardown.com.

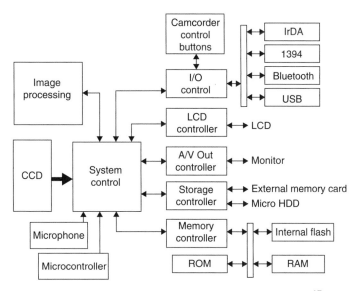

Figure 8.17: Block diagram of a digital camcorder[17]

8.4.2 Layout of a Digital Video Camera

Digital video cameras are often referred to as camcorders. These devices have used small form factor digital videotape for capturing the video images for many years and these are still the most common digital camcorders. However, camcorders that record digital video on optical discs, hard disk drives, and even flash memory are becoming more common and probably eventually will displace videotape.

Figure 8.17 is a block diagram of a digital camcorder showing the basic design elements in a digital camcorder.

Figure 8.18 shows a teardown of a digital video camera showing various labeled components. In this device the hard disk drive for video storage includes an accelerometer and appropriate software to detect when the camera is falling so the heads on the hard disk drive can be retracted in order to protect the drive from damage.

8.4.3 Storage Requirements for Digital Camcorders

Many modern digital video recorders allow capture of digital still images as well as video images. Hard disk drive based digital camcorders allow direct file transfer of data to a

---

[17]After Lattice Semiconductor block diagram.

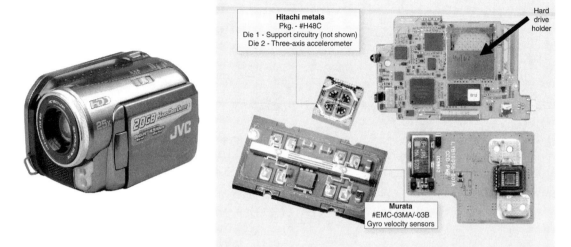

Figure 8.18: Teardown image of JVC hard-disk-drive based digital camcorder, JVC GZ-MG20[18]

computer. Since the disk drive in the camera appears as just another computer disk drive to the computer, it is attached using a USB, Firewire (IEEE 1394), or eSATA connection. Optical disc digital camcorders let the user remove the optical disc and play it back immediately on a DVD player or computer. Optical media for these applications has moved to blue laser discs for professional video cameras and it is likely that the market segment of digital video cameras that use optical disc media will move to blue laser products over the next few years.

Flash memory and hard disk drives show relatively steady increases in storage capacity while optical disc formats have remained static for many years. The resolution of video as well as optical content in camcorders will continue to grow for the next few years. Digital still image megapixel resolution has doubled about every four years. With the growth of high definition commercial video in the home I expect that for the next several years video resolution in camcorders will on average double about every four years as well.

The result is that multiple market niches are developing for digital camcorders. The very high end may continue to use some digital tape and hard disk drives, an endangered mid-range may use optical discs, and a very popular and rugged low end (lower resolution and lower cost optics camcorders) could offer lower resolution video at a very affordable price, storing the images on removable flash memory media.

---

[18]Courtesy of Portelligent, Inc.—teardown.com.

With the increasing storage capacity for an affordable price that flash offers, it may be possible to make sub-$150 and sub-$100 digital camcorders that can serve untapped markets. In general, the price of digital camcorders should decrease even though video resolution will likely increase as the expensive tape systems that are common today are replaced by less complex storage media such as optical discs, hard disk drives, and flash memory.

8.4.4 Road Maps for Camcorder Digital Storage

As the resolution of integrated or stand-alone cameras increases, the price drops, and more people start to share their content, greater digital storage capacity will be required along with greater bandwidth on networks to share this content. It is likely that cameras will be equipped with wireless capability, making it easy to transfer content to a nearby computer on a local network and possibly even to backup images and video to a wireless mobile storage device that you carry with you on your *Personal Area Network* (PAN).

Developments in camera optics could also be in for major changes as once-stodgy optical technology is transformed. Researchers are developing new devices using negative index meta-materials to create more complex and compact optical paths, allowing even more powerful and compact imaging systems. Multiple wavelength recording (not just visible wavelengths) is possible in a device that can be easily incorporated into other devices or turned into a lightweight and compact stand-alone imaging device. This multi-wavelength imager would allow the capture of much more than just surface images around us (imaging combining visible light, infrared light, and even higher or lower frequencies). New very small laser projection technologies could help turn digital cameras into their own projectors for playing back their content. All of these capabilities will encourage people to take more images and store more of those images in more places.

Increasing ubiquity of still and video cameras, GPS location devices, and other technologies could also lead to a new type of device discussed later in this section that allows the recording of events as they happen in a person's life and storing and even sharing parts of the content with others.

The visual image is one of the most powerful ways for humans to communicate with each other and to learn about the world around us. It is only natural that we should quickly incorporate these new advances in image capture, processing, and presentation to enhance the human experience and capture it for future generations.

8.5 Mobile Phones

There is no more ubiquitous portable consumer device than the cell phone. Over one billion of these versatile devices are shipping annually. With such a huge market there are naturally many market segments with many different capabilities and price points. For many people, particularly in the developing world, their first phone is a mobile phone. Thus mobile phones represent a very important development in creating personal and business networks. In this section we examine the different types of digital phones and the digital storage needs and requirements for these products.

8.5.1 Basic Layout of a Cell Phone

A modern cell phone provides two-way communications using one of several cellular standards (GSM, CDMA, TDMA, etc.). It often integrates one or more of the following functions:

- Vibrating ringer.

- Polyphonic ringer.

- Touch screen.

- Still and/or video camera.

- Broadcast radio receiver (FM, AM).

- MP3 player.

- PDA functions.

- Mobile TV capability.

- Location-based services.

- Digital storage and storage expansion.

- PC connection.

Figure 8.19 shows a reference block diagram for a cell phone including digital camera imaging capability. Note that the more features there are on the phone, the more memory, battery power, and processing power will be required. In particular, for capturing or storing higher definition still and video images, larger amounts of digital storage are required.

The RF section of the circuit block diagram includes antenna diversity, down-conversion, up-conversion, frequency synthesis, RF power amplifier as well as power ramp control, and other features. The baseband section of the circuit block diagram includes the DSP, *Voice Band Codec* (VBC), memory, and user interface controls for devices such as keypads, LCD, ringer, and so on.

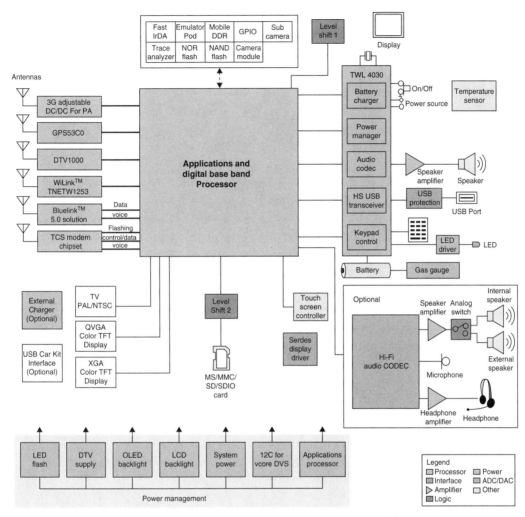

Figure 8.19: Block diagram of a mobile phone[19]

Figure 8.20 shows a teardown of a Sanyo S750 phone. This is a typical mass-market camera cell phone. The teardown of the main circuit board identifies the memory on the board, which includes some NAND flash and some SRAM.

Most phones use solid state memory. NOR is common for system program memory and NAND flash is common for small amounts of audio, still photo, and video capture or playback. A small number of cell phones including some built by Samsung have included

---

[19]Courtesy of Texas Instruments.

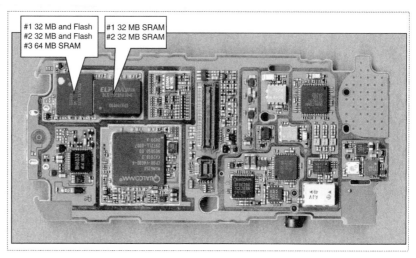

#1 32 MB and Flash
#2 32 MB and Flash
#3 64 MB SRAM

#1 32 MB SRAM
#2 32 MB SRAM

Figure 8.20: Teardown of Sanyo S750 cell phone[20]

1-inch and even the smaller form factor 0.85-inch disk drives. These small disk drives provide larger capacity for an affordable price for music or video files played on the cell phone.

8.5.2 Development of Cell Phone Technologies and the Vision of Convergence Devices

Cell phones differentiate themselves by offering various features in addition to the basic digital cell phone services. Cell phones can include still and video cameras, MP3 players, mobile TV viewers, and PDA capability including office productivity programs and files. At the time of the writing of this book, very few multiple-function phones did a very good job of handling multiple functions in a cell phone with acceptable performance and battery life and that were easy to use. There is a case to be made that multiple-function devices will seldom work as well as single or limited purpose devices.

My own multiple function Microsoft windows PDA phone does an amazing number of things but many of them don't work that well, and with so many buttons and ways to turn functions on, I never know what I will be pulling out of my pocket—a phone, a camera, a document, or a recording device. In addition, many modern phones will easily call back a prior called number resulting in what are called "pocket calls," where a phone in one's pocket calls someone up and they get to listen to whatever you are doing or saying at the time. Talk about a lack of privacy!

---

[20]Courtesy of Portelligent, Inc.—teardown.com.

There is interest in creating one mobile device that can serve for all one's communication, information access, and work needs. Such a device could make our lives much easier but we are probably a long way from that vision becoming a reality. Convergence devices with multiple capabilities need to be more stable and respond to something other than just touch to activate most functions—it is too easy to touch a button by mistake when it is in a pocket or a bag. Voice activated devices would probably be better, since, hopefully, one would not accidentally say the wrong function code word and activate the wrong function.

Voice activation and voice recognition, if easy to use and reliable, would be a big step forward in such devices. Today you can buy "dictation" software for your PC from multiple vendors that will recognize thousands of words, and that will capture, edit, and print a document by voice. You can call a voice activated assistant on the phone for banking, travel, ordering, and many other activities. The speech recognition algorithms used in these applications are quite large, consuming thousands of MIPs of processor performance and hundreds of megabytes of memory. Note that Dragon Naturally Speaking Professional can take as much as 2.5 GB of storage capacity depending upon the number of speech files installed.

With greater processing speed in mobile devices and higher storage capacity (combined with noise cancellation technologies) we will have the critical requirements for implanting usable voice recognition into mobile consumer devices. In recent years, algorithms based on modern embedded speech recognition have become available on many cell phones. They allow phone dialing by name or by number using your voice, and often allow voice activation of other cell phone functions. The application understands how names, numbers, and other words sound in a particular language, and can match your utterance to a name, number, or command in the phone. Users have found this new functionality straightforward to use and easy to remember.

Worldwide, there are more than 20 million phones that include these modern embedded speech applications. They work very well, calling the numbers listed by name from your phone book, allowing you to dial your phone by saying the phone number, and letting you look up a contact entry, launch a browser, start a game, and more. Many of these phones use low power ARM processors. A few companies, such as Voice-Signal and Advanced Recognition Technologies, are working on low-power voice recognition solutions. The speech recognition software is running on hardware similar to that of a decade ago (except that it is very power efficient and small), but the speech algorithms are almost state-of-the-art.[21] It is an easy step to see voice recognition technology becoming ubiquitous on mobile consumer devices.

---

[21]Source: IQ Online, Volume 3, Number 5, 2005.

The use of displays that have contrast characteristics more like paper in ambient light would help in viewing content. In 2006, Motorola introduced the 9-mm-thick Motofone, which targets emerging markets like India and Brazil. This product uses an electrophoretic display, or EPD—an ultrathin, low-power display often referred to as electronic paper. While the yield of these displays is not good at present and the resolution does not match that of other display technologies, it is very low power and can be seen easily with ambient light. As this technology improves it could point the way to future lower-power and easier-to-see display technologies for mobile devices such as mobile phones.

Adding additional functions to phones makes these devices more popular and allows offering subscription services that can help retain customers from going to competitors and competing technologies. This development, combined with the astounding growth of mobile phones of various sorts throughout the world, will result in a significant amount of development and experimentation on using phones as an element in personal networking and device convergence.

Most cell phones today work on cellular networks for their communication and downloading functions, but designs are now starting to appear that allow the device to use Wi-Fi, Bluetooth, and other wireless networking technologies as well. This is perhaps the best example of a mobile convergence device since the phone can act as a local media and data center for the user and perhaps for other devices that the user has. Incorporating IP wireless networks also opens up the possibility of creating PANs that allow communication between various devices that a person carries on him or her. Such PANs also open up the possibility of moving some digital storage and other functions off of a single device while still allowing for coordination and synchronization of content when needed. I believe that this may be the greatest means for near term convergence technology.

A viable storage and communications network and improving local power management and power availability will lead to amazing advances in mobile consumer products such as mobile phones in years to come.

8.5.3 Importance of Mobile Power Sources and Cell Phone Size

Power sources for mobile devices are a key factor in the usefulness of the device. For this reason, mobile products should always be designed to incorporate as many power-saving modes as are practical. This includes power-saving modes for storage products particularly where hard disk drives are used, because when these are actively spinning they consume a fair amount of power. New mobile phone designs incorporating media player functionality and other capabilities are allocating roughly 10 percent of their power budget to storage needs which, depending on the battery technology, translates to an average power consumption of about 110 mW.

New features such as the laser pocket projector from Novalux described in Section 8.3.3 combined with all the other possible functions that could be incorporated into mobile phones could require considerably more power. In that section we also covered expected developments of mobile power technology that will enable more of these mobile applications. Also, technologies to harvest ambient power or to use wireless recharging capabilities tied to specially designed furniture with this capability built-in will enable a richer and more satisfying consumer experience. In order to keep the mobile phone at an acceptable size, either more power needs to be available in small power sources, or these applications need to use less power.

8.6 Other Consumer Devices

There are several other types of consumer devices that utilize digital storage to enable their operation. We shall briefly discuss some of these as well as a short section speculating about future products enabled by very large mobile digital storage.

8.6.1 Fixed and Mobile Game Systems

Game systems use memory to store games. Home-based gaming systems that use a lot of graphics for their operation often use CD or DVD disks for game loading. Some of these systems also use hard disk drives to store game graphical video content so that it can be accessed quickly. The hard disk drives used for these applications are usually not very high capacity, but must be as inexpensive as possible in order to meet the bill of materials cost requirements.

Mobile games tend to use semiconductor memory for game storage. The graphical requirements for the smaller screens do not require large storage files as do home-based systems, and so lower capacity flash memory or other nonvolatile solid state memory can be used. Use of solid state memory in this application is wise since these mobile game systems often get rather harsh treatment by the children playing with them.

Multiplayer games played over the internet could change some of these storage requirements, particularly for the home-based game systems. This could lead to a greater amount of storage on a network and especially on the internet that gamers access. It is likely that resolution requirements and digital storage capacity requirements will continue to increase as other consumer devices increase in their video resolution and thus their storage capacity demands.

8.6.2 Handheld Navigation Devices

Handheld navigation devices run the gamut from inexpensive systems selling for less than $99 to more elaborate systems that can cost several hundred dollars. The latter navigation systems are often designed to be used for automobile or marine navigation as well as for handheld use while hiking and other outdoor activities. As GPS receivers get built in phones and other devices and the price and size of these capabilities decline, it is likely that, like DVRs, GPS will become a standard function that many consumer devices will have built-in to their mobile devices in the next several years. Stand-alone devices that are just GPS map systems may become rare just as most DVRs are now part of set-top boxes.

Regarding digital storage requirements for such devices, the storage needs are similar to those discussed in Section 8.2 of this chapter (on automobile navigation and entertainment systems). As the map details become more precise, the digital storage requirements increase. Also, if there is a need for real-time data updates to maps, terrain, or road conditions, a rewritable device such as flash memory or a smaller hard disk drive may be desirable.

Figure 8.2 in Section 8.2.2 shows a block diagram of a GPS receiver unit. This is the key to a GPS map system. Data stored on some storage medium, here an embedded flash or SmartCard, contains the map data to display the location.

8.6.3 Other Mobile Applications

Don Norman speculated about a ***Personal Life Recorder*** (PLR) type of device in his 1992 book *Turn Signals Are the Facial Expression of Automobiles*. He theorized that these PLRs would start out as a device given to young children, called the "Teddy." The Teddy would record all of a child's personal life moments, and as the child matured, the data could be transferred to new devices that matched his or her maturity level.

Projects such as "My Life Bits" from Microsoft are also exploring the requirements for such devices. Someday we may use such devices to share clips of our lives with our families or friends or to help recall past events and contacts. Combined with the capability of organizing and indexing the saved content, such devices could provide a very powerful personal database that could be accessed any time. In this book we shall refer to such devices as life-logs.

Such devices are today laboratory concepts, but in practice one could capture and sample one's life including audio, video, and GPS information linked together. The pieces are now available and all that remains is for a creative and imaginative company to find a market for such a product. In an age of social networking and cameras in every cell phone, such a function could become very popular. With volume and with technology development, such mobile recording devices could become quite affordable.

Depending upon the resolution, these devices would require huge amounts of information and the support system to store and organize the total aggregate content in the home could be even greater; a terabyte in the pocket and a petabyte in the home would not be out of the question. All together the proliferation of such devices could lead to the single biggest use of digital storage in the world, since there are a lot of people out there and thus a lot of people that may want to record their lives.

8.7 Chapter Summary

- Automobiles will use digital storage for entertainment and navigation purposes. These devices must be able to withstand very harsh environments and continue functioning. As the complexity of automobile navigation aids and automobile entertainment resolution increases, storage capacity requirements will increase. Automobiles will also be part of the home network, allowing them to obtain content over this network. This could significantly automate the acquisition of content and navigation information.

- Music and video player technology continues to develop. For highly compressed content (such as MP3 music) flash memory will probably provide all the digital storage needed in the future. However for less compressed (and higher quality) audio as well as video there will develop a lower end market for flash based devices as well as a higher resolution/content market for small form factor hard disk drive systems.

- The use of a larger screen on video players allows the use of 1.8-inch hard disk drives that could provide much larger capacity for the price than flash systems for several more years. Thus these small drives may not suffer the same fate that 1-inch disk drives used in MP3 players such as the mini-iPod suffered.

- New power source and wireless power technologies could provide extended usage life for mobile devices and enable the use of technologies that require more power, such as small projector systems for projecting images and video. Such systems could show higher resolution content than is required for smaller screens and thus require larger digital storage capacity.

- Digital still and video cameras are being incorporated into many other devices as well as continuing to exist as a stand-alone device category. Lower cost flash based video cameras will create a very popular market category. These products will require larger onboard storage as content resolution increases and they will also require larger storage capacities for off-loaded content—such as into home storage direct or network-attached storage devices.

•   Ultimately the growth in available capacity of mobile digital storage components will enable the creation of devices that can record a person's life experiences as they happen. Such a life-log will require significant digital storage capacity in the device as well as in the home network.

Bibliography

Simone, L. (2007). *If I Only Changed the Software Why Is the Phone on Fire?* New York: Elsevier.

Handy, J. (1993). *The Cache Memory*. San Diego, CA: Academic Press.

CHAPTER 9

Analog Low-Pass Filters

Marc T. Thompson

At first I had a good chuckle over the apparently self-cancelling title of Marc Thompson's book Intuitive Analog Circuit Design. *If there was ever a technology that's nonintuitive to anyone with an EE degree minted in the last 30 years, it's analog. Analog has long been the domain of old, bearded guys in back cubicles who handcraft the analog portions of your project and hand over black-box GDSII that you pray works with the rest of your design. Now that almost all new designs are analog/mixed-signal (AMS), you need to know more about this formerly black art. But making it "intuitive" is asking for a lot. Thompson pulls it off.*

Crafting analog circuits reliably in silicon is a tricky business, but modeling and constructing them from discrete components isn't. If you've spent years doing digital signal processing—and the word 'filter' means FIR, IIR and FFT to you—then this chapter will be a refresher course in the analog world of Butterworth, Chebyshev and Bessel filters. As Thompson points out, many digital filters begin as analog prototypes, and DSP systems generally have an analog low-pass filter front end; so getting reacquainted with these ubiquitous analog building blocks is a useful exercise.

Thompson starts off with a review of the three main types of analog low-pass filters, comparing the frequency and step responses, group delays, etc. The meat of the chapter is his description of a number of ways to implement your filter design, using both passive RLC ladders and active filters. For example, he shows how to implement a 5th-order Chebyshev filter with 0.5 dB passband ripple first using an RLC network, then as an active Sallen-Key circuit. He then takes this another step and shows how to implement an analog elliptic or "brick wall" filter with a very sharp cutoff. (And you thought that could only be done digitally!)

Thompson goes on to show how to design a 40-Hz Sallen-Key filter with adjustable Q; a 1-MHz low-pass filter; and a variation on a Butterworth—but you get the idea. The approaches, math and circuits in this chapter are both a useful refresher course in analog filter design and a handy reference to enlist when you have to design them yourself.

—John Donovan

9.1 Introduction

The low-pass filter is a ubiquitous component in many different kinds of signal-processing systems. Channel-separation, A/D antialiasing, and general signal processing are applications for low-pass filters, just to name a few. Even if you are a digital filter designer, it behooves you to know something about analog filter design, since many digital filters begin as analog prototypes, and are then transformed to the digital domain. Also, DSP systems generally have an analog front end that includes an analog low-pass filter for antialiasing purposes (Figure 9.1).

As with any kind of design, the "devil is in the details." Your specification will lead you to choices in filter topology and filter order, depending on the attenuation you need, the ripple that you can live with, and also the group delay variation that you can live with. In the following sections, we discuss design issues associated with low-pass filter design. The results can be extended without much trouble to band-pass and high-pass filters as well.

9.2 Review of Low-Pass Filter Basics

The magnitude response of the ideal low-pass filter is shown in Figure 9.2a. The gain of this filter is perfectly flat in the passband (for frequencies less than the filter cutoff[1] frequency ω_h), and the response drops to zero for frequencies higher than the cutoff frequency.

The magnitude response of a real-world low-pass filter is shown in Figure 9.2b. The non-ideal effects include:

- Possible ripple in the filter passband.

- Possible maximum attenuation floor in the stopband.

- Finite transition width between passband and stopband.

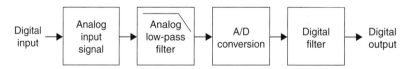

Figure 9.1: Typical digital signal-processing chain

---

[1] In general, the cutoff frequency of a filter is specified to be the frequency at which the gain through the filter has dropped to 0.707 of the DC value, or $-3\,\text{dB}$.

The magnitude response, however, only tells half the story. In addition, we must be concerned with the phase response of filters. As we'll see in the following sections the phase response (and by association the group delay[2] response) affects the transient response of filters. An ideal filter has a linear phase shift with frequency, and hence constant group delay as in Figures 9.2c and 9.2d. The following section discusses in detail several different low-pass filter types that to varying degrees approximate the ideal magnitude and phase of a low-pass filter.

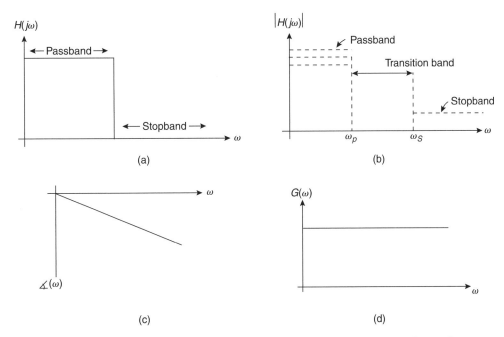

Figure 9.2: Response of low-pass filter. (a) Ideal low-pass filter magnitude |H(jω)|. (b) Real-world magnitude response |H(jω)| showing possible ripple in the passband and stopband, and a finite transition width between passband and stopband. (c) Ideal low-pass filter phase response ∠(ω) showing a negative phase shift that increases linearly with frequency. (d) Ideal low-pass filter group delay response G(ω) which is constant.

---

[2] The group delay of a filter is given by the derivative with respect to frequency of the phase response, or: $G(\omega) = -d\angle(\omega)/d\omega$. The group delay is a measure of how much a given frequency component is delayed passing through the filter. For low pulse distortion, you want all Fourier components to be delayed by the same amount of time, and hence you want constant group delay, and hence linear phase response.

9.3 Butterworth Filter

The Butterworth is a class of filters that provides maximally flat response in the passband. The pole locations for an Nth-order Butterworth filter are found equally spaced around a circle with radius equal to the filter cutoff frequency. A Butterworth filter with $-3\,\mathrm{dB}$ cutoff frequency $\omega_{3\,\mathrm{dB}}=1$ radian/second has poles at locations:

$$-\sin\frac{(2k-1)\pi}{2N} + j\cos\frac{(2k-1)\pi}{2N} \quad k = 1,2...N \tag{9.1}$$

The frequency response is given by:

$$\left|H(j\omega)\right| = \frac{1}{\sqrt{1 + \omega^{2N}}} \tag{9.2}$$

The pole transfer functions for Butterworth filters of varying order are shown in Table 9.1. Note that the transfer functions have been broken up into first-order and second-order

Table 9.1: Transfer function for Butterworth filters broken up into first-order and second-order factors. Transfer functions are shown for varying filter order N, with filter cutoff frequency $\omega_{3\,\mathrm{dB}}=1$ rad/sec

N	Transfer function
2	$\dfrac{1}{s^2 + 1.414s + 1}$
3	$\dfrac{1}{(s + 1)(s^2 + s + 1)}$
4	$\dfrac{1}{(s^2 + 0.7654s + 1)(s^2 + 1.8478s + 1)}$
5	$\dfrac{1}{(s + 1)(s^2 + 0.6180s + 1)(s^2 + 1.6180s + 1)}$
6	$\dfrac{1}{(s^2 + 0.5176s + 1)(s^2 + 1.4142s + 1)(s^2 + 1.9318s + 1)}$
7	$\dfrac{1}{(s + 1)(s^2 + 0.4550s + 1)(s^2 + 1.2480s + 1)(s^2 + 1.8019s + 1)}$
8	$\dfrac{1}{(s^2 + 0.3902s + 1)(s^2 + 1.1111s + 1)(s^2 + 1.6629s + 1)(s^2 + 1.9616s + 1)}$

factors.[3] Breaking up the transfer function in this fashion will help us to implement our filter, since first- and second-order sections are easily synthesized with op-amps. We also show the transfer function for a filter cutoff frequency of one radian per second.

The magnitude response, step response and group delay response for Butterworth filters are shown in Figure 9.3, Figure 9.4 and Figure 9.5, respectively. We note the following:

- As the filter order increases, the sharpness of the attenuation characteristic in the transition band increases. For instance, at 10 radians/second, an 8th order filter has an attenuation of roughly 160 dB, while lower-order filters have less attenuation.

- With regard to step response, the overshoot and delay through the filter increases as the filter order increases.

- With regard to group delay, the peak-peak variation in group delay increases as the filter order increases.

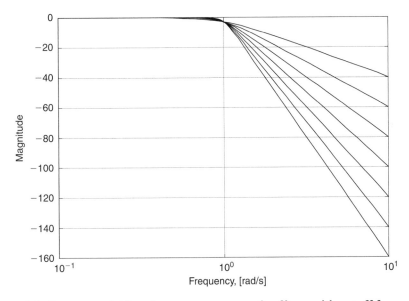

Figure 9.3: Butterworth filter frequency response for filters with cutoff frequency 1 radian/second, filter orders N = 2 to 8

---

[3] This is done to help facilitate the implementation of the transfer function using op-amps. For instance, first-order sections can be implemented with simple RC filters. Second-order transfer functions can be implemented using any number of op-amp circuits, including the Sallen-Key filter. More on this later on.

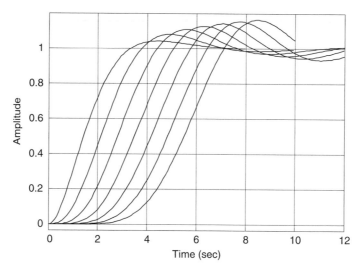

Figure 9.4: Butterworth filter step response for filters with cutoff frequency
1 radian/second, filter orders N=2 to 8

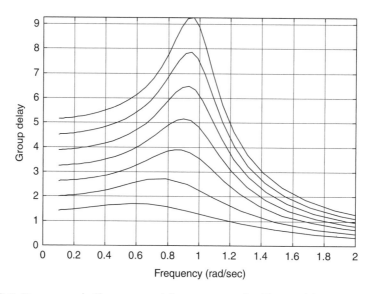

Figure 9.5: Butterworth filter group delay response for filters with cutoff frequency
1 radian/second, filter orders N=2 to 8

The preceding allows us to make some general statements regarding the relationship between filter frequency response, group delay, and transient response. A filter with a sharper cutoff will in general have more group delay variation and more overshoot and/or ringing in the frequency response. This is the reason why group delay variation is an important design parameter in filter design: more group delay variation results in more pulse distortion of a waveform passing through a filter. Conversely, filters with near-constant group delay pass pulses without significant distortion.

9.4 Chebyshev Filter

The Chebyshev filter has some ripple in the passband, and a sharper cutoff in the transition band than the Butterworth filter. As we'll see, the price one pays for using the Chebyshev design compared to the Butterworth is more group delay variation.

The pole locations of an Nth order Chebyshev are at locations $\sigma_k + j\omega_k$, with real parts and imaginary parts given by:

$$\sigma_k = -\sinh\left[\frac{1}{N}\sinh^{-1}\left(\frac{1}{\varepsilon}\right)\right]\sin(2k-1)\frac{\pi}{2N}, \quad k = 1,2\ldots N$$

$$\omega_k = \cosh\left[\frac{1}{N}\sinh^{-1}\left(\frac{1}{\varepsilon}\right)\right]\cos(2k-1)\frac{\pi}{2N}, \quad k = 1,2\ldots N \qquad (9.3)$$

The Chebyshev filter is characterized by two parameters: filter order N and allowable ripple in the passband. The ripple parameter ε is given by:

$$\varepsilon = \sqrt{10^{\frac{R_{db}-1}{10}}} \qquad (9.4)$$

where R_{db} is the allowable peak ripple in decibels. These pole locations yield a magnitude transfer function that is sharper than that of the Butterworth filter. Shown in Figure 9.6a is the Bode plot of an N=5, 0.20 dB ripple Chebyshev filter.

Looking at the passband detail of this filter, we plot the frequency response from 0.1 to 1 radian/second in Figure 9.6b, we see that the filter order (N=5) equals the number of up and down ripples in the passband. We also note that this Chebyshev formulation does not result in a filter cutoff frequency of exactly 1 radian/second. Rather, this formulation results in a filter where we leave the allowable ripple band at 1 radian/second; the -3 dB point is somewhat higher. We can show that the -3 dB point is given by:

$$\omega_{3db} = \cosh\left[\left[\frac{1}{N}\right]\cosh^{-1}\left(\frac{1}{\varepsilon}\right)\right] \qquad (9.5)$$

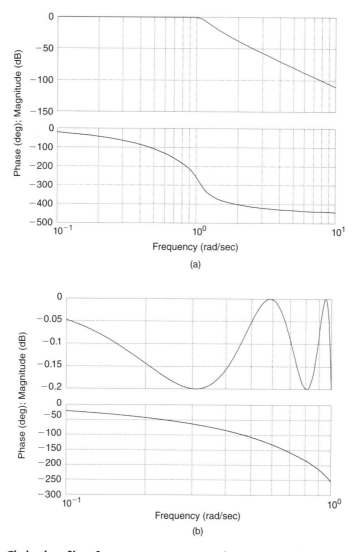

Figure 9.6: Chebyshev filter frequency response, order N=5, passband ripple 0.20 dB. (a) Overall response. (b) Detail of passband ripple. Note that the −3 dB cutoff frequency is slightly higher than 1 radian/second. The Chebyshev approximation shown here results in a passband ripple frequency of 1 radian/second.

For instance, for our N=5, 0.20 dB example, the -3 dB bandwidth is 1.10 radians/second.

A tabulation of transfer functions and -3 dB frequencies for three different types of Chebyshev filters are shown in Table 9.2, Table 9.3 and Table 9.4.

Table 9.2: Transfer functions[4] for Chebyshev filters of varying filter order N, with filter ripple frequency 1 rad/sec., passband ripple=0.1 dB. The -3 dB point ω_c is somewhat higher, as shown

N	Transfer function	$\omega_{3\,dB}$ (rad/s)
2	$\dfrac{3.314}{s^2 + 2.3724s + 3.314}$	1.94
3	$\dfrac{1.6381}{(s + 0.9694)(s^2 + 0.9694s + 1.6899)}$	1.38
4	$\dfrac{0.8285}{(s^2 + 0.5283s + 1.33)(s^2 + 1.2755s + 0.6229)}$	1.21
5	$\dfrac{0.4095}{(s + 0.5389)(s^2 + 0.3331s + 1.1949)(s^2 + 0.8720s + 0.6359)}$	1.13
6	$\dfrac{0.2071}{(s^2 + 0.2294s + 1.1294)(s^2 + 0.6267s + 0.6964)(s^2 + 0.8561s + 0.2634)}$	1.09
7	$\dfrac{0.1024}{(s + 0.3768)(s^2 + 0.1677s + 1.0924)(s^2 + 0.4698s + 0.7532)(s^2 + 0.6789s + 0.3302)}$	1.07
8	$\dfrac{0.0518}{(s^2 + 0.12805s + 1.0695)(s^2 + 0.3644s + 0.7989)(s^2 + 0.5454s + 0.4162)(s^2 + 0.6433s + 0.1456)}$	1.05

Table 9.3: Transfer functions for Chebyshev filters of varying filter order N, with filter ripple frequency 1 rad/sec., passband ripple=0.20 dB

N	Transfer function	$\omega_{3\,dB}$ (rad/s)
2	$\dfrac{2.3568}{s^2 + 1.9271s + 2.3568}$	1.67
3	$\dfrac{1.1516}{(s + 0.8146)(s^2 + 0.8146s + 1.4136)}$	1.28

(Continued)

---

[4]These transfer functions were generated using MATLAB and the CHEBY1 function that calculates the pole locations of Chebyshev filters with a given passband ripple.

Table 9.3: (Continued)

N	Transfer function	$\omega_{3\,dB}$ (rad/s)
4	$$\dfrac{0.5892}{(s^2 + 0.4496s + 1.1987)(s^2 + 1.0855s + 0.4916)}$$	1.16
5	$$\dfrac{0.2879}{(s + 0.4614)(s^2 + 0.2852s + 1.1174)(s^2 + 0.7466s + 0.5584)}$$	1.10
6	$$\dfrac{0.1473}{(s^2 + 0.1970s + 1.0779)(s^2 + 0.5383s + 0.6449)(s^2 + 0.7354s + 0.2119)}$$	1.07
7	$$\dfrac{0.0720}{(s + 0.3243)(s^2 + 0.1433s + 1.0557)(s^2 + 0.4044s + 0.7164)(s^2 + 0.5844s + 0.2934)}$$	1.05
8	$$\dfrac{0.0368}{(s^2 + 0.1103s + 1.0418)(s^2 + 0.3141s + 0.7712)(s^2 + 0.4700s + 0.3886)(s^2 + 0.5544s + 0.1180)}$$	1.04

Table 9.4: Transfer functions for Chebyshev filters of varying filter order N, with filter ripple frequency=1 rad/sec., passband ripple=0.50 dB.

N	Transfer function	$\omega_{3\,dB}$ (rad/s)
2	$$\dfrac{1.5162}{s^2 + 1.4256s + 1.5162}$$	1.39
3	$$\dfrac{0.7157}{(s + 0.6265)(s^2 + 0.6265s + 1.1424)}$$	1.17
4	$$\dfrac{0.3791}{(s^2 + 0.3507s + 1.0635)(s^2 + 0.8467s + 0.3564)}$$	1.09
5	$$\dfrac{0.1789}{(s + 0.3623)(s^2 + 0.2239s + 1.0358)(s^2 + 0.5762 + 0.4768)}$$	1.06
6	$$\dfrac{0.0948}{(s^2 + 0.1553s + 1.0230)(s^2 + 0.4243s + 0.5900)(s^2 + 0.5796s + 0.1570)}$$	1.04
7	$$\dfrac{0.0447}{(s + 0.2562)(s^2 + 0.1140s + 1.0161)(s^2 + 0.3194s + 0.6769)(s^2 + 0.4616s + 0.2539)}$$	1.03
8	$$\dfrac{0.0237}{(s^2 + 0.0872s + 1.0119)(s^2 + 0.2484s + 0.7413)(s^2 + 0.3718s + 0.3587)(s^2 + 0.4386s + 0.0881)}$$	1.02

A comparison of group delay responses for Chebyshev filters[5] of varying order are shown in Figure 9.7. Note the peaking in the group delay near the cutoff frequency. This peaking, and group delay ripple in the passband, become more pronounced as we increase the passband ripple of the Chebyshev filter.

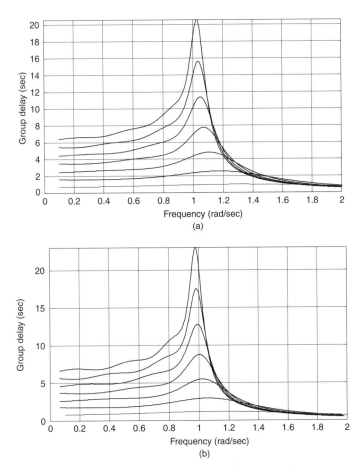

Figure 9.7: Comparison of group delay responses for Chebyshev filters with passband ripple 0.1 dB, 0.2 dB and 0.5 dB and cutoff frequencies near 1 radian/sec, and filter orders N=2 to 8. Comparison of group delay responses for Chebyshev filters with passband ripple 0.1 dB, 0.2 dB and 0.5 dB and cutoff frequencies near 1 radian/sec.

[5] In each case, the filters are normalized for a frequency at which the ripple band is first exceeded of 1 radian/second. Therefore, the cutoff frequency for each filter is approximately 1 radian/sec., but not exactly.

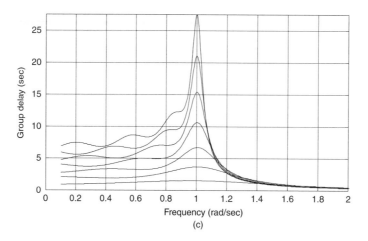

Figure 9.7: (Continued)

9.5 Bessel Filter

The Bessel filter is optimized to provide constant group delay in the filter passband, while sacrificing sharpness in the frequency response. We note these effects with reference to Figure 9.8, Figure 9.9 and Figure 9.10. The frequency response is gradual in the transition band, and there is little or no overshoot in the step response. The group delay exhibits very flat response in the passband.

The resultant transfer functions of the Bessel filters are shown in Table 9.6.

9.6 Comparison of Responses of Different Filter Types

A comparison of frequency responses for three different types of an $N=5$ filter is shown in Figure 9.11. Note that, as expected, the Chebyshev filter has the sharpest cutoff characteristics, while the Bessel response is more gradual in the stopband.

A comparison of step responses for the three different filters is shown in Figure 9.12. Note that the Bessel has minimal overshoot, while the overshoots of the Butterworth and Chebyshev filters are comparable. A Chebyshev filter with more passband ripple would have higher overshoot.

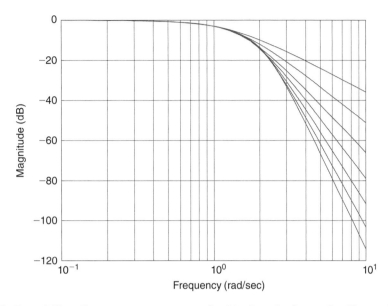

Figure 9.8: Bessel filter frequency response, order N=2 to 8, shown for filters with −3 dB cutoff frequency of 1 radian/second

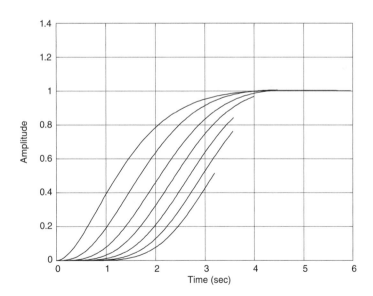

Figure 9.9: Bessel filter step response, order N=2 to 8, shown for filters with −3 dB cutoff frequency of 1 radian/second

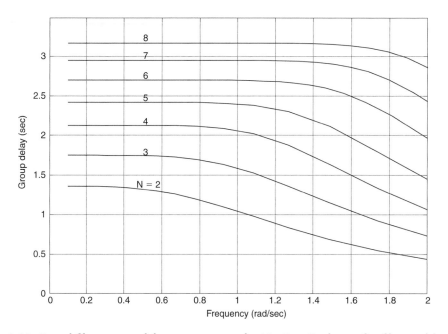

Figure 9.10: Bessel filter group delay response, order N=2 to 8, shown for filters with −3 dB cutoff frequency of 1 radian/second. Note that as the filter order increases, the flatness of the group delay response in the passband improves.

Table 9.5: Pole locations for Bessel filters of varying filter order N, with filter cutoff frequency ω_c=1 rad/sec

N	Real Part ($-\sigma$)	Imaginary Part ($\pm j\omega$)
2	1.1030	0.6368
3	1.0509	1.0025
	1.3270	
4	1.3596	0.4071
	0.9877	1.2476
5	1.3851	0.7201
	0.9606	1.4756
	1.5069	

(Continued)

Table 9.5: (Continued)

N	Real Part ($-\sigma$)	Imaginary Part ($\pm j\omega$)
6	1.5735	0.3213
	1.3836	0.9727
	0.9318	1.6640
7	1.6130	0.5896
	1.3797	1.1923
	0.9104	1.8375
	1.6853	
8	1.7627	0.2737
	0.8955	2.0044
	1.3780	1.3926
	1.6419	0.8253

Table 9.6: Transfer function broken up into first-order and second-order quadratic factors for Bessel filters of varying filter order N, with filter cutoff frequency $\omega_c = 1$ rad/sec

N	Transfer function
2	$\dfrac{1.6221}{s^2 + 2.206s + 1.6221}$
3	$\dfrac{2.7992}{(s + 1.3270)(s^2 + 2.1018s + 2.1094)}$
4	$\dfrac{5.1002}{(s^2 + 2.7192s + 2.0142)(s^2 + 1.9754s + 2.5321)}$
5	$\dfrac{11.3845}{(s + 1.5069)(s^2 + 2.7702s + 2.4370)(s^2 + 1.9212s + 3.1001)}$
6	$\dfrac{26.8328}{(s^2 + 3.1470s + 2.5791)(s^2 + 2.7672s + 2.8605)(s^2 + 1.8636s + 3.6371)}$
7	$\dfrac{69.5099}{(s + 1.6853)(s^2 + 3.2262s + 2.9497)(s^2 + 2.7594s + 3.3251)(s^2 + 1.8208s + 4.2052)}$
8	$\dfrac{198.7746}{(s^2 + 3.5254s + 3.1820)(s^2 + 1.7910s + 4.1895)(s^2 + 2.7560s + 3.8382)(s^2 + 3.2838s + 3.3770)}$

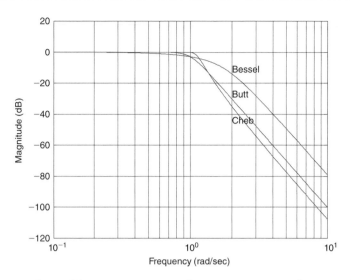

Figure 9.11: Comparison of frequency response for N=5 Butterworth, N=5 Chebyshev with 0.1 dB ripple and N=5 Bessel filter, each with a cutoff frequency of 1 radian/second

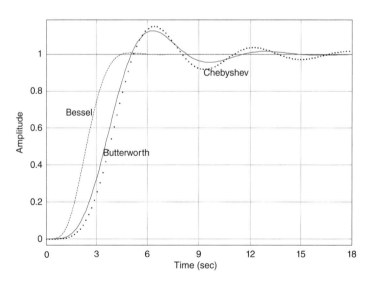

Figure 9.12: Comparison of step response for N=5 Butterworth, Chebyshev (0.1 dB ripple) and Bessel filter, each with a cutoff frequency of 1 radian/second. Legend: Butterworth (solid), Bessel (dashed), Chebyshev (dotted). Note that there is minimal overshoot in the Bessel response, while the Butterworth and Chebyshev responses are comparable.

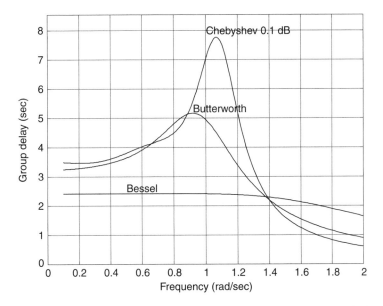

Figure 9.13: Comparison of group delay response for N=5 Butterworth, Chebyshev (0.1 dB ripple) and Bessel filter, each with a cutoff frequency of 1 radian/second

In Figure 9.13 we compare the group delay responses of the three filters. We note that the Bessel filter has flat group delay response in the passband, while the Butterworth has moderate variation in group delay in the passband. The Chebyshev exhibits pronounced group delay peaking in the passband.

9.7 Filter Implementation

So, you've done all this hard work to figure out the filter type and order, but how do you build it in practice? There are several methods and we'll discuss a few of them in this next section.

9.7.1 Ladder

For high-frequency filters, one option is to build a passive ladder using resistors, inductors and capacitors. The topology for an Nth-order ladder filter suitable for implementing Butterworth, Chebyshev and Bessel filters is shown in Figure 9.14a. The filter is made with alternating inductors and capacitors, with source and termination resistors. We note that, for an even-order filter, the filter terminates with a resistor and capacitor and for an odd-order filter the filter terminates with an inductor and resistor. Ladders for N=4 and N=5 filters are shown in Figure 9.14b and Figure 9.14c, respectively.

Values have been extensively tabulated for filters with cutoff frequency 1 radian per second. The termination resistor for this normalized filter is 1.0 ohms, and the filter can have any in-put resistance R_s that you desire. Tabulated ladder element values for Butterworth, Bessel and Chebyshev filters of various orders are shown in Table 9.7 through Table 9.11.

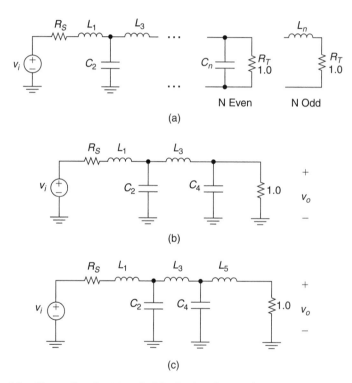

Figure 9.14: Ladder filter of order N, suitable for implementing Butterworth, Chebyshev and Bessel filters. (a) Generic filter. (b) N=4 ladder. (c) N=5 ladder.

Table 9.7: Butterworth inductor[6] and capacitor values of varying filter order N, with filter cutoff frequency $\omega_c = 1$ rad/sec., and termination resistance $R_T = 1.0\,\Omega$. Note that for this case, all values of the source resistance are $R_s = 1.0\,\Omega$

N	R_s	L_1	C_2	L_3	C_4	L_5	C_6	L_7	C_8
2	1.0	1.4142	1.4142						
3	1.0	1.0000	2.0000	1.0000					
4	1.0	0.7654	1.8478	1.8478	0.7654				
5	1.0	0.6180	1.6180	2.0000	1.6180	0.6180			
6	1.0	0.5176	1.4142	1.9319	1.9319	1.4142	0.5176		
7	1.0	0.4450	1.2470	1.8019	2.0000	1.8019	1.2470	0.4450	
8	1.0	0.3902	1.1111	1.6629	1.9616	1.9616	1.6629	1.1111	0.3902

Table 9.8: Bessel inductor and capacitor values of varying filter order N, with filter cutoff frequency $\omega_c = 1$ rad/sec., and termination resistance $R_T = 1.0\,\Omega$. Note that for this case, all values of the source resistance are $R_s = 1.0\,\Omega$

N	R_s	L_1	C_2	L_3	C_4	L_5	C_6	L_7	C_8
2	1.0	0.5755	2.1478						
3	1.0	0.3374	0.9705	2.2034					
4	1.0	0.2334	0.6725	1.0815	2.2404				
5	1.0	0.1743	0.5072	0.8040	1.1110	2.2582			
6	1.0	0.1365	0.4002	0.6392	0.8538	1.1126	2.2645		
7	1.0	0.1106	0.3259	0.5249	0.7020	0.8690	1.1052	2.2659	
8	1.0	0.0919	0.2719	0.4409	0.5936	0.7303	0.8695	1.0956	2.2656

Table 9.9: 0.1 dB Chebyshev filter inductor and capacitor values, with filter cutoff frequency $\omega_c = 1$ rad/sec., and termination resistance $R_T = 1.0\,\Omega$. Note that the values of the source resistance vary from filter order to filter order

N	R_s	L_1	C_2	L_3	C_4	L_5	C_6	L_7	C_8
2	1.3554	1.2087	1.6382						
3	1.0	1.4328	1.5937	1.4328					
4	1.3554	0.9924	2.1476	1.5845	1.3451				
5	1.0	1.3013	1.5559	2.2411	1.5559	1.3013			
6	1.3554	0.9419	2.0797	1.6581	2.2473	1.5344	1.2767		
7	1.0	1.2615	1.5196	2.2392	1.6804	2.2392	1.5196	1.2615	
8	1.3554	0.9234	2.0454	1.6453	2.2826	1.6841	2.2300	1.5091	1.2515

---

[6]All inductor and capacitor value charts adopted from A. Zverev, Handbook of Filter Synthesis, John Wiley, 1967. For other source resistances Rs, the reader is invited to visit this reference.

Table 9.10: 0.25 dB Chebyshev filter inductor and capacitor values, with filter cutoff frequency
$\omega_c = 1$ rad/sec., and termination resistance $R_T = 1.0\,\Omega$

N	R_s	L_1	C_2	L_3	C_4	L_5	C_6	L_7	C_8
2	2.0	0.6552	2.7632						
3	1.0	1.6325	1.4360	1.6325					
4	2.0	0.6747	3.6860	1.0247	1.8806				
5	1.0	1.5046	1.4436	2.4050	1.4436	1.5046			
6	2.0	0.6867	3.2074	0.9308	3.8102	1.2163	1.7088		
7	1.0	1.5120	1.4169	2.4535	1.5350	2.4535	1.4169	1.5120	

Table 9.11: 0.5 dB Chebyshev filter inductor and capacitor values, with filter cutoff frequency
$\omega_c = 1$ rad/sec., and termination resistance $R_T = 1.0\,\Omega$

N	R_s	L_1	C_2	L_3	C_4	L_5	C_6	L_7	C_8
2	1.9841	0.9827	1.9497						
3	1.0	1.8636	1.2804	1.8636					
4	1.9841	0.9202	2.5864	1.3036	1.8258				
5	1.0	1.8068	1.3025	2.6912	1.3025	1.8068			
6	1.9841	0.9053	2.5774	1.3675	2.7133	1.2991	1.7961		
7	1.0	1.7896	1.2961	2.7177	1.3848	2.7177	1.2961	1.7896	
8	1.9841	0.8998	2.5670	1.3697	2.7585	1.3903	2.7175	1.2938	1.7852

Example 9.1   Design example: Fifth-order Chebyshev filter with 0.5 dB passband ripple

We'll use the filter charts to design a fifth-order 1-MHz low-pass Chebyshev filter with 0.5 dB ripple in the passband. From the filter chart (Table 9.11) we find the corresponding values for a filter with cutoff frequency 1 radian/second as:

$$R_s = R_T = 1\Omega$$
$$L_1 = 1.8068$$
$$C_2 = 1.3025$$
$$L_3 = 2.6914$$
$$C_4 = 1.3025$$
$$L_5 = 1.8068$$

We next need to pick more reasonable values for source and termination resistors (instead of the $1.0\,\Omega$ normalized values). For this design example we'll choose

$R_s = R_T = 50\,\Omega$. We now make use of an unnormalization process to transform the filter with cutoff of 1 radian per second to our desired frequency of 1 MHz. The unnormalization process is:

$$C = \frac{C_n}{2\pi f_c R}$$

$$L = \frac{L_n R}{2\pi f_c} \tag{9.6}$$

where C_n and L_n are the normalized values found from the filter charts, f_c is the desired new cutoff frequency, and R is the resistor value used in the new filter. Applying this process to our filter results in:

$$L_1' = \frac{L_1 R}{2\pi f_c} = \frac{(1.8068)(50)}{(2\pi)(10^6)} = 14.378\,\mu\text{H}$$

$$C_2' = \frac{C_2}{2\pi f_c R} = \frac{(1.3025)}{(2\pi)(10^6)(50)} = 4146\,\text{pF}$$

$$L_3' = \frac{L_3 R}{2\pi f_c} = \frac{(2.6914)(50)}{(2\pi)(10^6)} = 23.407\,\mu\text{H}$$

$$C_4' = \frac{C_4}{2\pi f_c R} = \frac{(1.3025)}{(2\pi)(10^6)(50)} = 4146\,\text{pH}$$

$$L_5' = \frac{L_5 R}{2\pi f_c} = \frac{(1.8068)(50)}{(2\pi)(10^6)} = 14.378\,\mu\text{H} \tag{9.7}$$

The resultant circuit and frequency response for the 1-MHz filter is shown in Figure 9.15.

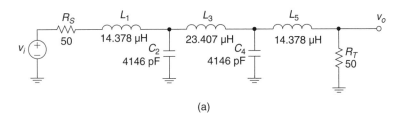

(a)

Figure 9.15: 1-MHz, 0.5 dB Chebyshev low-pass filter implemented as a ladder. (a) PSPICE circuit. (b) Frequency response. Note that the DC gain of this filter is $-6\,\text{dB}$ (or a factor of $1/2$) due to the $R_s - R_T$ resistive divider.

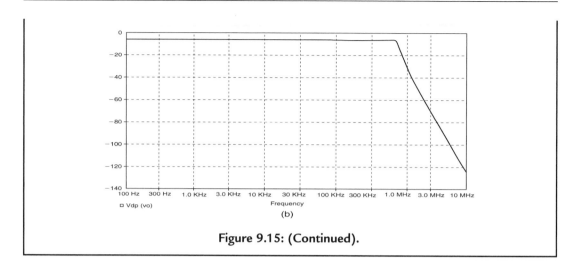

Figure 9.15: (Continued).

9.7.2 Filter Implementation—Active

We can also directly implement the filter transfer functions using active filters, such as the Sallen-Key filter. The Sallen-Key filter (Figure 9.16) generates the transfer function

$$\frac{v_o}{v_i} = \frac{1}{R_1 R_2 C_1 C_2 S^2 + [R_2 C_2 + R_1 C_2]s + 1} \tag{9.8}$$

For instance, to implement a fourth-order filter, you could cascade two Sallen-Key circuits, provided you know where all the filter poles are.

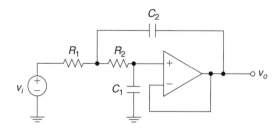

Figure 9.16: Sallen-Key circuit (with DC gain=1)

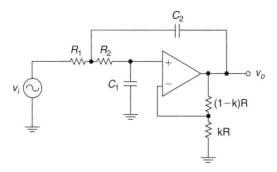

Figure 9.17: Sallen-Key Circuit (with adjustable Q). The adjustment can be made by implementing the $(1-k)R$ and kR resistor with a potentiometer.

A variation on the theme is the Sallen-Key circuit with adjustable DC gain, which also adjusts the damping of the filter, as shown in Figure 9.17. The transfer function of this filter is:

$$\frac{v_{out}}{v_{in}} = \left(\frac{1}{k}\right)\frac{1}{R_1R_2C_1C_2s^2 + \left[R_2C_2 + R_1C_2 + R_1C_1\left(1 - \frac{1}{k}\right)\right]s + 1} \tag{9.9}$$

9.7.3 Some Comments on Elliptic (or "Brick Wall") Filters

The detailed design of elliptic filters is beyond the scope of this book, but a few comments are in order. Elliptic filters, also called "brick wall" filters, have very sharp filter cutoff characteristics. Again, this is done at the expense of very nonlinear group delay. One flavor of elliptic filter has zero ripple in the passband but finite ripple in the stopband. This is accomplished by having zeroes in the transfer function. A 1 radian/second elliptic low-pass filter is shown in Figure 9.18. Note that the filter has parallel LC sections. These sections generate zeros in the transfer function.

We can now unnormalize this filter as before to generate a filter with a cutoff frequency of 10 MHz (Figure 9.19a). The frequency response (Figure 9.19b) shows a very fast rolloff in the transition band, and has a minimum attenuation floor.

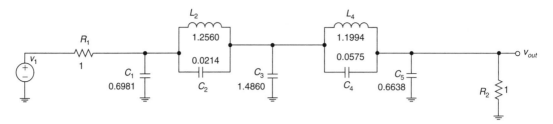

Figure 9.18: Elliptic low-pass filter prototype (1 r/s cutoff)

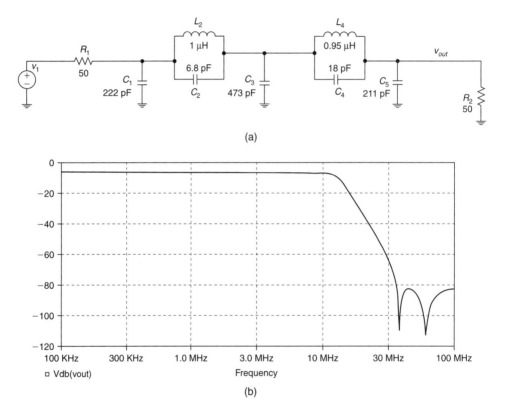

(a)

(b)

Figure 9.19: Unnormalized elliptic low-pass filter prototype (10-MHz cutoff). (a) PSPICE Circuit. (b) Frequency response.

Example 9.2   Design Example: 40-Hz Sallen-Key with Adjustable Q

Following is a design of a 40-Hz low-pass filter with adjustable Q:

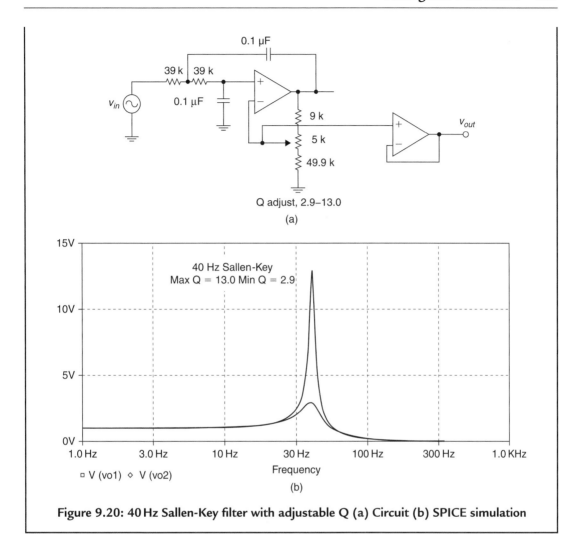

Figure 9.20: 40 Hz Sallen-Key filter with adjustable Q (a) Circuit (b) SPICE simulation

9.7.4 All-pass Filters

An all-pass filter is a filter that has a magnitude response of unity, but which provides phase shift. You can use all-pass filters to tailor group delay responses in your filters. You may find that you will need to cascade your filter with an all-pass filter in order to meet the group delay specification. A first-order all-pass circuit is shown in Figure 9.21. Note that this all-pass provides a DC gain of -1. If you want, you can cascade an inverting op-amp stage with the all-pass to take care of this phase inversion.

The transfer function, angle and group delay for a first-order all-pass filter are:

$$H(s) = \frac{RCs - 1}{RCs + 1} = \frac{s - \dfrac{1}{RC}}{s + \dfrac{1}{RC}} = \frac{s - a}{s + a}$$

$$\angle H(s) = -2\tan^{-1}\frac{\omega}{a}$$

$$D(j\omega) = \frac{2a}{a^2 + \omega^2} \tag{9.10}$$

The group delay characteristics for a=1 r/s is shown in Figure 9.22. Note that the DC delay is twice the value of RC.

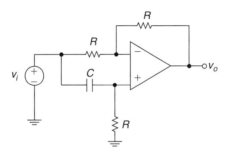

Figure 9.21: First-order all-pass filter

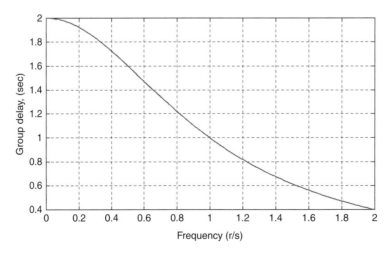

Figure 9.22: First-order all-pass filter group delay response

---

Example 9.3   Design case study: 1-MHz low-pass filter

In this case study, we'll design and simulate an analog low-pass filter meeting the following specifications (Table 9.12):

Table 9.12: Design case study specification

Filter type	Low-pass
Nominal $-3\,dB$ bandwidth	1 MHz
Passband gain	0 dB nominal, within 0.25 dB of nominal up to 750 kHz
Attenuation	$>50\,dB$ at 2.5 MHz
Group delay response	Group delay variation from the DC value less than 1000 nanoseconds up to 1 MHz

This design could be implemented with either a ladder filter with passive elements (resistors, inductors and capacitor) or a cascade of Sallen-Key active second-order sections. If you at-tempt to use a Bessel filter, the filter order will be quite high. If you look at elliptic filters the group delay specification may be very difficult to meet.

In this design $\omega_s/\omega_c = 2.5$ and minimum attenuation at ω_s is $-50\,dB$. A Bessel filter was not implemented due to the high filter order required. In order to meet the specification for $-50\,dB$ gain at a normalized frequency of 2.5, a Bessel filter of order N > 10 would be needed. A sixth-order Butterworth filter would almost make the specification, since

$$20\log_{10}\left(\frac{1}{\sqrt{1 + 2.5^{12}}}\right) = -47.8\,dB$$

So, if we use a Butterworth, we'll need at least N=7. The advantage of the Butterworth is that there is no magnitude ripple in the passband.

Another alternative will be to try a Chebyshev design with N=7 or less and some passband ripple. From our previous work on the Chebyshev filter, it looks like an N=6 Chebyshev with 0.1 dB ripple in the passband will meet the specification. An N=5 Chebyshev with 0.25 dB passband ripple barely misses the attenuation spec. So, let's go with the N=6, 0.1 dB Chebyshev design (Figure 9.23).

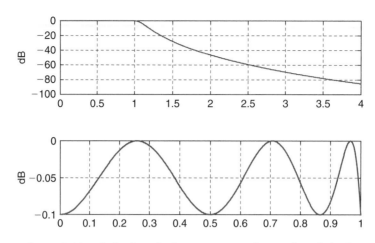

Figure 9.23: Chebyshev design, N=6, 0.1 dB passband ripple

Using a ladder filter topology for an N=6 Chebyshev with 0.1 dB ripple, normalized component values for $\omega_c = 1$ r/s and $R_T = 1\,\Omega$ are:

$$
\begin{aligned}
R_s &= 1.3554 \\
L_1 &= 0.9419 \\
C_2 &= 2.0797 \\
L_3 &= 1.6581 \\
C_4 &= 2.2473 \\
L_5 &= 1.5344 \\
C_6 &= 1.2767
\end{aligned}
$$

$$
L = \frac{L_{norm} R}{\omega_c}
$$

$$
C = \frac{C_{norm}}{\omega_c R}
$$

Remember, we scale the normalized filter by:

From this, we calculate the unnormalized component values for a filter with $-3\,dB$ point of 1 MHz (Figure 9.24), and using a termination resistor value of $75\,\Omega$. The overall magnitude of the frequency response of the Chebyshev ladder filter is shown in Figure 9.25.

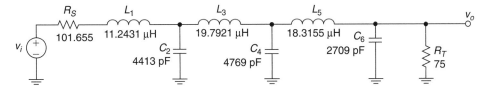

Figure 9.24: Chebyshev ladder filter design

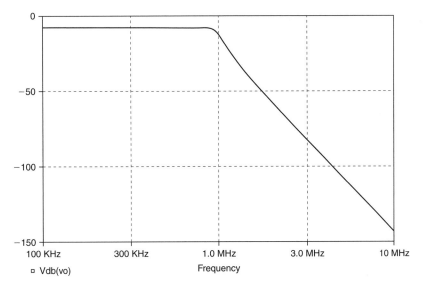

Figure 9.25: Chebyshev ladder filter magnitude of the frequency response (from SPICE). Vertical scale in dB

The stopband detail shows that the magnitude is attenuated >60 dB at 2.5 MHz, as expected (Figure 9.26).

Note that there is slightly more than expected ripple in the passband. This may be due to roundoff error in the ladder components (Figure 9.27).

SPICE results show that the variation in group delay in the DC to 1-MHz range is approximately 1200 nanoseconds, which violates the group delay specification (Figure 9.28). From the above, we see that the delay variation in the DC to 1-MHz range is over 1000 nanoseconds. So, let's cascade a first-order all-pass network in an attempt to fill in the group delay hole below 1 MHz (Figure 9.29). We'll assume that we have ideal op-amps at our disposal. What this means is that we need to choose op-amps with gain-bandwidth product much higher than our frequency range

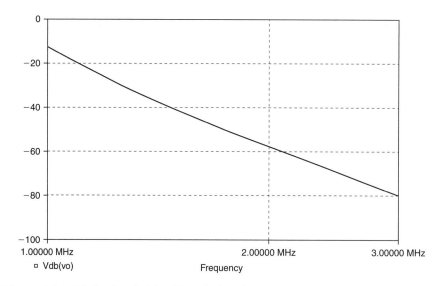

Figure 9.26: Chebyshev ladder filter design frequency response, stopband detail
(Vertical scale in dB)

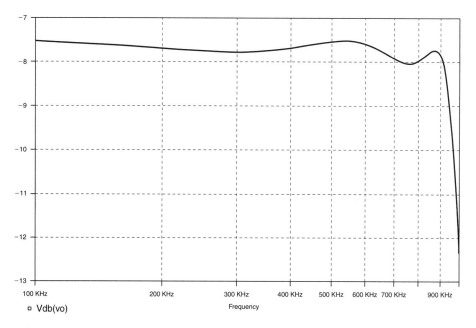

Figure 9.27: Chebyshev ladder filter design frequency response, passband detail
(Vertical scale in dB)

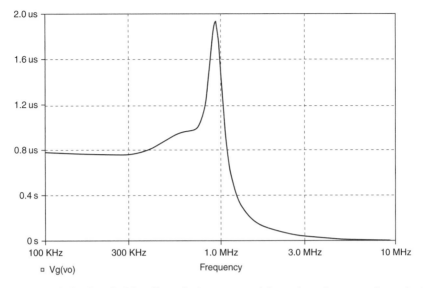

Figure 9.28: Chebyshev ladder filter design group delay. There is approximately 1200
nanoseconds of group delay variation in the passband

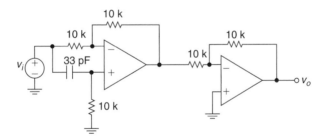

Figure 9.29: Group delay equalizer for filter example

of interest, or greater than 1 MHz. The low-frequency delay of a first-order all-pass is 2RC, so in this case we've chosen a DC delay of 660 nanoseconds.

The delay-equalized filter meets the group delay specification. The peak-to-peak delay variation in the passband is approximately 900 nanoseconds (Figure 9.30), and hence we meet the group delay specification.

The step response of original and delay equalized circuits (Figure 9.31) shows that the equalized circuit has less overshoot in the step response, as expected. However, the delay through the filter is increased (due to the all-pass network).

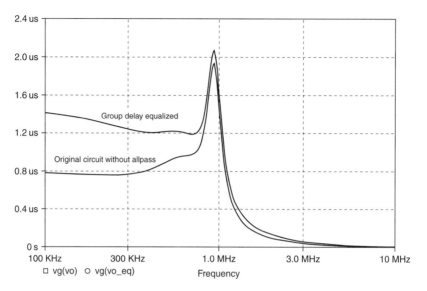

Figure 9.30: Chebyshev ladder filter design group delay equalized, compared to original circuit

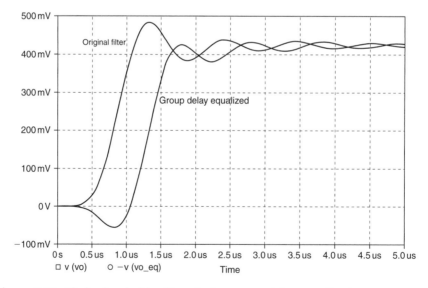

Figure 9.31: Chebyshev ladder filter design group delay equalized, step response

Example 9.4   Alternate design using Butterworth filter

Here's a seventh-order Butterworth design, which also meets the group delay specification without any further all-pass filtering. In this case, the Butterworth turns out to be the simpler design, even though it has a higher filter order, since a delay equalizer is not needed.

Let's choose a design with $R_s=1$, $L_1=0.445$, $C_2=1.247$, $L_3=1.8019$, $C_4=2.0$, $L_5=1.8019$, $C_6=1.247$, $L_7=0.445$ and $R_T=1$. The unnormalized values are shown in the circuit of Figure 9.32a. We note that we meet the gain specification as well as the group delay specification (Figure 9.32b) using this filter, without any additional group delay equalization.

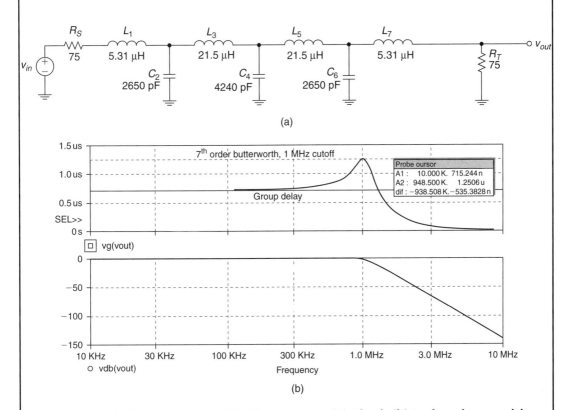

(a)

(b)

Figure 9.32: N=7 Butterworth ladder filter response (a) Circuit (b) Peak-peak group delay variation is less than 1000 ns (approximately 535 ns) (Note that a gain of +2 to compensate for the resistive divider is assumed following the filter.)

Bibliography

Balch, Brent (1988). A simple technique boosts performance of active Filters. *EDN, November 10*, 277–286.

Blinchikoff, H., & Zverev, A. (1976). *Filtering in the time and frequency domains*. John Wiley.

Burton, L.T., & Treleaven, D. (1973). Active filter design using generalized impedance converters. *EDN, February 5*, 68–75.

Chambers, William (1991). Know your options and requirements when designing filters. *EDN, August 5*, 129–138.

Corral, C. (2000). Designing elliptic filters with maximum selectivity. *EDN, May 25*, 101–109.

Corrington, Murlan S. (September 1949). Transient response of filters. *RCA Review, 10*(3), 397–429.

Downs, Rick (1991). Vintage filter scheme yields low distortion in new audio designs. *EDN, November 7*, 267–272.

Steer, Robert, Jr. (1989) Antialiasing Filters Reduce Errors in A/D Converters. *EDN, March 30*, 171–186.

Tow, J. (1969). A step-by-step active-filter design. *IEEE Spectrum, December*, 64–68.

Williams, A., & Taylor, F. (1988). *Electronic filter design handbook*. McGraw-Hill.

Yager, Charles, & Laber, Carlos (1989). Create a high-frequency complex filter. *Electronic Design, April 13*, 123–127.

Zverev, Anatol (1967). *Handbook of filter synthesis*. John Wiley.

CHAPTER 10

Class-A Amplifiers

Douglas Self

In the beginning there was Class A, and it was good. It was simple, linear, and had no problem with crossover distortion, switch-off phenomena, nonlinear voltage-amplifier stage (VAS) loading, injection of supply rails, induction from supply currents, erroneous feedback or Beta mismatch in the output devices. OK, so efficiency was horrendous—get a bigger heatsink!—but the sound quality was spectacular, so we learned to live with that.

Class A amplifiers will always be around—and they'll always be a challenge, particularly for battery-powered portable devices. But when your application needs the lowest possible distortion, they're hard to avoid.

In his classic Audio Power Amplifier Design Handbook *Douglas Self spends a lot of time on Class A, AB and B amplifiers—including the detailed features, upsides and downsides of dozens of permutations, including several of his own design. No matter how much you may have worked with these circuits, his chapter on Class A amplifiers will provide new insights and clever workarounds.*

There are numerous ways to construct a Class A amplifier, from single-ended/resistively-loaded to constant current to a variety of push-pull configurations. Most commercial amplifiers use some variation of push-pull, since going that route doubles your theoretical efficiency from 12.5% for a single-ended design to 25–50% at maximum output, depending on the configuration (the operative word here being theoretical). Self comes down in favor of an increased-bias Class B topology, which operates effectively in Class A if biased well beyond the AB transition level. It has the additional advantage, if the load impedance suddenly drops, of falling back to Class AB instead of just melting down.

Close quiescent current control is important in Class B to minimize distortion; in Class A, where you're always drawing current, it's needed to prevent thermal runaway. Self critiques the usual methods of a Vbe-multiplier bias generator and feedback current control before introducing his own somewhat complex but effective way of solving the problem.

Self makes his own contribution to the state of the art, describing the design of a quasi-complementary trimodal (Class A/AB/B) amplifier that is ultralinear most of the time (Class A), very linear all the time (well-designed Class B) with occasional excursions into AB. With voltage efficiency of approximately 93% positive/negative and total harmonic distortion (THD) almost too low to measure, this design alone is worth the price of the book.

All told, "Class-A Amplifiers" is a thorough treatment of an important topic, with some clever ways of addressing some long-standing, vexing problems.

—**John Donovan**

10.1 An Introduction to Class A

The two salient facts about Class-A amplifiers are that they are inefficient, and that they give the best possible distortion performance. They will never supplant Class-B amplifiers; but they will always be around.

The quiescent dissipation of the classic Class-A amplifier is equal to twice the maximum output power, making massive power outputs impractical, if only because of the discomfort engendered in the summer months. However, the nature of human hearing means that the power of an amplifier must be considerably increased to sound significantly louder. Doubling the sound pressure level (SPL) is not the same as doubling subjective loudness, the latter being measured in Sones rather than dB above threshold, and it appears that doubling subjective loudness requires nearer a 10 dB rather than 6 dB rise in SPL[1]. This implies amplifier power must be increased something like tenfold, rather than merely quadrupled, to double subjective loudness. Thus a 40 W Class-B amplifier does not sound much larger than its 20 W Class-A cousin.

There is an attractive simplicity and purity about Class A. Most of the distortion mechanisms stem from Class B, and we can thankfully forget crossover and switch-off phenomena, nonlinear VAS loading, injection of supply-rail signals, induction from supply currents, and erroneous feedback connections. Beta-mismatch in the output devices can also be ignored.

The only real disadvantage of Class-A is inefficiency, so inevitably efforts have been made to compromise between A and B. As compromises go, traditional Class-AB is not a happy one because when the AB region is entered the step-change in gain generates significantly greater high-order distortion than that from optimally biased Class-B. However, a well-designed AB amplifier does give pure Class-A performance below the AB threshold, something a Class-B amp cannot do.

Another possible compromise is the so-called non-switching amplifier, with its output devices clamped to always pass a minimum current. However, it is not obvious that a sudden halt in current-change as opposed to complete turn-off makes a better crossover region. Those residual oscillograms that have been published seem to show that some kind of discontinuity still exists at crossover[2].

One potential problem is the presence of maximum ripple on the supply-rails at zero signal output; the PSRR must be taken seriously if good noise and ripple figures are to be obtained.

10.2 Class-A Configurations and Efficiency

There is a canonical sequence of efficiency in Class-A amplifiers. The simplest version is single-ended and resistively loaded, as at Figure 10.1a. When it sinks output current, there is an inevitable voltage drop across the emitter resistance, limiting the negative output capability, and resulting in an efficiency of 12.5% (erroneously quoted in at least one textbook as 25%, apparently on the grounds that power not dissipated in silicon does not count). This would be of purely theoretical interest—and not much of that—except that a single-ended design by Fuller Audio has recently appeared. This reportedly produces a 10 W output for a dissipation of 120 W, with output swing predictably curtailed in one direction[3].

A better method—Constant-current Class-A—is shown in Figure 10.1b. The current sunk by the lower constant-current source is no longer related to the voltage across it, and so the output voltage can approach the negative rail with a practicable quiescent current (hereafter shortened to Iq). Maximum efficiency is doubled to 25% at maximum output; for an example with 20 W output (and a big fan) see Nelson[4]. Some versions (Krell) make the current-source value switchable, controlling it with a kind of noise-gate.

Push-pull operation once more doubles full-power efficiency, getting us to a more practical 50%; most commercial Class-A amplifiers have been of this type. Both output halves now swing from zero to twice the Iq, and least voltage corresponds with maximum current, reducing dissipation. There is also the intriguing prospect of cancelling the even-order harmonics generated by the output devices.

Push-pull action can be induced in several ways. Figure 10.1c, d show the lower constant current-source replaced by a voltage-controlled current-source (VCIS). This can be driven directly by the amplifier forward path, as in Figure 10.1c[5], or by a current-control negative-feedback loop, as at Figure 10.1d[6]. The first of these methods has the drawback that the stage generates gain, phase-splitter TR1 doubling as the VAS; hence there is no circuit node that can be treated as the input to a unity-gain output stage, making the circuit hard to analyze, as

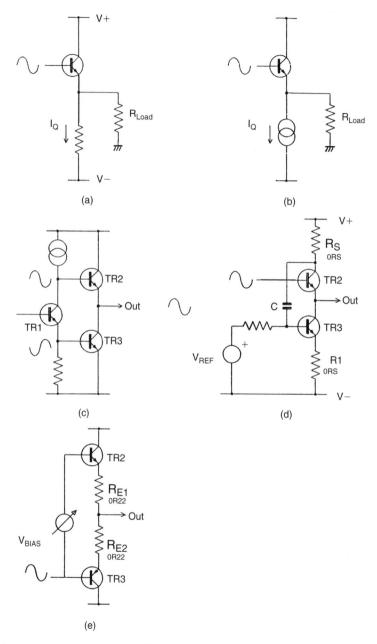

(a)

(b)

(c)

(d)

(e)

Figure 10.1: The canonical sequence of Class-A configurations; c, d and e are push-pull variants, and achieve 50% efficiency; e is simply a Class-B stage with higher Vbias.

VAS distortion cannot be separated from output stage nonlinearity. There is also no guarantee that upper and lower output devices will be driven appropriately for Class-A; in Linsley-Hood[5] the effective quiescent varies by more than 40% over the cycle.

The second push-pull method in Figure 10.1d is more dependable, and I have designed several versions that worked well. The disadvantage with the simple version shown is that a regulated supply is required to prevent rail ripple from disrupting the current-loop control. Designs of this type have a limited current-control range—in Figure 10.1d TR3 cannot be turned on any further once the upper device is fully off—so the lower VCIS will not be able to respond to an unforeseen increase in the output loading. In this event there is no way of resorting to Class-AB to keep the show going and the amplifier will show some form of asymmetrical hard clipping.

The best push-pull stage seems to be that in Figure 10.1e, which probably looks rather familiar. Like all the conventional Class-B stages, this one will operate effectively in pure push-pull Class-A if the quiescent bias voltage is sufficiently increased; the increment over Class-B is typically 700 mV, depending on the value of the emitter resistors. For an example of high-biased Class B see Nelson-Jones[7]. This topology has the great advantage that, when confronted with an unexpectedly low load impedance, it will operate in Class-AB. The distortion performance will be inferior not only to Class-A but also to optimally biased Class-B, once above the AB transition level, but can still be made very low by proper design.

The push-pull concept has a maximum efficiency of 50%, but this is only achieved at maximum sinewave output; due to the high peak/average ratio of music, the true average efficiency probably does not exceed 10%, even at maximum volume before obvious clipping.

Other possibilities are signal-controlled variation of the Class-A amplifier rail voltages, either by a separate Class-B amplifier or by a modulated switch-mode supply. Both approaches are capable of high power output, but involve extensive extra circuitry, and present some daunting design problems.

A Class-B amplifier has a limited voltage output capability, but is flexible about load impedances; more current is simply turned on when required. However, Class-A has also a current limitation, after which it enters Class AB, and so loses its raison d'être. The choice of quiescent value has a major effect on thermal design and parts cost; so Class-A design demands a very clear idea of what load impedance is to be driven in pure A before we begin. The calculations to determine the required Iq are straightforward, though lengthy if supply ripple, Vce(sat)s, and Re losses, etc. are all considered, so I just give the results here. (An unregulated supply with 10,000 μF reservoirs is assumed.)

A 20 W/8Ω amplifier will require rails of approximately ±24 V and a quiescent of 1.15 A. If this is extended to give roughly the same voltage swing into 4Ω, then the output power becomes 37 W, and to deliver this in Class-A the quiescent must increase to 2.16 A, almost doubling dissipation. If however full voltage swing into 6Ω will do (which it will for many reputable speakers) then the quiescent only needs to increase to 1.5 A; from here on I assume a quiescent of 1.6 A to give a margin of safety.

10.3 Output Stages in Class A

I consider here only the increased-bias Class-B topology, because it is probably the best approach, effectively solving the problems presented by the other methods. Figure 10.2 shows a Spice simulation of the collector currents in the output devices versus output voltage, and also the sum of these currents. This sum of device currents is in principle constant in Class-A,

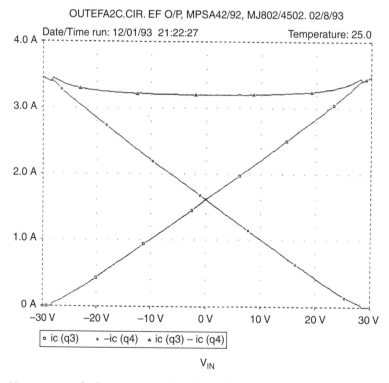

Figure 10.2: How output device current varies in push-pull Class-A. The sum of the currents is near-constant, simplifying biasing.

though it need not be so for low THD; the output signal is the difference of device currents, and is not inherently related to the sum. However, a large deviation from this constant-sum condition means increased inefficiency, as the stage must be conducting more current than it needs to for some part of the cycle.

The constancy of this sum-of-currents is important because it shows that the voltage measured across Re1 and Re2 together is also effectively constant so long as the amplifier stays in Class-A. This in turn means that quiescent current can be simply set with a constant-voltage bias generator, in very much the same way as Class-B.

Figures 10.3, 10.4 and 10.5, show Spice gain plots for open-loop output stages, with 8Ω loading and 1.6 A quiescent. The upper traces show Class-A gain, and the lower traces optimal-bias Class-B gain for comparison. Figure 10.3 shows an emitter-follower output, Figure 10.4 a simple quasi-complementary stage, and Figure 10.5 a CFP output.

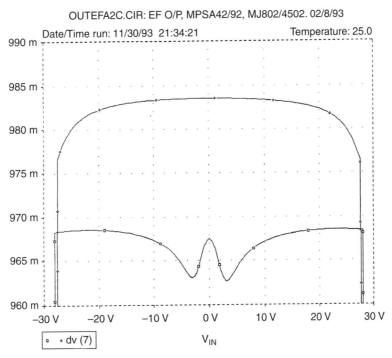

Figure 10.3: Gain linearity of the Class-A emitter-follower output stage. Load is 8Ω, and quiescent current (Iq) is 1.6 A.

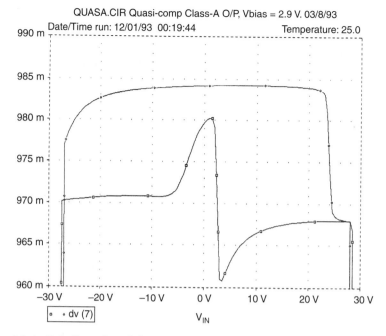

Figure 10.4: Gain linearity of the Class-A quasi-complementary output stage. Conditions as Figure 10.3

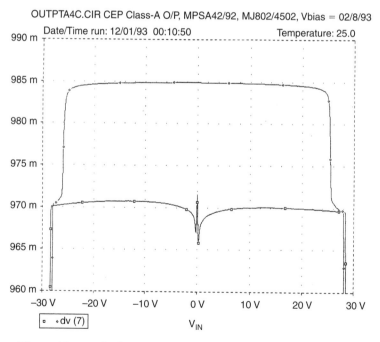

Figure 10.5: Gain linearity of the Class-A CFP output stage

Table 10.1: Harmonic and THD figures for the three configurations

Harmonic	Emitter Follower (%)	Quasi-Comp (%)	CFP Output (%)
Second	0.00012	0.0118	0.00095
Third	0.0095	0.0064	0.0025
Fourth	0.00006	0.0011	0.00012
Fifth	0.00080	0.00058	0.00029
THD	0.0095	0.0135	0.0027
(THD is calculated from the first nine harmonics, though levels above the fifth are very small)			

We would expect Class-A stages to be more linear than B, and they are. (Harmonic and THD figures for the three configurations, at 20 V Pk, are shown in Table 10.1.) There is absolutely no gain wobble around 0 V, as in Class-B, and push-pull Class-A really can and does cancel even-order distortion.

It is at once clear that the emitter-follower has more gain variation, and therefore worse linearity, than the CFP, while the quasi-comp circuit shows an interesting mix of the two. The more curved side of the quasi gain plot is on the −ve side, where the CFP half of the quasi circuit is passing most of the current; however, we know by comparing Figure 10.3 and Figure 10.5 that the CFP is the more linear structure. Therefore it appears that the shape of the gain curve is determined by the output half that is turning off, presumably because this shows the biggest gm changes. The CFP structure maintains gm better as current decreases, and so gives a flatter gain curve with less rounding of the extremes.

The gain behavior of these stages is reflected in their harmonic generation; Table 10.1 reveals that the two symmetrical topologies give mostly odd-order harmonics, as expected. The asymmetry of the quasi-comp version causes a large increase in even-order harmonics, and this is reflected in the higher THD figure. Nonetheless all the THD figures are still 2 to 3 times lower than for their Class-B equivalents.

This modest factor of improvement may seem a poor return for the extra dissipation of Class-A, but not so. The crucial point about the distortion from a Class-A output stage is not just that it is low in magnitude, but that it is low-order, and so benefits much more from the typical NFB factor that falls with frequency than does high-order crossover distortion.

The choice of Class-A output topology is now simple. For best performance, use the CFP; apart from greater basic linearity, the effects of output device temperature on Iq are servoed-out by local feedback, as in Class B. For utmost economy, use the quasi-complementary with two NPN devices; these need only a low Vce(max) for a typical Class-A amp, so here is an opportunity to recoup some of the money spent on heatsinking. The rules here are somewhat different from Class-B; the simple quasi-complementary configuration gives first-class results with moderate NFB, and adding a Baxandall diode to simulate a complementary emitter-follower stage gives little improvement in linearity. See however Nelson-Jones[7] for an example of its use.

It is sometimes assumed that the different mode of operation of Class-A makes it inherently short-circuit proof. This may be true with some configurations, but the high-biased type studied here will continue delivering current in time-honored Class-B fashion until it bursts, and overload protection seems to be no less essential.

10.4 Quiescent Current Control Systems

Unlike Class-B, precise control of quiescent current is not required to optimize distortion; for good linearity there just has to be enough of it. However, the Iq must be under some control to prevent thermal runaway, particularly if the emitter-follower output is used. A badly designed quiescent controller can ruin the linearity, and careful design is required. There is also the point that a precisely held standing-current is considered the mark of a well-bred Class-A amplifier; a quiescent that lurches around like a drunken sailor does not inspire confidence.

Straightforward thermal compensation with a Vbe-multiplier bias generator works[8], and will prevent thermal runaway. However, unlike Class-B, Class-A gives the opportunity of tightly controlling Iq by negative feedback. This is profoundly ironic because now that we can precisely control Iq, it is no longer critical. Nevertheless it seems churlish to ignore the opportunity, and so feedback quiescent control will be examined.

There are two basic methods of feedback current-control. In the first, the current in one output device is monitored, either by measuring the voltage across one emitter-resistor (Rs in Figure 10.6a), or by a collector sensing resistor; the second method monitors the sum of the device currents, which as described above, is constant in Class-A.

The first method as implemented in Figure 10.6a[7] compares the Vbe of TR4 with the voltage across Rs, with filtering by RF, CF. If quiescent is excessive, then TR4 conducts more, turning on TR5 and reducing the bias voltage between points A and B. In Figure 10.6b, which uses the VCIS approach, the voltage across collector sensing resistor Rs is compared with Vref by TR4, the value of Vref being chosen to allow for TR4 Vbe[9]. Filtering is once more by RF, CF.

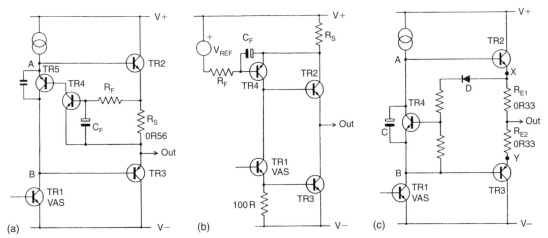

Figure 10.6: Current-control systems. Only that at c avoids the need to low-pass filter the control signal; C simply provides feedforward to speed up signal transfer to TR2.

For either Figure 10.6a or b, the current being monitored contains large amounts of signal, and must be low-pass filtered before being used for control purposes. This is awkward as it adds one more time-constant to worry about if the amplifier is driven into asymmetrical clipping. In the case of collector-sensing there are unavoidable losses in the extra sense resistor. It is also my experience that imperfect filtering causes a serious rise in LF distortion.

The Better Way is to monitor current in both emitter resistors; as explained above, the voltage across both is very nearly constant, and in practice filtering is unnecessary. An example of this approach is shown in Figure 10.6c, based on a concept originated by Nelson Pass[10]. Here TR4 compares its own Vbe with the voltage between X and B; excessive quiescent turns on TR4 and reduces the bias directly. Diode D is not essential to the concept, but usefully increases the current-feedback loop-gain; omitting it more than doubles Iq variation with TR7 temperature in the Pass circuit.

The trouble with this method is that TR3 Vbe directly affects the bias setting, but is outside the current-control loop. A multiple of Vbe is established between X and B, when what we really want to control is the voltage between X and Y. The temperature variations of TR4 and TR3 Vbe partly cancel, but only partly. This method is best used with a CFP or quasi output so that the difference between Y and B depends only on the driver temperature, which can be kept low. The reference is TR4 Vbe, which is itself temperature-dependent; even if it is kept away from the hot bits it will react to ambient temperature changes, and this explains the poor performance of the Pass method for global temp changes (Table 10.2).

Table 10.2: Iq change per degree C

	Changing TR7 temp only (%)	Changing Global temp (%)
Quasi+Vbe-mult	+0.112	−0.43
Pass: as Fig. 10.6c	+0.0257	−14.1
Pass: no diode D	+0.0675	−10.7
New system:	+0.006	−0.038
(assuming OR22 emitter resistors and 1.6 A Iq)		

10.5 A Novel Quiescent Current Controller

To solve this problem, I would like to introduce the novel control method in Figure 10.7. We need to compare the floating voltage between X and Y with a fixed reference, which sounds like a requirement for two differential amplifiers. This can be reduced to one by sitting the reference Vref on point Y; this is a very low-impedance point and can easily swallow a reference current of 1 mA or so. A simple differential pair TR15, 16 then compares the reference voltage with that at point Y; excess quiescent turns on TR16, causing TR13 to conduct more and reducing the bias voltage.

The circuitry looks enigmatic because the high-impedance of TR13 collector would seem to prevent the signal from reaching the upper half of the output stage; this is in essence true, but the vital point is that TR13 is part of an NFB loop that establishes a voltage at A that will keep the bias voltage between A and B constant. This comes to the same thing as maintaining a constant Vbias across TR5. As might be imagined, this loop does not shine at transferring signals quickly, and this duty is done by feed-forward capacitor C4. Without it, the loop (rather surprisingly) works correctly, but HF oscillation at some part of the cycle is almost certain. With C4 in place the current-loop does not need to move quickly, since it is not required to transfer signal but rather to maintain a DC level.

The experimental study of Iq stability is not easy because of the inaccessibility of junction temperatures. Professional SPICE implementations like PSpice allow both the global circuit temperature and the temperature of individual devices to be manipulated; this is another aspect where simulators shine. The exact relationships of component temperatures in an amplifier is hard to predict, so I show here only the results of changing the global temperature of all devices, and changing the junction temp of TR7 alone (Figure 10.7) with different current-controllers. TR7 will be one of the hottest transistors and unlike TR9 it is not in a local NFB loop, which would greatly reduce its thermal effects.

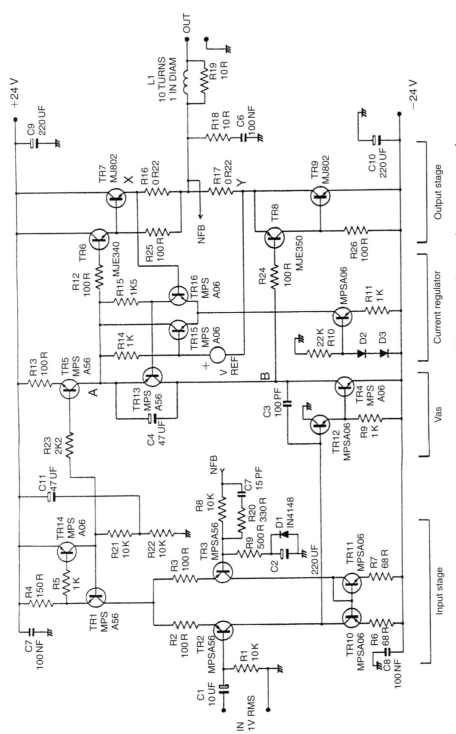

Figure 10.7: A Blameless 20W Class-A power amplifier, using the novel current-control system

10.6 A Class-A design

A design example of a Blameless 20 W/8Ω Class-A power amplifier is shown in Figure 10.7. This is as close as possible in operating parameters to the previous Class-B design, to aid comparison; in particular the NFB factor remains 30 dB at 20 kHz. The front-end is as for the Class-B version, which should not be surprising as it does exactly the same job, input Distortion 1 being unaffected by output topology. As before the input pair uses a high tail current, so that R2, 3 can be introduced to linearize the transfer characteristic and set the transconductance. Distortion 2 (VAS) is dealt with as before, the beta-enhancer TR12 increasing the local feedback through Cdom. There is no need to worry about Distortion 4 (nonlinear loading by output stage) as the input impedance of a Class-A output, while not constant, does not have the sharp variations shown by Class-B.

Figure 10.7 uses a standard quasi output. This may be replaced by a CFP stage without problems. In both cases the distortion is extremely low, but gratifyingly the CFP proves even better than the quasi, confirming the simulation results for output stages in isolation.

The operation of the current regulator TR13, 15, 16 has already been described. The reference used is a National LM385/1.2. Its output voltage is fixed at 1.223V nominal; this is reduced to approximately 0.6 V by a 1 k–1 k divider (not shown). Using this band-gap reference, a 1.6 A Iq is held to within ±2 mA from a second or two after switch-on. Looking at Table 10.2, there seems no doubt that the new system is effective.

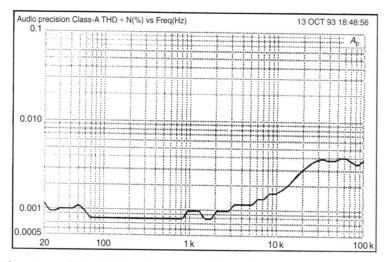

Figure 10.8: Class-A amplifier THD performance with quasi-comp output stage. The steps in the LF portion of the trace are measurement artifacts.

As before, a simple unregulated power supply with 10,000 μF reservoirs was used, and despite the higher prevailing ripple, no PSRR difficulties were encountered once the usual decoupling precautions were taken.

The closed-loop distortion performance (with conventional compensation) is shown in Figure 10.8 for the quasi-comp output stage, and in Figure 10.9 for a CFP output version. The THD residual is pure noise for almost all of the audio spectrum, and only above 10 kHz do small amounts of third-harmonic appear. The expected source is the input pair, but this so far remains unconfirmed.

The distortion generated by the Class-B and A design examples is summarized in Table 10.3, which shows a pleasing reduction as various measures are taken to deal with it. As a final

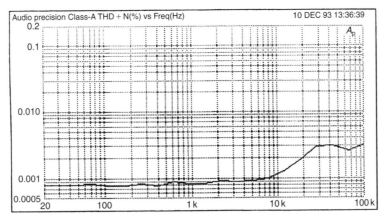

Figure 10.9: Class-A distortion performance with CFP output stage

Table 10.3: Summary of distortion generated by Class-B and Class-A design examples

	1 kHz (%)	10 kHz (%)	20 kHz (%)	Power (W)
Class B EF	<0.0006	0.0060	0.012	50
Class B CFP	<0.0006	0.0022	0.0040	50
Class B EF 2-pole	<0.0006	0.0015	0.0026	50
Class A quasi	<0.0006	0.0017	0.0030	20
Class A CFP	<0.0006	0.0010	0.0018	20
Class A CFP 2-pole	<0.0006	0.0010	0.0012	20
(All for 8 Ω loads and 80 kHz bandwidth. Single-pole compensation unless otherwise stated.)				

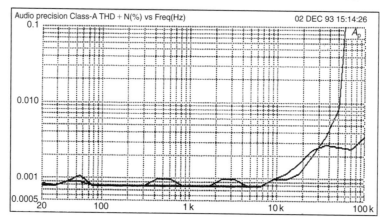

Figure 10.10: Distortion performance for CFP output stage with 2-pole compensation. The THD drops to 0.0012% at 20 kHz, but the extra VAS loading has compromised the positive-going slew capability.

fling two-pole compensation was applied to the most linear (CFP) of the Class-A versions, reducing distortion to a rather small 0.0012% at 20 kHz, at some cost in slew-rate. (Figure 10.10). While this may not be the fabled Distortionless Amplifier, it must be a near relation.

10.7 The Trimodal Amplifier

I present here my own contribution to global warming in the shape of an improved Class-A amplifier; it is believed to be unique in that it not only copes with load impedance dips by means of the most linear form of Class-AB possible, but will also operate as a Blameless Class-B engine. The power output in pure Class-A is 20 to 30 W into 8 Ω, depending on the supply-rails chosen.

This amplifier uses a Complementary-Feedback Pair (CFP) output stage for best possible linearity, and some incremental improvements have been made to noise, slew-rate and maximum DC offset. The circuit naturally bears a very close resemblance to a Blameless Class-B amplifier, and so it was decided to retain the Class-B Vbe-multiplier, and use it as a safety-circuit to prevent catastrophe if the relatively complex Class-A current-regulator failed. From this the idea arose of making the amplifier instantly switchable between Class-A and Class-B modes, which gives two kinds of amplifier for the price of one, and permits of some interesting listening tests. Now you really can do an A/B comparison.

In the Class-B mode the amplifier has the usual negligible quiescent dissipation. In Class-A the thermal dissipation is naturally considerable, as true Class-A operation is extended down to 6 Ω resistive loads for the full output voltage swing, by suitable choice of the quiescent current; with heavier loading the amplifier gracefully enters Class-AB, in which it will give full output down to 3Ω before the Safe-Operating-Area (SOAR) limiting begins to act. Output into 2Ω is severely curtailed, as it must be with one output pair, and this kind of load is definitely not recommended.

In short, the amplifier allows a choice between:

1. being very linear all the time (Blameless Class-B) and

2. ultra-linear most of the time (Class-A) with occasional excursions into Class-AB. The AB mode is still extremely linear by current standards, though inherently it can never be quite as good as properly handled Class-B. Since there are three classes of operation I have decided to call the design a trimodal power amplifier.

It is impossible to be sure that you have read all the literature; however, to the best of my knowledge this is the first ever trimodal amplifier.

As previously said, designing a low-distortion Class-A amplifier is in general a good deal simpler than the same exercise for Class-B, as all the difficulties of arranging the best possible crossover between the output devices disappear. Because of this it is hard to define exactly what Blameless means for a Class-A amplifier. In Class-B the situation is quite different, and Blameless has a very specific meaning; when each of the eight or more distortion mechanisms has been minimized in effect, there always remains the crossover distortion inherent in Class-B, and there appears to be no way to reduce it without departing radically from what might be called the generic Lin amplifier configuration. Therefore the Blameless state appears to represent some sort of theoretical limit for Class-B, but not for Class-A.

However, Class-B considerations cannot be ignored, even in a design intended to be Class-A only, because if the amplifier does find itself driving a lower load impedance than expected, it will move into Class-AB, and then all the additional Class-B requirements are just as significant as for a Class-B design proper. Class-AB can never give distortion as low as optimally biased Class-B, but it can be made to approach it reasonably closely, if the extra distortion mechanisms are correctly handled.

In a class-A amplifier, certain sacrifices are made in the name of quality, and so it is reasonable not to be satisfied with anything less than the best possible linearity. The amplifier described here therefore uses the Complementary-Feedback Pair (CFP) type of output stage, which has the lowest distortion due to the local feedback loops wrapped around the output devices. It also has the advantage of better output efficiency than the emitter-follower (EF) version, and inherently superior quiescent current stability. It will shortly be seen that these are both important for this design.

Half-serious thought was given to labelling the Class-A mode Distortionless as the THD is completely unmeasurable across most of the audio band.

However, detectable distortion products do exist above 10 kHz, so this provocative idea was regretfully abandoned.

It seemed appropriate to take another look at the Class-A design, to see if it could be inched a few steps nearer perfection. The result is a slight improvement in efficiency, and a 2 dB improvement in noise performance. In addition the expected range of output DC offset has been reduced from ±50 mV to ±15 mV, still without any adjustment.

10.8 Load Impedance and Operating Mode

The amplifier is 4 Ω capable in both A/AB and B operating modes, though it is the nature of things that the distortion performance is not quite so good. All solid-state amplifiers (without qualification, as far as I am aware) are much happier with an 8 Ω load, both in terms of linearity and efficiency; loudspeaker designers please note. With a 4 Ω load, Class-B operation gives better THD than Class-A/AB, because the latter will always be in AB mode, and therefore generating extra output stage distortion through gm-doubling. (Which should really be called gain-deficit-halving, but somehow I do not see this term catching on.) These not entirely obvious relationships are summarized in Table 10.4.

Figure 10.11 attempts to show diagrammatically just how power, load resistance, and operating mode are related. The rails have been set to ±20 V, which just allows 20 W into 8 Ω in Class-A. The curves are lines of constant power (i.e., V × I in the load), the upper horizontal line represents maximum voltage output, allowing for Vce(sat)s, and the sloping line on the right is the SOAR protection locus; the output can never move outside this area in either mode. The intersection between the load resistance lines sloping up from the origin and the ultimate limits of voltage-clip and SOAR protection define which of the curved constant-power lines is reached.

Table 10.4: Distortion and dissipation for different output stage classes

Load (Ω)	Mode	Distortion	Dissipation
8	A/AB	Very low	High
4	A/AB	High	High
8	B	Low	Low
4	B	Medium	Medium
(Note: High distortion in the context of this sort of amplifier means about 0.002% THD at 1 kHz and 0.01% at 10 kHz)			

In A/AB mode, the operating point must be left of the vertical push-pull current-limit line (at 3 A, twice the quiescent current) for Class-A. If we move to the right of this limit along one of the impedance lines, the output devices will begin turning off for part of the cycle; this is the AB operation zone. In Class-B mode, the 3 A line has no significance and the amplifier remains in optimal Class-B until clipping or SOAR limiting occurs. Note that the diagram axes represent instantaneous power in the load, but the curves show sinewave RMS power, and that is the reason for the apparent factor-of-two discrepancy between them.

10.9 Efficiency

Concern for efficiency in Class-A may seem paradoxical, but one way of looking at it is that Class-A watts are precious things, wrought in great heat and dissipation, and so for a given quiescent power it makes sense to ensure that the amplifier approaches its limited theoretical

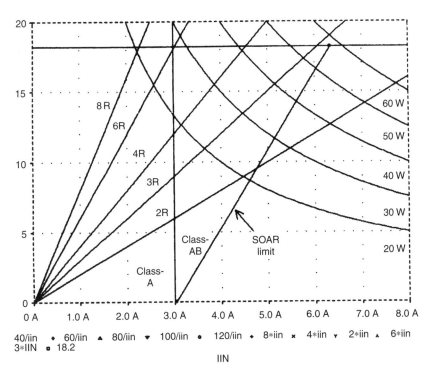

Figure 10.11: The relationships between load, mode, and power output. The intersection between the sloping load resistance lines and the ultimate limits of voltage-clipping and SOAR protection define which of the curved constant-power lines is reached. In A/AB mode, the operating point must be to the left of the vertical push-pull current-limit line for true Class-A.

efficiency as closely as possible. I was confirmed in this course by reading of another recent design[11] which seems to throw efficiency to the winds by using a hybrid BJT/FET cascode output stage. The voltage losses inherent in this arrangement demand ± 50 V rails and six-fold output devices for a 100 W Class-A capability; such rail voltages would give 156 W from a 100% efficient amplifier.

The voltage efficiency of a power amplifier is the fraction of the supply-rail voltage which can actually be delivered as peak-to-peak voltage swing into a specified load; efficiency is invariably less into 4Ω due to the greater resistive voltage drops with increased current.

The Class-B amplifier described earlier has a voltage efficiency of 91.7% for positive swings, and 92.5% for negative, into 8Ω. Amplifiers are not in general completely symmetrical, and so two figures need to be quoted; alternatively the lower of the two can be given as this defines the maximum undistorted sinewave. These figures above are for an emitter-follower output stage, and a CFP output does better, the positive and negative efficiencies being 94.0% and 94.7%, respectively. The EF version gives a lower output swing because it has two more Vbe drops in series to be accommodated between the supply-rails; the CFP is always more voltage-efficient, and so selecting it over the EF for the current Class-A design is the first step in maximising efficiency.

Figure 10.12 shows the basic CFP output stage, together with its two biasing elements. In Class-A the quiescent current is rigidly controlled by negative-feedback; this is possible because in Class-A the total voltage across both emitter resistors Re is constant throughout the cycle. In Class-B this is not the case, and we must rely on thermal feedback from the output stage, though to be strictly accurate this is not feedback at all, but a kind of feedforward. Another big advantage of the CFP configuration is that Iq depends only on driver temperature, and this is important in the Class-B mode, where true feedback control of quiescent current is not possible, especially if low-value Re's such as 0.1Ω, are chosen, rather than the more usual 0.22Ω; the motivation for doing this will soon become clear.

The voltage efficiency for the quasi-complementary Class-A circuit of the circuit in Figure 10.7 into 8Ω is 89.8% positive and 92.2% negative. Converting this to the CFP output stage increases this to 92.9% positive and 93.6% negative. Note that a Class-A quiescent current (Iq) of 1.5 A is assumed throughout; this allows 31 W into 8Ω in push-pull, if the supply-rails are adequately high. However the assumption that loudspeaker impedance never drops below 8Ω is distinctly doubtful, to put it mildly, and so as before this design allows for full Class-A output voltage swing into loads down to 6Ω.

So how else can we improve efficiency? The addition of extra and higher supply-rails for the small-signal section of the amplifier surprisingly does not give a significant increase in

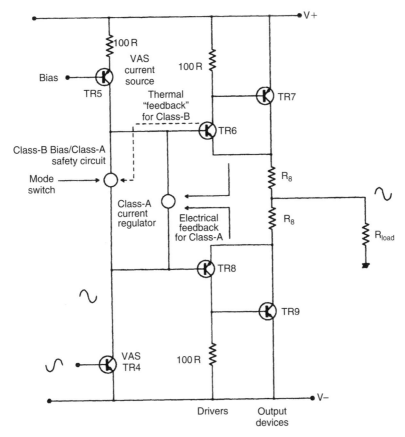

Figure 10.12: The basic CFP output stage, equally suited to operating Class B, AB and A, depending on the magnitude of Vbias. The emitter resistors Re may be from 0.1 to 0.47Ω.

output; examination of Figure 10.13 shows why. In this region, the output device TR6 base is at a virtually constant 880 mV from the V+ rail, and as TR7 driver base rises it passes this level, and keeps going up; clipping has not yet occurred. The driver emitter follows the driver base up, until the voltage difference between this emitter and the output base (i.e., the driver Vce) becomes too small to allow further conduction; this choke point is indicated by the arrows A–A. At this point the driver base is forced to level off, although it is still about 500 mV below the level of V+. Note also how the voltage between V+ and TR5 emitter collapses. Thus a higher rail will give no extra voltage swing, which I must admit came as something of a surprise. Higher sub-rails for small-signal sections only come into their own in FET amplifiers, where the high Vgs for FET conduction (5 V or more) makes their use almost mandatory.

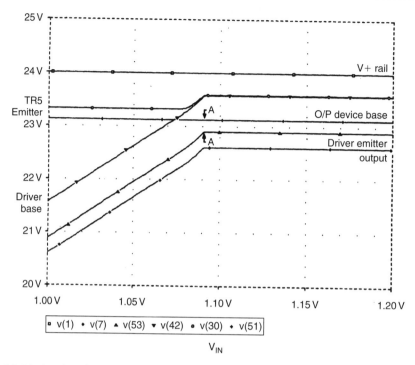

Figure 10.13: PSpice simulation showing how positive clipping occurs in the CFP output. A higher subrail for the VAS cannot increase the output swing, as the limit is set by the minimum driver Vce, and not the VAS output swing.

The efficiency figures given so far are all greater for negative rather than positive voltage swings. The approach to the rail for negative clipping is slightly closer because there is no equivalent to the 0.6 V bias established across R13; however, this advantage is absorbed by the need to lose a little voltage in the RC filtering of the V− supply to the current-mirror and VAS. This is essential if really good ripple/hum performance is to be obtained.

In the quest for efficiency, an obvious variable is the value of the output emitter resistors Re. The performance of the current-regulator described, especially when combined with a CFP output stage, is more than good enough to allow these resistors to be reduced while retaining first-class Iq stability. I took $0.1\,\Omega$ as the lowest practicable value, and even this is comparable with PCB track resistance, so some care in the exact details of physical layout is essential; in particular the emitter resistors must be treated as four-terminal components to exclude unwanted voltage drops in the tracks leading to the resistor pads.

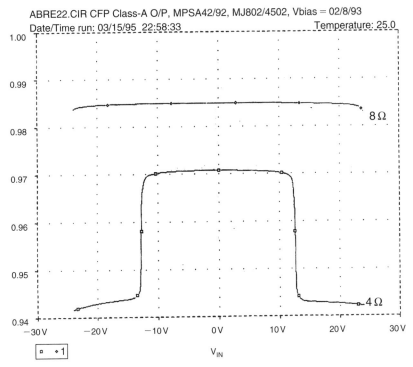

Figure 10.14: CFP output stage linearity with Re = OR22. Upper trace is Class-A into 8Ω, lower is Class-AB operation into 4Ω, showing step changes in gain of 0.024 units.

If Re is reduced from 0.22Ω to 0.1Ω then voltage efficiency improves from 92.9%/93.6%, to 94.2%/95.0%. Is this improvement worth having? Well, the voltage-limited power output into 8Ω is increased from 31.2 to 32.2 W with ±24 V rails, at zero cost, but it would be idle to pretend that the resulting increase in SPL is highly significant; it does however provide the philosophical satisfaction that as much Class-A power as possible is being produced for a given dissipation; a delicate pleasure.

The linearity of the CFP output stage in Class-A is very slightly worse with 0.1Ω emitter resistors, though the difference is small and only detectable open-loop; the simulated THD for 20 V pk–pk into 8Ω is only increased from 0.0027% to 0.0029%. This is probably due simply to the slightly lower total resistance seen by the output stage.

However, at the same time, reducing the emitter resistors to 0R1 provides much lower distortion when the amplifier runs out of Class-A; it halves the size of the step gain changes inherent in Class-AB, and so effectively reduces distortion into 4Ω loads. See Figures 10.14

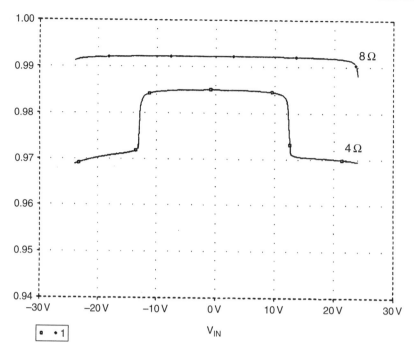

Figure 10.15: CFP output linearity with Re = OR1, rebiased to keep Iq at 1.5 A. There is slightly poorer linearity in the flat-topped Class-A region than for Re = OR22, but the 4Ω AB steps are halved in size at 0.012 units. Note that both gains are now closer to unity; same scale as Figure 10.14.

and 10.15 for output linearity simulations; the measured results from a real and Blameless Trimodal amplifier are shown in Figure 10.16, where it can be clearly seen that THD has been halved by this simple change. To the best of my knowledge this is a new result; if you must work in Class-AB, then keep the emitter resistors as low as possible, to minimize the gain changes.

Having considered the linearity of Class-A and AB, we must not neglect what effect this radical Re change has on Class-B linearity. The answer is, not very much; see Figure 10.17, where crossover distortion seems to be slightly higher with Re = 0.2Ω than for either 0.1 or 0.4Ω. Whether this is a consistent effect (for CFP stages anyway) remains to be seen.

The detailed mechanisms of bias control and mode-switching are described in the following sections.

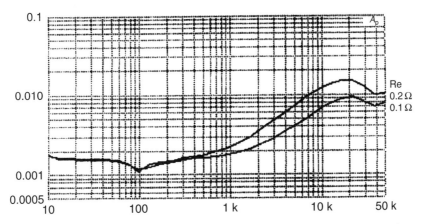

Figure 10.16: Distortion in Class-AB is reduced by lowering the value of Re.

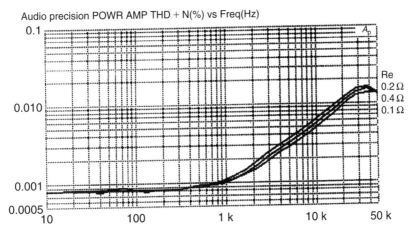

Figure 10.17: Proving that emitter resistors matter much less in Class-B. Output was 20 W 8Ω, with optimal bias. Interestingly, the bias does not need adjusting as the value of Re changes.

10.10 On Trimodal Biasing

Figure 10.18 shows a simplified rendering of the Trimodal biasing system; the full version appears in Figure 10.19. The voltage between points A and B is determined by one of two controller systems, only one of which can be in command at a time. Since both are basically shunt voltage regulators sitting between A and B, the result is that the lowest voltage wins. The novel Class-A current-controller introduced in section 10.5 is used here adapted for 0.1Ω emitter resistors, mainly by reducing the reference voltage to 300 mV, which gives a quiescent current (Iq) of 1.5 A when established across the total emitter resistance of 0.2Ω.

In parallel with the current-controller is the Vbe-multiplier TR13. In Class-B mode, the current-controller is disabled, and critical biasing for minimal crossover distortion is provided in the usual way by adjusting preset PR1 to set the voltage across TR13. In Class-A/AB mode, the voltage TR13 attempts to establish is increased (by shorting out PR1) to a value greater than that required for Class-A. The current-controller therefore takes charge of the voltage between X and Y, and unless it fails TR13 does not conduct. Points A, B, X, and Y are the same circuit nodes as in the simple Class-A design (see Figure 10.6c).

10.11 Class-A/AB Mode

In Class-A/AB mode, the current-controller (TR14, 15, 16 in Figure 10.18) is active and TR13 is off, as TR20 has shorted out PR1. TR15, 16 form a simple differential amplifier that compares the reference voltage across R31 with the Vbias voltage across output emitter resistors R16 and R17; as explained above, in Class-A this voltage remains constant despite

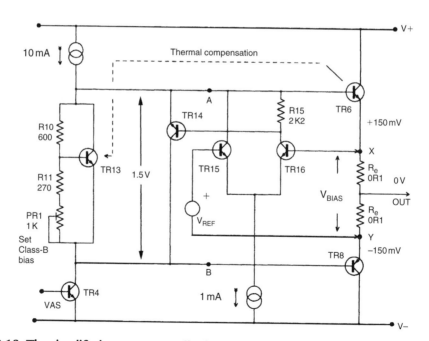

Figure 10.18: The simplified current-controller in action, showing typical DC voltages in Class-A. Points A, B, X and Y are in Figure 10.6.

delivery of current into the load. If the voltage across R16, 17 tends to rise, then TR16 conducts more, turning TR14 more on and reducing the voltage between A and B. TR14, 15 and 16 all move up and down with the amplifier output, and so a tail current-source (TR17) is used.

I am very aware that the current-controller is more complex than the simple Vbe-multiplier used in most Class-B designs. There is an obvious risk that an assembly error could cause a massive current that would prompt the output devices to lay down their lives to save the rail fuses. The tail-source TR17 is particularly vulnerable because any fault that extinguishes the tail-current removes the drive to TR14, the controller is disabled, and the current in the output stage will be very large. In Figure 10.18 the Vbe-multiplier TR13 acts as a safety-circuit which limits Vbias to about 600 mV rather than the normal 300 mV, even if the current-controller is completely non-functional and TR14 fully off. This gives a quiescent of 3.0 A, and I can testify this is a survivable experience for the output devices in the short-term; however they may eventually fail from overheating if the condition is allowed to persist.

There are some important points about the current-controller. The entire tail-current for the error-amplifier, determined by TR17, is syphoned off from VAS current source TR5, and must be taken into account when ensuring that the upper output half gets enough drive current.

There must be enough tail-current available to turn on TR14, remembering that most of TR16 collector-current flows through R15, to keep the pair roughly balanced. If you feel moved to alter the VAS current, remember also that the base current for driver TR6 is higher in Class-A than Class-B, so the positive slew-rate is slightly reduced in going from Class-A to B.

The original Class-A amplifier used a National LM385/1.2, its output voltage fixed at 1.223 V nominal; this was reduced to approximately 0.6 V by a 1k–1k divider. The circuit also worked well with Vref provided by a silicon diode, 0.6 V being an appropriate Vbias drop across two 0.22 Ω output emitter resistors. This is simple, and retains the immunity of Iq to heatsink and output device temperatures, but it does sacrifice the total immunity to ambient temperature that a band-gap reference gives.

The LM385/1.2 is the lowest voltage band-gap reference commonly available; however, the voltages shown in Figure 10.18 reveal a difficulty with the new lower Vbias value and the CFP stage; points A and Y are now only 960 mV apart, which does not give the reference room to work in if it is powered from node A, as in the original circuit. The solution is to power the reference from the V+ rail, via R42 and R43. The mid-point of these two resistors is bootstrapped from the amplifier output rail by C5, keeping the voltage across R43 effectively constant. Alternatively, a current-source could be used, but this might reduce positive headroom. Since there is no longer a strict upper limit on the reference voltage, a more easily obtainable 2.56 V device could be used providing R30 is suitably increased to 5k to maintain Vref at 300 mV across R31.

In practical use, Iq stability is very good, staying within 1% for long periods. The most obvious limitation on stability is differential heating of TR15, 16 due to heat radiation from the main heatsink. TR14 should also be sited with this in mind, as heating it will increase its beta and slightly imbalance TR15, 16.

10.12 Class-B Mode

In Class-B mode, the current-controller is disabled, by turning off tail-source TR17 so TR14 is firmly off, and critical biasing for minimal crossover distortion is provided as usual by Vbe-multiplier TR13. With 0.1Ω emitter resistors Vbias (between X and Y) is approximately 10 mV. I would emphasize that in Class-B this design, if constructed correctly, will be as Blameless as a purpose-built Class-B amplifier. No compromises have been made in adding the mode-switching.

As in the previous Class-B design, the addition of R14 to the Vbe-multiplier compensates against drift of the VAS current-source TR5. To make an old but much-neglected point, the preset should always be in the bottom arm of the Vbe divider R10, 11, because when presets fail it is usually by the wiper going open; in the bottom arm this gives minimum Vbias, but in the upper it would give maximum.

In Class-B, temperature compensation for changes in driver dissipation remains vital. Thermal runaway with the CFP is most unlikely, but accurate quiescent setting is the only way to minimize cross-over distortion. TR13 is therefore mounted on the same small heatsink as driver TR6. This is often called thermal feedback, but it is no such thing as TR13 in no way controls the temperature of TR6; thermal feedforward would be a more accurate term.

10.13 The Mode-switching System

The dual nature of the biasing system means Class-A/Class-B switching can be implemented fairly simply. A Class-A amplifier is an uneasy companion in hot weather, and so I have been unable to resist the temptation to subtitle the mode switch Summer/Winter, by analogy with a car air intake.

The switchover is DC-controlled, as it is not desirable to have more signal than necessary running around inside the box, possibly compromising interchannel crosstalk. In Class-A/AB mode, SW1 is closed, so TR17 is biased normally by D5, 6, and TR20 is held on via R33, shorting out preset PR1 and setting TR13 to safety mode, maintaining a maximum Vbias limit of 600 mV. For Class-B, SW1 is opened, turning off TR17 and therefore TR15, 16 and 14. TR20 also ceases to conduct, protected against reverse-bias by D9, and reduces the voltage set by TR13 to a suitable level for Class-B. The two control pins of a stereo amplifier can be connected together, and the switching performed with a single-pole switch, without interaction or increased crosstalk.

The mode-switching affects the current flowing in the output devices, but not the output voltage, which is controlled by the global feedback loop, and so it is completely silent in operation. The mode may be freely switched while the amplifier is handling audio, which allows some interesting A/B listening tests.

It may be questioned why it is necessary to explicitly disable the current controller in Class-B; TR13 is establishing a lower voltage than the current controller which latter subsystem will therefore turn TR14 off as it strives futilely to increase Vbias. This is true for 8 Ω loads, but 4 Ω impedances increase the currents flowing in R16, 17 so they are transiently greater than the Class-A Iq, and the controller will therefore intermittently take control in an attempt to reduce the average current to 1.5 A. Disabling the controller by turning off TR17 via R44 prevents this.

If the Class-A controller is enabled, but the preset PR1 is left in circuit (e.g., by shorting TR20 base-emitter) we have a test mode which allows suitably cautious testing; Iq is zero with the preset fully down, as TR13 overrides the current-controller, but increases steadily as PR1 is advanced, until it suddenly locks at the desired quiescent current. If the current-controller is faulty, then Iq continues to increase to the defined maximum of 3.0 A.

10.14 Thermal Design

Class-A amplifiers are hot almost by definition, and careful thermal design is needed if they are to be reliable, and not take the varnish off the Sheraton. The designer has one good card to play; since the internal dissipation of the amplifier is maximal with no signal, simply turning on the prototype and leaving it to idle for several hours will give an excellent idea of worst-case component temperatures. In Class-B the power dissipation is very program-dependent, and estimates of actual device temperatures in realistic use are notoriously hard to make.

Table 10.5 shows the output power available in the various modes, with typical transformer regulation, etc.; the output mode diagram in Figure 10.11 shows exactly how the amplifier changes mode from A to AB with decreasing load resistance. Remember that in this context

Table 10.5: Power capability

	Load resistance			
	8Ω	**6Ω**	**4Ω**	**Distortion**
Class-A	20W	27W	15W	Low
Class-AB	n/a	n/a	39W	High
Class-B	21W	28W	39W	Medium

high distortion means 0.002% at 1 kHz. This diagram was produced in the analysis section of PSpice simply by typing in equations, and without actually simulating anything at all.

The most important thermal decision is the size of the heatsink; it is going to be expensive, so there is a powerful incentive to make it no bigger than necessary. I have ruled out fan cooling as it tends to make concern for ultra-low electrical noise look rather foolish; let us rather spend the cost of the fan on extra cooling fins and convect in ghostly silence. The exact thermal design calculations are simple but tedious, with many parameters to enter—the perfect job for a spreadsheet. The final answer is the margin between the predicted junction temperatures and the rated maximum. Once power output and impedance range are decided, the heatsink thermal resistance to ambient is the main variable to manipulate; and this is a compromise between coolness and cost, for high junction temperatures always reduce semiconductor reliability. Looking at it very roughly:

	Thermal resistance (°C)/W	Heat flow (W)	Temp rise (°C)	Temp (°C)
Juncn to TO3 Case	0.7	36	25	100 junction
Case to Sink	0.23	36	8	75 TO3 case
Sink to air	0.65	72	47	67 Heatsink
Total			80	20 Ambient

This shows that the transistor junctions will be 80° above ambient, i.e., at around 100°C; the rated junction maximum is 200°C, but it really is not wise to get anywhere close to this very real limit. Note the Case-Sink thermal washers were high-efficiency material, and standard versions have a slightly higher thermal resistance.

The heatsinks used in the prototype had a thermal resistance of 0.65°C/W per channel. This is a substantial chunk of metal, and since aluminum is basically congealed electricity, it's bound to be expensive.

10.15 A Complete Trimodal Amplifier Circuit

The complete Class-A amplifier is shown in Figure 10.19, complete with optional input bootstrapping. It may look a little complex, but we have only added four low-cost transistors to realize a high-accuracy Class-A quiescent controller, and one more for mode-switching. Since the biasing system has been described above, only the remaining amplifier subsystems are dealt with here.

The input stage follows my design methodology by using a high tail current to maximize transconductance, and then linearizing by adding input degeneration resistors R2, 3 to reduce the final transconductance to a suitable level. Current-mirror TR10, 11 forces the collector currents of the two input devices TR2, 3 to be equal, balancing the input stage to prevent the generation of second-harmonic distortion. The mirror is degenerated by R6, 7 to eliminate the effects of Vbe mismatches in TR10, 11. With some misgivings I added the input network R9, C15, which is definitely not intended to define the system bandwidth, unless fed from a buffer stage; with practical values the HF roll-off could vary widely with the source impedance driving the amplifier. It is intended rather to give the possibility of dealing with RF interference without having to cut tracks. R9 could be increased for bandwidth definition if the source impedance is known, fixed, and taken into account when choosing R9; bear in mind that any value over 47Ω will measurably degrade the noise performance. The values given roll off above 150 MHz to keep out UHF.

The input-stage tail current is increased from 4 to 6 mA, and the VAS standing current from 6 to 10 mA over the original circuit. This increases maximum positive and negative slew-rates from +21, –48 V/μsec to +37, –52 V/μsec; this amplifier architecture is bound to slew asymmetrically. One reason is feedthrough in the VAS current source; in the original circuit an unexpected slew-rate limit was set by fast edges coupling through the current-source c-b capacitance to reduce the bias voltage during positive slewing. This effect is minimized here by using the negative-feedback type of current source bias generator, with VAS collector current chosen as the controlled variable. TR21 senses the voltage across R13, and if it attempts to exceed Vbe, turns on further to pull up the bases of TR1 and TR5. C11 filters the DC supply to this circuit and prevents ripple injection from the V+ rail. R5, C14 provide decoupling to prevent TR5 from disturbing the tail-current while controlling the VAS current.

The input tail-current increase also slightly improves input-stage linearity, as it raises the basic transistor gm and allows R2, 3 to apply more local NFB.

The VAS is linearized by beta-enhancing stage TR12, which increases the amount of local NFB through Miller dominant-pole capacitor C3 (i.e., Cdom). R36 has been increased to 2k2

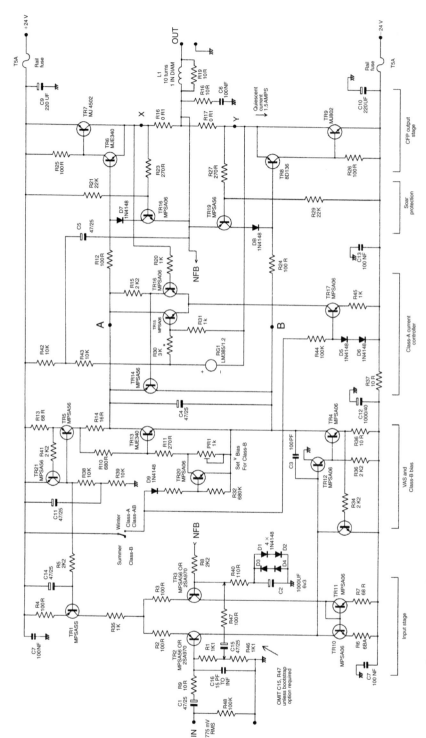

Figure 10.19: The complete circuit diagram of Trimodal amplifier, including the optional bootstrapping components, R47 and C15

to minimize power dissipation, as there seems no significant effect on linearity or slewing. Do not omit it altogether, or linearity will be affected and slewing much compromised.

The simplest way to prevent ripple from entering the VAS via the V− rail is old-fashioned RC decoupling, with a small R and a big C. We have some 200 mV in hand in the negative direction, compared with the positive, and expending this as the voltage-drop through the RC decoupling will give symmetrical clipping. R37 and C12 perform this function; the low rail voltages in this design allow the 1000 μF C12 to be a fairly compact component.

The output stage is of the Complementary-Feedback Pair (CFP) type, which as previously described, gives the best linearity and quiescent stability, due to the two local negative feedback loops around driver and output device. Quiescent stability is particularly important with R16, 17 at 0.1 Ω, and this low value might be rather dicey in a double emitter-follower (EF) output stage. The CFP voltage efficiency is also higher than the EF version. R25, 26 define a suitable quiescent collector current for the drivers TR6, 8, and pull charge carriers from the output device bases when they are turning off. The lower driver is now a BD136; this has a higher fT than the MJE350, and seems to be more immune to odd parasitics at negative clipping.

The new lower values for the output emitter resistors R16, 17 halve the distortion in Class-AB. This is equally effective when in Class-A with too low a load impedance, or in Class-B but with Iq maladjusted too high. It is now true in the latter case that too much Iq really is better than too little—but not much better, and AB still comes a poor third in linearity to Classes A and B.

SOAR (Safe Operating ARea) protection is given by the networks around TR18, TR19. This is a single-slope SOAR system that is simpler than two-slope SOAR, and therefore somewhat less efficient in terms of getting the limiting characteristic close to the true SOAR of the output transistor. In this application, with low rail voltages, maximum utilization of the transistor SOAR is not really an issue; the important thing is to observe maximum junction temperatures in the A/AB mode.

The global negative-feedback factor is 32 dB at 20 kHz, and this should give a good margin of safety against Nyquist-type oscillation. Global NFB increases at 6 dB/octave with decreasing frequency to a plateau of around 64 dB, the corner being at a rather ill-defined 300 Hz; this is then maintained down to 10 Hz. It is fortunate that magnitude and frequency here are non-critical, as they depend on transistor beta and other doubtful parameters.

It is often stated in hi-fi magazines that semiconductor amplifiers sound better after hours or days of warm-up. If this is true (which it certainly is not in most cases) it represents truly spectacular design incompetence. This sort of accusation is applied with particular venom to Class-A designs, because it is obvious that the large heatsinks required take time to reach final temperature, so I thought it important to state that in Class-A this design stabilizes its electrical operating conditions in less than a second, giving the full intended performance. No warm-up time beyond this is required; obviously the heatsinks take time to reach thermal equilibrium, but as described above, measures have been taken to ensure that component temperature has no significant effect on operating conditions or performance.

10.16 The Power Supply

A suitable unregulated power supply must be designed for continuous operation at maximum current, so the bridge rectifier must be properly heat-sunk, and careful consideration given to the ripple-current ratings of the reservoirs. This is one reason why reservoir capacitance has been doubled to 20,000 μF per rail, over the 10,000 μF that was adequate for the Class-B design; the ripple voltage is halved, which improves voltage efficiency as it is the ripple troughs that determine clipping onset, but in addition the ripple current, although unchanged in total value, is now split between two components. (The capacitance was not increased to reduce ripple injection, which is dealt with far more efficiently and economically by making the PSRR high.) Do not omit the secondary fuses; even in these modern times rectifiers do fail, and transformers are horribly expensive.

10.17 The Performance

The performance of a properly designed Class-A amplifier challenges the ability of even the Audio Precision measurement system. To give some perspective on this, Figure 10.20 shows the distortion of the AP oscillator driving the analyzer section directly for various bandwidths.

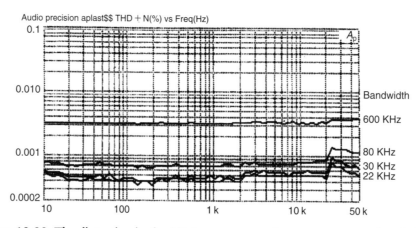

Figure 10.20: The distortion in the AP-1 system at various measurement bandwidths

There appear to be internal mode changes at 2 kHz and 20 kHz, causing step increases in oscillator distortion content; these are just visible in the THD plots for Class-A mode.

Figure 10.21 shows Class-B distortion for 20 W into 8 and 4 Ω, while Figure 10.22 shows the same in Class-A/AB. Figure 10.23 shows distortion in Class-A for varying measurement

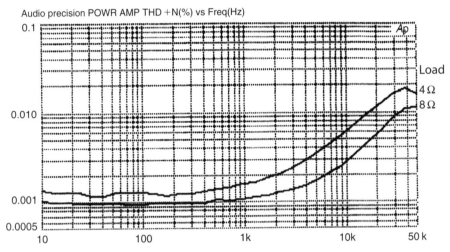

Figure 10.21: Distortion in Class-B (Summer) mode. Distortion into 4Ω is always worse. Power was 20 W in 8Ω and 40 W in 4Ω, bandwidth 80 kHz.

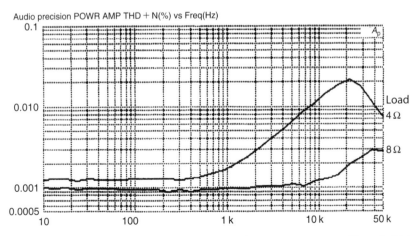

Figure 10.22: Distortion in Class-A/AB (Winter) mode, same power and bandwidth of Figure 10.21. The amplifier is in AB mode for the 4Ω case, and so distortion is higher than for Class-B into 4Ω. At 80 kHz bandwidth, the Class-A plot below 10 kHz merely shows the noise floor.

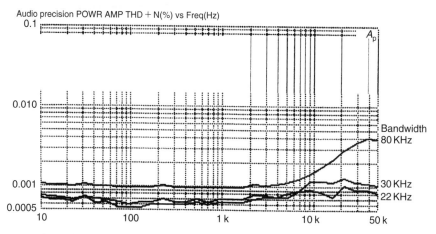

Figure 10.23: Distortion in Class-A only (20 W/8Ω) for varying measurement bandwidths. The lower bandwidths ignore HF distortion, but give a much clearer view of the excellent linearity below 10 kHz.

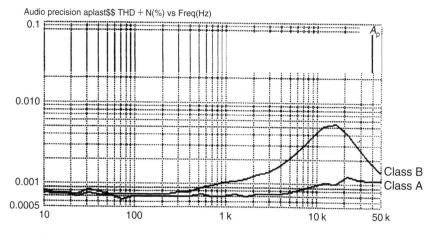

Figure 10.24: Direct comparison of Classes A and B (20 W/8Ω) at 30 kHz bandwidth. The HF rise for B is due to the inability of negative feedback that falls with frequency to linearize the high-order crossover distortion in the output stage.

bandwidths. The lower bandwidths misleadingly ignore the HF distortion, but give a much clearer view of the excellent linearity below 10 kHz. Figure 10.24 gives a direct comparison of Classes A and B. The HF rise for B is due to high-order crossover distortion being poorly linearized by negative feedback that falls with frequency.

10.18 Further Possibilities

One interesting extension of the ideas presented here is the Adaptive Trimodal Amplifier. This would switch into Class-B on detecting device or heatsink over-temperature, and would be a unique example of an amplifier that changed mode to suit the operating conditions. The thermal protection would need to be latching; flipping from Class-A to Class-B every few minutes would subject the output devices to unnecessary thermal cycling.

References

1. Moore BJ: An Introduction to the Psychology of Hearing. Academic Press; 1982 pp. 48–50.

2. Tanaka S. A New Biasing Circuit for Class-B Operation. *JAES* Jan/Feb 1981:27.

3. Fuller S, Private communication.

4. Nelson Pass. Build a Class-A Amplifier. *Audio* Feb 1977:28. Constant-current.

5. Linsley-Hood J. Simple Class-A Amplifier. *Wireless World* April 1969:148.

6. Self D. High-Performance Preamplifier. *Wireless World* Feb 1979:41.

7. Nelson-Jones L. Ultra-Low Distortion Class-A Amplifier. *Wireless World* March 1970:98.

8. Giffard T. Class-A Power Amplifier. *Elektor* Nov 1991:37.

9. Linsley-Hood J. High-Quality Headphone Amp. *HiFi News and RR* Jan 1979:81.

10. Nelson Pass. The Pass/A40 Power Amplifier. *The Audio Amateur* 1978:4. (Push-pull).

11. Thagard N, Build a 100W Class-A Mono Amp. *Audio* Jan 95, p. 43.

CHAPTER 11

MPEG-4 and H.264

Keith Jack

Now that cell phones have morphed into portable media players (PMPs), consumers expect to be able to carry dozens of videos, hundreds of pictures and thousands of songs along with them. The previous two chapters dealt with the storage problems this presents. This chapter deals with your first line of defense: codecs.

Codecs compress audio and video not only to conserve storage but also to reduce the bandwidth required to transport these data streams. Successive generations of codecs have achieved higher compression ratios resulting in smaller streams and file sizes. Still, the trade-off is the computational requirements to perform the compression. Application processors typically offload audio and graphics processing from the CPU, but you still don't get performance without burning clock cycles, which impacts battery life.

The MPEG series of codecs have been very successful at addressing the storage problem in portable devices. MPEG-1 Layer III—also known as MP3—is almost universally used to store music on PMPs. It's not lossless, but it gives CD quality playback with a compression ratio of 12:1, enabling you to store dozens of albums on the flash memory card in your iPod. Get the hard disk version and the sky's the limit.

Video is another matter. MPEG-2 uses inter-frame encoding—interspersing occasional full frames with frames that just convey the changes from the previous screen—to achieve 30:1 compression for general video. MPEG-4 AVC (for Advanced Video Coding)—also known as MPEG-4 Part 10 or H.264—achieves a compression ratio of 50:1 with better fidelity and a lower bit rate than MPEG-2; on panoramic scenes, it can achieve up to 1000:1. And unlike MPEG-2, H.264 was developed to be a single-chip solution that can be implemented cheaply in silicon—at least its Simple and Advanced Simple profiles.

MPEG-4 takes a very different approach than its predecessors. Instead of just coding a bitstream, MPEG-4 takes an object-oriented approach to a scene; instead of sending frames, it provides a standardized way to represent audio, video, and still image media objects using descriptive *elements instead of pixels. MPEG-4 breaks down the video stream into natural and synthetic visual objects (VOPs). Each VOP can be encoded in a scalable (mult-layer, 3D) or nonscalable (single layer, 2D) form, depending on the application. Data for each object is carried in a separate elementary stream. The scene description is carried in a separate stream by the Binary Format for Scenes (BIFS), a format for two- or three-dimensional audiovisual content. BIFS is based on the virtual reality modeling language (VRML) and is detailed in MPEG-4 Part 11.*

If all this seems complicated, the details are actually a lot more complicated. However, Keith Jack in this chapter from Video Demystified *takes you through a detailed description of MPEG-4 and H.264 without being mystifying. If you're even considering multimedia in your next portable design, this chapter is a must read.*

—**John Donovan**

MPEG-4 builds upon the success and experience of MPEG-2. It is best known for:

- Lower bit-rates than MPEG-2 (for the same quality of video)

- Use of natural or synthetic objects that can be rendered together to make a scene

- Support for interactivity

For *authors*, MPEG-4 enables creating content that is more reusable and flexible, with better content protection capabilities.

For *consumers*, MPEG-4 can offer more interactivity and, due to the lower bit-rate over MPEG-2, the ability to enjoy content over new networks (such as DSL) and mobile products.

MPEG-4 is an ISO standard (ISO/IEC 14496), and currently consists of 19 parts:

systems	ISO/IEC 14496–1
visual	ISO/IEC 14496–2
audio	ISO/IEC 14496–3
conformance testing	ISO/IEC 14496–4
reference software	ISO/IEC 14496–5
DMIF	ISO/IEC 14496–6
reference software	ISO/IEC 14496–7
carriage over IP networks	ISO/IEC 14496–8

reference hardware	ISO/IEC 14496–9
advanced video (H.264)	ISO/IEC 14496–10
scene description	ISO/IEC 14496–11
ISO file format	ISO/IEC 14496–12
IPMP extensions	ISO/IEC 14496–13
MP4 file format	ISO/IEC 14496–14
H.264 file format	ISO/IEC 14496–15
animation extension	ISO/IEC 14496–16
streaming text format	ISO/IEC 14496–17
font compression	ISO/IEC 14496–18
synthesize texture stream	ISO/IEC 14496–19

MPEG-4 provides a standardized way to represent audio, video, or still image media objects using descriptive elements (instead of actual bits of an image, for example). A media object can be natural or synthetic (computer-generated) and can be represented independent of its surroundings or background.

It also describes how to merge multiple media objects to create a scene. Rather than sending bits of picture, the media objects are sent, and the receiver composes the picture. This allows:

- An object to be placed anywhere

- Geometric transformations on an object

- Grouping of objects

- Modifying attributes and transform data

- Changing the view of a scene dynamically

11.1 Audio Overview

MPEG-4 audio supports a wide variety of applications, from simple speech to multi-channel high-quality audio.

Audio objects (audio codecs) use specific combinations of tools to efficiently represent different types of audio objects. Profiles use specific combinations of audio object types to efficiently service a specific market segment. Levels specify size, rate, and complexity limitations within a profile to ensure interoperability.

Currently, most solutions support a few of the most popular audio codecs (usually AAC-LC and HE-AAC) rather than one or more profiles/levels.

11.1.1 General Audio Object Types

This category supports a wide range of quality, bit-rates, and number of channels. For natural audio, MPEG-4 supports the AAC (Advanced Audio Coding), BSAC (Bit Sliced Arithmetic Coding), and TwinVQ (Transform Domain Weighted Interleave Vector Quantization) algorithms. The following audio objects are available:

11.1.1.1 AAC-Main Objects

AAC-Main objects add the Perceptual Noise Shaping (PNS) tool to MPEG-2 AAC-Main.

11.1.1.2 AAC-LC Objects

AAC-LC (Low Complexity) objects add the PNS tool to MPEG-2 AAC-LC. There is also an Error Resilient version, ER AAC-LC.

11.1.1.3 AAC-SSR Objects

AAC-SSR (Scalable Sampling Rate) objects add the PNS tool to MPEG-2 AAC-SSR.

11.1.1.4 AAC-LTP Objects

AAC-LTP (Long Term Predictor) objects are similar to AAC-LC objects, with the long term predictor replacing the AAC-LC predictor. This gives the same efficiency with significantly lower implementation cost. There is also an Error Resilient version, ER AAC-LTP.

11.1.1.5 AAC-Scalable Objects

AAC-Scalable objects allow a large number of scalable combinations. They support only mono or 2-channel stereo sound. There is also an Error Resilient version, ER AAC-Scalable.

11.1.1.6 ER AAC-LD Objects

Error Resilient AAC-LD (Low Delay) is derived from AAC and all the capabilities for coding of two or more sound channels are supported. They support sample rates up 48 kHz and use frame lengths of 512 or 480 samples (compared to 1024 or 960 samples used by AAC) to enable a maximum algorithmic delay of 20 ms.

11.1.1.7 ER BSAC Objects

Error Resilient BSAC objects replace the noiseless coding of AAC quantized spectral data and the scale factors. One base layer bitstream and many small enhancement layer bitstreams are used, enabling real-time adjustments to the quality of service.

11.1.1.8 HE-AAC Objects

HE-AAC (High Efficiency), a combination of AAC and Spectral Band Replication (SBR) technology, is designed for ultra-low bit-rate coding, as low as 32 kbps for stereo.

11.1.1.9 TwinVQ Objects

TwinVQ objects are based on fixed rate vector quantization instead of the Huffman coding used in AAC. They operate at lower bit-rates than AAC, supporting mono and stereo sound. There is also an Error Resilient version, ER TwinVQ.

11.1.2 Speech Object Types

Speech coding can be done using bit-rates from 2–24 kbps. Lower bit-rates, such as an average of 1.2 kbps, are possible when variable rate coding is used. The following audio objects are available:

11.1.2.1 CELP Objects

CELP (Code Excited Linear Prediction) objects support 8 and 16 kHz sampling rates at bit-rates of 4–24 kbps. There is also an Error Resilient version, ER CELP.

11.1.2.2 HVXC Objects

HVXC (Harmonic Vector eXcitation Coding) objects support 8 kHz mono speech at fixed bit-rates of 2–4 kbps (below 2 kbps using a variable bit-rate mode), along with the ability to change the pitch and speed during decoding. There is also an Error Resilient version, ER HVXC.

11.1.3 Synthesized Speech Object Types

Scalable TTS (Text-to-Speech) objects offer a low bit-rate (200–1.2 kbps) phonemic representation of speech. Content with narration can be easily created without recording natural speech. The TTS Interface allows speech information to be transmitted in the International Phonetic Alphabet (IPA) or in a textual (written) form of any language. The synthesized speech can also be synchronized with a facial animation object.

11.1.4 Synthesized Audio Object Types

Synthetic Audio support is provided by a Structured Audio Decoder implementation that allows the application of score-based control information to musical instruments described in a special language. The following audio objects are available:

11.1.4.1 Main Synthetic Objects

Main Synthetic objects allow the use of the all MPEG-4 Structured Audio tools. They support synthesis using the Structured Audio Orchestra Language (SAOL) music-synthesis language and wavetable synthesis using Structured Audio Sample-Bank Format (SASBF).

11.1.4.2 Wavetable Synthesis Objects

Wavetable Synthesis objects are a subset of Main Synthetic, making use of SASBF and MIDI (Musical Instrument Digital Interface) tools. They provide relatively simple sampling synthesis.

11.1.4.3 General MIDI Objects

General MIDI objects provide interoperability with existing content.

11.2 Visual Overview

MPEG-4 visual is divided into two sections. MPEG-4.2 includes the original MPEG-4 video codecs discussed in this section. MPEG-4.10 specifies the "advanced video codec," also known as H.264, and is discussed at the end of this chapter.

The visual specifications are optimized for three primary bit-rate ranges:

- less than 64 kbps

- 64–384 kbps

- 0.384–4 Mbps

For high-quality applications, higher bit-rates are possible, using the same tools and bitstream syntax as those used for lower bit-rates.

With MPEG-4, visual objects (video codecs) use specific combinations of tools to efficiently represent different types of visual objects. Profiles use specific combinations of visual object types to efficiently service a specific market segment. Levels specify size, rate, and complexity limitations within a profile to ensure interoperability.

Currently, most solutions support only a couple of the MPEG-4.2 video codecs (usually Simple and Advanced Simple) due to silicon cost issues. Interest in MPEG-4.2 video codecs also dropped dramatically with the introduction of the MPEG-4.10 (H.264) and SMPTE 421M (VC-1) video codecs, which offer about $2\times$ better performance.

11.2.1 YCbCr Color Space

The 4:2:0 YCbCr color space is used for most objects. Each component can be represented by a number of bits ranging from 4 to 12 bits, with 8 bits being the most commonly used.

MPEG-4.2 Simple Studio and Core Studio objects may use 4:2:2, 4:4:4, 4:2:2:4, and 4:4:4:4:4:4 YCbCr or RGB sampling options, to support the higher picture quality required during the editing process.

Like H.263 and MPEG-2, the MPEG-4.2 video codecs are also macroblock, block, and DCT-based.

11.2.2 Visual Objects

Instead of the video frames or pictures used in earlier MPEG specifications, MPEG-4 uses natural and synthetic visual objects. Instances of video objects at a given time are called visual object planes (VOPs).

Much like MPEG-2, there are I (intra), P (predicted), and B (bi-directional) VOPs. The S-VOP is a VOP for a sprite object. The S(GMC)-VOP is coded using prediction based on global motion compensation from a past reference VOP.

Arbitrarily shaped video objects, as well as rectangular objects, may be used. An MPEG-2 video stream can be a rectangular video object, for example.

Objects may also be scalable, enabling the reconstruction of useful video from pieces of a total bitstream. This is done by using a base layer and one or more enhancement layers.

Only natural visual object types are discussed since they are currently of the most interest in the marketplace.

11.2.3 MPEG-4.2 Natural Visual Object Types

MPEG-4.2 supports many natural visual object types (video codecs), with several interesting ones shown in Table 11.1. The more common object types are:

Table 11.1: Available tools for common MPEG-4.2 natural visual object types

Tools	Object Type						
	Main	Core	Simple	Advanced Simple	Advanced Real Time Simple	Advanced Coding Efficiency	Fine Granularity Scalable
VOP types	I, P, B	I, P, B	I, P	I, P, B	I, P	I, P, B	I, P, B
chroma format	4:2:0	4:2:0	4:2:0	4:2:0	4:2:0	4:2:0	4:2:0
interlace	×	−	−	×	−	×	×
global motion compensation (GMC)	−	−	−	×	−	×	−
quarter-pel motion compensation (QPEL)	−	−	−	×	−	×	−
slice resynchronization	×	×	×	×	×	×	×
data partitioning	×	×	×	×	×	×	×
reversible VLC	×	×	×	×	×	×	×
short header	×	×	×	×	×	×	×
method 1 and 2 quantization	×	×	−	×	−	×	×
shape adaptive DCT	−	−	−	−	−	×	−
dynamic resolution conversion	−	−	−	−	×	×	−
NEWPRED	−	−	−	−	×	×	−
binary shape	×	×	−	−	−	×	−
grey shape	×	−	−	−	−	×	−
sprite	×	−	−	−	−	−	−
fine granularity scalability (FGS)	−	−	−	−	−	−	×
FGS temporal scalability	−	−	−	−	−	−	×

11.2.3.1 Main Objects

Main objects provide the highest video quality. Compared to Core objects, they also support grayscale shapes, sprites, and both interlaced and progressive content.

11.2.3.2 Core Objects

Core objects use a subset of the tools used by Main objects, although B-VOPs are still supported. They also support scalability by sending extra P-VOPs. Binary shapes can include a constant transparency but cannot do the variable transparency offered by grayscale shape coding.

11.2.3.3 Simple Objects

Simple objects are low bit-rate, error resilient, rectangular natural video objects of arbitrary aspect ratio. Simple objects use a subset of the tools used by Core objects.

11.2.3.4 Advanced Simple Objects

Advanced Simple objects looks much like Simple objects in that only rectangular objects are supported, but adds a few tools to make it more efficient: B-frames, ¼-pixel motion compensation (QPEL), and global motion compensation (GMC).

11.2.3.5 Fine Granularity Scalable Objects

Fine Granularity Scalable objects can use up to eight scalable layers so delivery quality can easily adapt to transmission and decoding circumstances.

11.2.4 MPEG-4.2 Natural Visual Profiles

MPEG-4.2 supports many visual profiles and levels. Only natural visual profiles (Tables 11.2 and 11.3) are discussed since they are currently of the most interest in the marketplace. The more common profiles are:

11.2.4.1 Main Profile

Main profile was created for broadcast applications, supporting both progressive and interlaced content. It combines highest quality video with arbitrarily shaped objects.

11.2.4.2 Core Profile

Core profile is useful for higher quality interactive services, combining good quality with limited complexity and supporting arbitrary shape objects. Mobile broadcast services can also be supported by this profile.

Table 11.2a: MPEG-4.2 natural vision profiles and levels

MPEG-4.2 Profile	Supported Shapes	Notes	Level	Typical Resolution	Maximum Number of Objects	Maximum Bit-Rate
Main	arbitrary	additional tools and functionality	L4	BT.709	32	38.4 Mbps
			L3	BT.601	32	15 Mbps
			L2	CIF	16	2 Mbps
Core	arbitrary	additional tools and functionality	L2	CIF	16	2 Mbps
			L1	QCIF	4	384 kbps
Advanced Core	arbitrary	higher coding efficiency	L2	CIF	16	2 Mbps
			L1	QCIF	4	384 kbps
N-Bit	arbitrary		L2	CIF	16	2 Mbps
Simple	rectangular		L3	CIF	4	384 kbps
			L2	CIF	4	128 kbps
			L1	QCIF	4	64 kbps
Advanced Simple	rectangular	higher coding efficiency	L5	BT.601	4	8 Mbps
			L4	352×576	4	3 Mbps
			L3b	CIF	4	1.5 Mbps
			L3	CIF	4	768 kbps
			L2	CIF	4	384 kbps
			L1	QCIF	4	128 kbps
			L0	QCIF	1	128 kbps
Advanced Real Time Simple	rectangular	higher error resilience	L4	CIF	16	2 Mbps
			L3	CIF	4	384 kbps
			L2	CIF	4	128 kbps
			L1	QCIF	4	64 kbps
Core Scalable	arbitrary	spatial and temporal scalability	L3	BT.601	16	4 Mbps
			L2	CIF	8	1.5 Mbps
			L1	CIF	4	768 kbps
Simple Scalable	rectangular	spatial and temporal scalability	L2	CIF	4	256 kbps
			L1	CIF	4	128 kbps
			L0	QCIF	1	128 kbps
Fine Granularity Scalable	rectangular	SNR and temporal scalability	L5	BT.601	4	8 Mbps
			L4	352×576	4	3 Mbps
			L3	CIF	4	768 kbps

(Continued)

Table 11.2a: (Continued)

MPEG-4.2 Profile	Supported Shapes	Notes	Level	Typical Resolution	Maximum Number of Objects	Maximum Bit-Rate
			L2	CIF	4	384 kbps
			L1	QCIF	4	128 kbps
			L0	QCIF	1	128 kbps
Advanced Coding Efficiency	arbitrary	higher coding efficiency	L4	BT.709	32	38.4 Mbps
			L3	BT.601	32	15 Mbps
			L2	CIF	16	2 Mbps
			L1	CIF	4	384 kbps

Table 11.2b: MPEG-4.2 Natural vision profiles and levels. 4:4:4:4:4:4 means 4:4:4 RGB + 3 auxiliary channels. 4:2:2:4 means 4:2:2 YCbCr + 1 auxiliary channel

MPEG-4.2 Profile	Supported Shapes	Notes	Level	Typical Resolution	Maximum Number of Objects	Maximum Bit-Rate
Core Studio (uses 10-bit pixel data)	arbitrary	additional tools and functionality	L4	BT.709, 60P, 4:4:4 BT.709, 30I, 4:4:4:4:4:4	16	900 Mbps
			L3	BT.709, 30I, 4:4:4 BT.601, 4:2:2:4	8	450 Mbps
			L2	BT.709, 30I, 4:2:2 BT.601, 4:4:4:4:4:4	4	300 Mbps
			L1	BT.601, 4:2:2:4 BT.601, 4:4:4	4	90 Mbps
Simple Studio (uses 10- or 12-bit pixel data)	arbitrary		L4	BT.709, 60P, 4:4:4 BT.709, 30I, 4:4:4:4:4:4	1	1800 Mbps
			L3	BT.709, 30I, 4:4:4 BT.709, 30I, 4:2:2:4	1	900 Mbps
			L2	BT.709, 30I, 4:2:2 BT.601, 4:4:4:4:4:4	1	600 Mbps
			L1	BT.601, 4:2:4 BT.601, 4:4:4	1	180 Mbps

Table 11.3: Objects supported by common MPEG-4.2 profiles

MPEG-4.2 Object Type	MPEG-4.2 Profile						
	Main	Core	Simple	Advanced Simple	Advanced Real Time Simple	Advanced Coding Efficiency	Fine Granularity Scalable
Main	✕	–	–	–	–	–	–
Core	✕	✕	–	–	–	✕	–
N-Bit	–	–	–	–	–	–	–
Simple	✕	✕	✕	✕	✕	✕	✕
Advanced Simple	–	–	–	✕	–	–	✕
Advanced Real Time Simple	–	–	–	–	✕	–	–
Advanced Coding Efficiency	–	–	–	–	–	✕	–
Fine Granularity Scalable	–	–	–	–	–	–	✕

11.2.4.3 Simple Profile

Simple profile was created with low complexity applications in mind. Primary applications are mobile and the Internet.

11.2.4.4 Advanced Simple Profile

Advanced Simple profile provides the ability to distribute single-layer frame-based video at a wide range of bit-rates.

11.2.4.5 Fine Granularity Scalable Profile

Fine Granularity Scalable profile was created with Internet streaming and wireless multimedia in mind.

11.3 Graphics Overview

Graphics profiles specify which graphics elements of the BIFS tool can be used to build a scene. Although it is defined in the Systems specification, graphics is really just another media profile like audio and video, so it is discussed here.

Four hierarchical graphics profiles are defined: Simple 2D, Complete 2D, Complete and 3D Audio Graphics. They differ in the graphics elements of the BIFS tool to be supported by the decoder, as shown in Table 11.4.

Simple 2D profile provides the basic features needed to place one or more visual objects in a scene.

Complete 2D profile provides 2D graphics functions and supports features such as arbitrary 2D graphics and text, possibly in conjunction with visual objects.

Complete profile provides advanced capabilities such as elevation grids, extrusions, and sophisticated lighting. It enables complex virtual worlds to exhibit a high degree of realism.

3D Audio Graphics profile may be used to define the acoustical properties of the scene (geometry, acoustics absorption, diffusion, material transparency). This profile is useful for applications that do environmental equalization of the audio signals.

11.4 Visual Layers

An MPEG-4 visual scene consists of one or more video objects. Currently, the most common video object is a simple rectangular frame of video.

Each video object may have one or more layers to support temporal or spatial scalable coding. This enables the reconstruction of video in a layered manner, starting with a base layer and adding a number of enhancement layers. Where a high degree of scalability is needed, such as when an image is mapped onto a 2D or 3D object, a wavelet transform is available.

The visual bitstream provides a hierarchical description of the scene. Each level of hierarchy can be accessed through the use of unique start codes in the bitstream.

11.4.1 Visual Object Sequence (VS)

This is the complete scene which contains all the 2D or 3D, natural or synthetic, objects and any enhancement layers.

11.4.2 Video Object (VO)

A video object corresponds to a particular object in the scene. In the most simple case this can be a rectangular frame, or it can be an arbitrarily shaped object corresponding to an object or background of the scene.

Table 11.4: Graphics elements (BIFS tools) supported by MPEG-4 graphics profiles

Graphics Element of BIFS Tool	Graphics Profile			Graphics Tool (BIFS node)	Graphics Profile		
	Simple 2D	Complete 2D	Complete		Simple 2D	Complete 2D	Complete
appearance	×	×	×	fog	−	−	×
box	−	−	×	font style	−	×	×
bitmap	×	×	×	indexed face set	−	−	×
background	−	−	×	indexed face set 2D	−	×	×
background 2D	−	×	×	indexed line set	−	−	×
circle	−	×	×	indexed line set 2D	−	×	×
color	−	×	×	line properties	−	×	×
cone	−	−	×	material	−	−	×
coordinate	−	−	×	material 2D	−	×	×
coordinate 2D	−	×	×	normal	−	−	×
curve 2D	−	×	×	pixel texture	−	×	×
cylinder	−	−	×	point light	−	−	×
directional light	−	−	×	point set	−	−	×
elevation grid	−	−	×	point set 2D	−	×	×
expression	−	−	×	rectangle	−	×	×
extrusion	−	−	×	shape	×	×	×
face	−	−	×	sphere	−	−	×
face def mesh	−	−	×	spot light	−	−	×
face def table	−	−	×	text	−	×	×
face def transform	−	−	×	texture coordinate	−	×	×
FAP	−	−	×	texture transform	−	×	×
FDP	−	−	×	viseme	−	−	×
FIT	−	−	×				

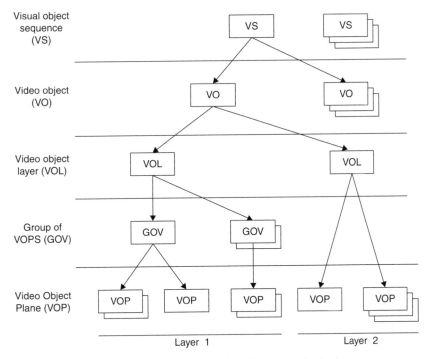

Figure 11.1: Example MPEG-4 video bitstream logical structure

11.4.3 Video Object Layer (VOL)

Each video object can be encoded in scalable (multi-layer) or nonscalable form (single layer), depending on the application, represented by the video object layer (VOL). The VOL provides support for scalable coding. A video object can be encoded using spatial or temporal scalability, going from coarse to fine resolution. Depending on parameters such as available bandwidth, computational power, and user preferences, the desired resolution can be made available to the decoder.

There are two types of video object layers, the video object layer that provides full MPEG-4 functionality, and a reduced functionality video object layer, the video object layer with short headers. The latter provides bitstream compatibility with baseline H.263.

11.4.4 Group of Video Object Plane (GOV)

Each video object is sampled in time; each time sample of a video object is a video object plane. Video object planes can be grouped together to form a group of video object planes.

The GOV groups together video object planes. GOVs can provide points in the bitstream where video object planes are encoded independently from each other, and can thus provide random access points into the bitstream. GOVs are optional.

11.4.5 Video Object Plane (VOP)

A VOP is a time sample of a video object. VOPs can be encoded independently of each other, or dependent on each other by using motion compensation. A conventional video frame can be represented by a VOP with rectangular shape.

11.5 Object Description Framework

Unlike MPEG-2, MPEG-4 does not multiplex multiple elementary streams together into a single transport or program stream.

Data for each object (audio, one layer of one visual object, etc.), scene description information (to declare the spatial-temporal relationship of objects), and object control information are carried in separate elementary streams. Synthetic objects may be generated using BIFS to provide the graphics and audio. BIFS is more than a scene description language—it integrates natural and synthetic objects into the same composition space.

The object description framework is a set of object descriptors used to identify, describe, and associate elementary streams to each other, and to objects used in the scene description, as illustrated in Figure 11.2.

An initial object descriptor, a derivative of the object descriptor, contains two descriptors. One descriptor points to the scene description [elementary] stream; the other points to the corresponding object descriptor [elementary] stream.

11.5.1 Object Descriptor (OD) Stream

The object descriptors are transported in a dedicated elementary stream, called the object descriptor stream.

The object descriptor effectively associates sets of related elementary streams so they are seen as a single entity by the decoder. Each object descriptor contains other descriptors that typically point to one or more elementary streams associated to a single node and a single audio or visual object. This allows support for multiple alternative streams, such as different languages.

In addition, an object descriptor can point to auxiliary data such as object content information (OCI) and intellectual property rights management and protection (IPMP).

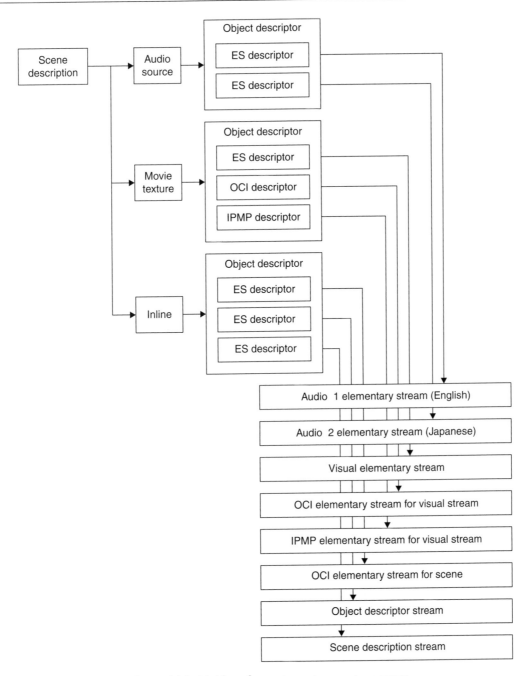

Figure 11.2: Linking elementary streams to a scene

Object descriptors are not simply present in an object descriptor stream one after the other. Rather, they are encapsulated in object descriptor commands. These commands enable object descriptors to be dynamically conveyed, updated, or removed at a specific point in time. This allows new elementary streams for an object to be advertised as they become available, or to remove references to elementary streams that are no longer available. Updates are time stamped to indicate when they are to take effect. The time stamp is placed on the sync layer, as with any other elementary stream.

11.5.2 Object Content Information (OCI)

The OCI elementary stream conveys OCI events. Each OCI event consists of OCI descriptors.

OCI descriptors communicate a number of features of the associated object, such as keywords, text description of the content, language, parental rating, creation date, authors, etc.

If the OCI information will never change, it may instead be conveyed using CCI descriptors within the object descriptor stream.

11.5.3 Intellectual Property Management and Protection (IPMP)

The IPMP elementary stream conveys IPMP messages to one or more IPMP systems. The IPMP system provides intellectual property management and content protection functions in the receiver.

If the IPMP information will rarely change, it may instead be conveyed using IPMP descriptors within the object descriptor stream.

11.6 Scene Description

To assemble a multimedia scene at the receiver, it is not sufficient to simply send just the multiple streams of data. For example, objects may be located in 2D or 3D space, and each has its local coordinate system. Objects are positioned within a scene by transforming each of them to the scene's coordinate system. Therefore, additional data is required for the receiver to assemble a meaningful scene for the user. This additional data is called scene description.

Scene graph elements (which are BIFS tools) describe audiovisual primitives and attributes. These elements, and any relationship between them, form a hierarchical scene graph, as illustrated in Figure 11.3. The scene graph is not necessarily static; elements may be added, deleted, or modified as needed.

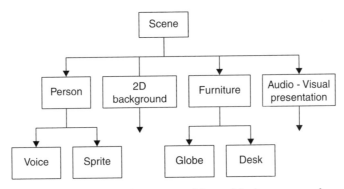

Figure 11.3: Example MPEG-4 hierarchical scene graph

The scene graph profile defines the allowable set of scene graph elements that may be used.

11.6.1 BIFS

BIFS (BInary Format for Scenes) is used to not only describe the scene composition information, but also graphical elements. A fundamental difference between the BIFS and VRML is that BIFS is a binary format, whereas VRML is a textual format. BIFS supports the elements used by VRML and several that VRML does not, including compressed binary format, streaming, streamed animation, 2D primitives, enhanced audio, and facial animation.

11.6.1.1 Compressed Binary Format

BIFS supports an efficient binary representation of the scene graph information. The coding may be either lossless or lossy. Lossy compression is possible due to context knowledge: if some scene graph data has been received, it is possible to anticipate the type and format of subsequent data.

11.6.1.2 Streaming

BIFS is designed so that a scene may be transmitted as an initial scene, followed by modifications to the scene.

11.6.1.3 Streamed Animation

BIFS includes a low-overhead method for the continuous animation of changes to numerical values of the elements in a scene. This provides an alternative to the interpolator elements supported in BIFS and VRML.

11.6.1.4 2D Primitives

BIFS has native support for 2D scenes to support low-complexity, low-cost solutions such as traditional television. Rather than partitioning the world into 2D vs. 3D, BIFS allows both 2D and 3D elements in a single scene.

11.6.1.5 Enhanced Audio

BIFS improves audio support through the use of an audio scene graph, enabling audio sources to be mixed or the generation of sound effects.

11.6.1.6 Facial Animation

BIFS exposes the animated face properties to the scene level. This enables it to be a full member of a scene that can be integrated with any other BIFS functionality, similar to other audiovisual objects.

11.7 Synchronization of Elementary Streams

11.7.1 Sync Layer

The sync layer (Figure 11.4) partitions each elementary stream into a sequence of access units, the smallest entity to which timing information can be associated. It then encodes all the relevant properties using a flexible syntax, generating SL packets. These SL packets are then passed on to a delivery (transport) layer. A sequence of SL packets from a single elementary stream is called an SL-packetized stream.

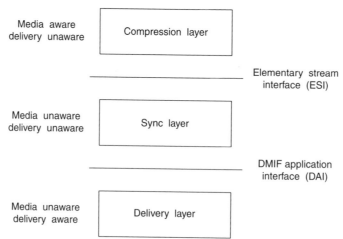

Figure 11.4: Relationship between the MPEG-4 compression, sync, and delivery layers

Unlike the MPEG-2 PES, the sync layer is not a self-contained stream. Instead, it is an intermediate format that is mapped to a specific delivery layer, such as IP, MPEG-2 transport stream, etc. For this reason, there is no need to include start codes, SL packet lengths, etc., in the sync layer—these are already included in the various delivery layer protocols.

SL packets serve two purposes. First, access units can be fragmented in any way during adaptation to a specific delivery layer. Second, it makes sense to have the encoder guide the fragmentation when it knows about delivery layer characteristics, such as the size of the maximum transfer unit (MTU).

Synchronization of multiple elementary streams is done by conveying object clock reference (OCR), decoding time stamps (DTS), composition time stamps (CTS), and clock references within the sync layer.

The sync layer syntax is flexible in that it can be configured individually for each elementary stream. For example, low bit-rate audio streams may desire time stamps with minimum overhead; high bit-rate video streams may need very precise time stamps.

11.7.2 DMIF Application Interface

Unlike MPEG-2, MPEG-4 supports multiple simultaneous usage scenarios (local retrieval, remote interaction, broadcast, multicast, etc.), and multiple simultaneous delivery technologies.

The DMIF (Delivery Multimedia Integration Framework) Application Interface, or DAI, is the interface that controls the data exchanged between a sync layer and a delivery layer (Figure 11.4) during both transmission and reception. It enables accessing, presenting, and synchronizing MPEG-4 content transmitted or received using different technologies, such as MPEG-2 transport stream, IP multicast, RTP, etc., even simultaneously.

As a result, a specification for control and data mapping to a specific delivery or storage protocol (also called a payload format specification) has to be done jointly with the organization that manages the delivery layer specification. In the example of MPEG-4 transport over IP, development work was done jointly with the Internet Engineering Task Force (IETF).

11.8 Multiplexing of Elementary Streams

Delivery of MPEG-4 content is a task that is usually dealt with outside the specification.

However, an analysis of existing delivery layers indicated a need for an additional layer of multiplexing. The occasionally bursty and low bit-rate MPEG-4 streams sometimes have to map to a delivery layer that uses fixed packet sizes (such as MPEG-2 transport streams) or high multiplexing overhead (such as RTP/UDP/IP). The potentially large number of delivery streams may also have a burden in terms of management and cost.

To address this situation, FlexMux, a very simple multiplex packet syntax, was defined. It allows multiplexing a number of SL-packetized streams into a self-contained FlexMux stream with low overhead.

In addition, the specifications for the encapsulation of SL-packetized streams into common delivery layer protocols, including MPEG-2 transport and program streams, IP, and MPEG-4 file format, have already been done.

11.8.1 FlexMux

FlexMux multiplexes one or more SL-packetized streams with varying instantaneous bit-rates into FlexMux packets that have variable lengths.

Identification of SL packets originating from different elementary streams is through FlexMux channel numbers. Each SL-packetized stream is mapped into one FlexMux channel. FlexMux packets with data from different SL-packetized streams can therefore be arbitrarily interleaved. The sequence of FlexMux packets that are interleaved into one stream is called a FlexMux stream.

11.8.2 MPEG-4 over MPEG-2

The MPEG-2 PES is the common denominator for encapsulating content. MPEG-4 defines the encapsulation of SL-packetized and FlexMux streams within PES packets.

One SL-packetized stream is mapped to one PID or stream_id of the MPEG-2 multiplex. Each SL packet is mapped to one PES packet. PES and SL packet header redundancy is reduced by conveying information only in the PES header, removing duplicate data from the SL packet header.

An integer number of FlexMux packets can also be conveyed in a PES packet to further reduce multiplex overhead. Several SL-packetized streams can then be mapped to one MPEG-2 PID or stream_id. The PES header is not used at all, since synchronization can be done with the time stamp information conveyed in the SL packet headers.

11.8.3 MP4 File Format

A file format for the exchange of MPEG-4 content has also been defined. The file format supports metadata in order to allow indexing, fast searches, and random access.

11.9 Intellectual Property Management and Protection (IPMP)

IPMP, also called digital rights management (DRM), provides an interface and tools, rather than a complete system, for implementing intellectual property rights management.

The level and type of management and protection provided is dependent on the value of the content and the business model. For this reason, the complete design of the IPMP system is left to application developers.

The architecture enables both open and proprietary solutions to be used, while enabling interoperability, supporting the use of more than one type of protection (i.e., decryption, watermarking, rights management, etc.) and supporting the transferring of content between devices using a defined inter-device message (reflecting the issue of content distribution over home networks).

For protected content, the IPMP tool requirements are communicated to the decoder before the presentation starts. Tool configuration and initialization information is conveyed by the IPMP Descriptor or IPMP elementary stream. Needed tools can be embedded, downloaded, or acquired by other means.

Control point and ordering sequence information in the IPMP Descriptor allows different tools to function at different places in the system. IPMP data, carried in either an IPMP Descriptor or IPMP elementary stream, includes rights containers, key containers, and tool initialization data.

11.10 MPEG-4.10 (H.264) Video

Previously known as "H.26L," "JVT," "JVT codec," "AVC," and "Advanced Video codec," ITU-T H.264 is one of two new video codecs, the other being SMPTE 421M (VC-1), which is based on Microsoft Windows Media Video 9 codec. H.264 is incorporated into the MPEG-4 specifications as Part 10.

Rather than a single major advancement, H.264 employs many new tools designed to improve performance. These include:

- Support for 8-, 10-, and 12-bit 4:2:2 and 4:4:4 YCbCr

- Integer transform

- UVLC, CAVLC, and CABAC entropy coding

- Multiple reference frames

- Intra prediction

- In-loop deblocking filter

- SP and SI slices

- Many new error resilience tools

11.10.1 Profiles and Levels

Similar to other video codecs, profiles specify the syntax (i.e., algorithms) and levels specify various parameters (resolution, frame rate, bit-rate, etc.). The various levels are described in Table 11.5.

11.10.1.1 Baseline Profile (BP)

Baseline profile is designed for progressive video such as video conferencing, video-over-IP, and mobile applications. Tools used by Baseline profile include:

- I and P slice types

- ¼-pixel motion compensation

- UVLC and CAVLC entropy coding

- Arbitrary slice ordering (ASO)

- Flexible macroblock ordering (FMO)

- Redundant slices (RS)

- 4:2:0 YCbCr format

Note that Baseline profile is not a subset of Main profile. Many solutions implement a subset of Baseline profile, without ASO or FMO; this is a subset of Main profile (and much easier to implement).

11.10.1.2 Extended Profile (XP)

Extended profile is designed for mobile and Internet streaming applications. Additional tools over Baseline profile include:

- B, SP, and SI slice types

- Slice data partitioning

- Weighted prediction

Table 11.5: MPEG-4.10 (H.264) Levels ("MB" = macroblock, "MV" = motion vector)

Level	Maximum MB per Second	Maximum Frame Size (MB)	Typical Frame Resolution	Typical Frames per Second	Maximum MVs per Two Consecutive MBs	Maximum Reference Frames	Maximum Bit-Rate
1	1,485	99	176×144	15	–	4	64 kbps
1.1	3,000	396	176×144	30	–	9	192 kbps
			320×240	10		3	
			352×288	7.5		3	
1.2	6,000	396	352×288	15	–	6	384 kbps
1.3	11,880	396	352×288	30	–	6	768 kbps
2	11,880	396	352×288	30	–	6	2 Mbps
2.1	19,800	792	352×480	30	–	6	4 Mbps
			352×576	25			
2.2	20,250	1,620	720×480	15	–	5	4 Mbps
			720×576	12.5			
3	40,500	1,620	720×480	30	32	5	10 Mbps
			720×576	25			
3.1	108,000	3,600	1280×720	30	16	5	14 Mbps
3.2	216,000	5,120	1280×720	60	16	4	20 Mbps
4	245,760	8,192	1920×1080	30	16	4	20 Mbps
			1280×720	60			
4.1	245,760	8,192	1920×1080	30	16	4	50 Mbps
			1280×720	60			
4.2	491,520	8,192	1920×1080	60	16	4	50 Mbps
5	589,824	22,080	2048×1024	72	16	5	135 Mbps
5.1	983,040	36,864	2048×1024	120	16	5	240 Mbps
			4096×2048	30			

11.10.1.3 Main Profile (MP)

Main profile is designed for a wide range of broadcast applications. Additional tools over Baseline profile include:

- Interlaced coding

- B slice type

- CABAC entropy coding

- Weighted prediction

- 4:2:2 and 4:4:4 YCbCr, 10- and 12-bit formats

- ASO, FMO, and RS are not supported

11.10.1.4 High Profiles (HP)

After the initial specification was completed, the Fidelity Range Extension (FRExt) amendment was added. This resulted in four additional profiles being added to the specification:

- **High Profile (HP)**: adds support for adaptive selection between 4×4 and 8×8 block sizes for the luma spatial transform and encoder-specified frequency-dependent scaling matrices for transform coefficients

- **High 10 Profile (Hi10P)**: adds support for 9- or 10-bit 4:2:0 YCbCr

- **High 4:2:2 Profile (Hi422P)**: adds support for 4:2:2 YCbCr

- **High 4:4:4 Profile (Hi444P)**: adds support for 11- or 12-bit samples, 4:4:4 YCbCr or RGB, residual color transform and predictive lossless coding

11.10.2 Supplemental Enhancement Information (SEI) Messages

Supplemental enhancement information (SEI) messages assist in processes related to decoding, display, or other purposes. SEI messages include:

- Buffering period

- Picture timing

- Pan-scan rectangle

- Filler payload

- User data registered

- User data unregistered

- Recovery point

- Decoded reference picture marking repetition

- Spare picture

- Scene information

- Sub-sequence information

- Sub-sequence layer characteristics

- Sub-sequence characteristics

- Full-frame freeze

- Full-frame freeze release

- Full-frame snapshot

- Progressive refinement segment start

- Progressive refinement segment end

- Motion-constrained slice group set

The Fidelity Range Extension (FRExt) amendment added three new supplemental enhancement information (SEI) messages:

- Film grain characteristics

- Deblocking filter display preference

- Stereo video

11.10.3 Video Coding Layer

11.10.3.1 YCbCr Color Space

H.264 uses the YCbCr color space, supporting 4:2:0, 4:2:2, and 4:4:4 sampling. The 4:2:2 and 4:4:4 sampling options increase the chroma resolution over 4:2:0, resulting in better picture quality. In addition to 8-bit YCbCr data, H.264 supports 10- and 12-bit YCbCr data to further improve picture quality.

11.10.3.2 Macroblocks

With H.264, the partitioning of the 16×16 macroblocks has been extended, as illustrated in Figure 11.5.

Such fine granularity leads to a potentially large number of motion vectors per macroblock (up to 32) and number of blocks that must be interpolated (up to 96). To constrain encoder/decoder complexity, there are limits on the number of motion vectors used for two consecutive macroblocks.

Error concealment is improved with Flexible Macroblock Ordering (FMO), which assigns macroblocks to another slice so they are transmitted in a non-scanning sequence.

This reduces the chance that an error will affect a large spatial region, and improves error concealment by being able to use neighboring macroblocks for prediction of a missing macroblock.

11.10.3.3 In-loop Deblocking Filter

H.264 adds an in-loop deblocking filter. It removes artifacts resulting from adjacent macroblocks having different estimation types and/or different quantizer scales. The filter also removes artifacts resulting from adjacent blocks having different transform/quantization and motion vectors.

The loop filter uses a content adaptive nonlinear filter to modify the two samples on either side of a block or macroblock boundary.

11.10.3.4 Slices

The slice has greater importance in H.264 since it is now the basic independent spatial element. This prevents an error in one slice from affecting other slices. This flexibility allows

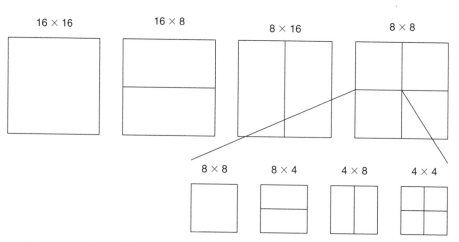

Figure 11.5: Segmentations of H.264 macroblocks for motion compensation. Top: segmentation of macroblocks. Bottom: segmentation of 8 × 8 partitions

extending the I-, P-, and B-picture types down to the slice level, resulting in I-, P-, and B-slice types.

Arbitrary Slice Ordering (ASO) enables slices to be transmitted and received out of order. This can improve low-delay performance in video conferencing and Internet applications.

Redundant slices are also allowed for additional error resilience. This alternative data can be used to recover any corrupted macroblocks.

SP- and SI-Slices

In addition to I-, P-, and B-slices, H.264 adds support for SP-slices (Switching P) and SI-slices (Switching I). SP-slices use motion compensated prediction, taking advantage of temporal redundancy to enable reconstruction of a slice even when different reference slices are used. SI-slices take advantage of spatial prediction to enable identically reconstructing a corresponding SP-slice.

Use of S-slices enable efficient bitstream switching, random access, fast-forward, and error resilience/recovery, as illustrated in Figures 11.6 and 11.7.

11.10.3.5 Intra Prediction

When motion estimation is not efficient, intra prediction can be used to eliminate spatial redundancies. This technique attempts to predict the current block based on adjacent blocks. The difference between the predicted block and the actual block is then coded. This tool is very useful in flat backgrounds where spatial redundancies often exist.

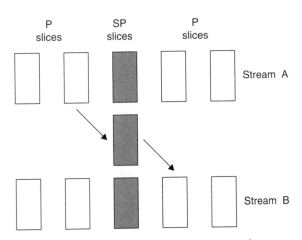

Figure 11.6: Using SP slices to switch to another stream

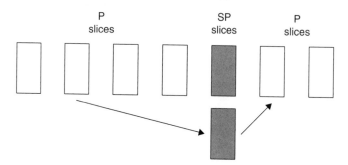

Figure 11.7: Using SP slices to fast-forward

11.10.3.6 Motion Compensation

¼-Pixel Motion Compensation

Motion compensation accuracy is improved from the ½-pixel accuracy used by most earlier video codecs. H.264 supports the same ¼-pixel accuracy that is used on the latest MPEG-4 video codec.

Multiple Reference Frames

H.264 adds supports for multiple reference frames. This increases compression by improving the prediction process and increases error resilience by being able to use another reference frame in the event that one was lost.

A single macroblock can use up to 8 reference frames (up to 3 for HDTV), with a total limit of 16 reference frames used within a frame.

To compensate for the different temporal distances between current and reference frames, predicted blocks are averaged with configurable weighting parameters. These parameters can either be embedded within the bitstream or the decoder may implicitly derive them from temporal references.

Unrestricted Motion Search

This allows for reference frames that are outside the picture. Missing data is spatially predicted from boundary data.

11.10.3.7 Transform, Scaling, and Quantization

H.264 uses a simple 4×4 integer transform. In contrast, older video codecs use an 8×8 DCT that operates on floating-point coefficients. An additional 2×2 transform is applied to the four CbCr DC coefficients. Intra 16×16 macroblocks have an additional 4×4 transform performed for the sixteen Y DC coefficients.

Blocking and ringing artifacts are reduced as a result of the smaller block size used by H.264. The use of integer coefficients eliminates rounding errors that cause drifting artifacts common with DCT-based video codecs.

For quantization, H.264 uses a set of 52 uniform scalar quantizers, with a step increment of about 12.5% between each.

The quantized coefficients are then scanned, from low frequency to high frequency, using one of two scan orders.

11.10.3.8 Entropy Coding

After quantization and zig-zag scanning, H.264 uses two types of entropy encoding: variable-length coding (VLC) and Context Adaptive Binary Arithmetic Coding (CABAC).

For everything but the transform coefficients, H.264 uses a single Universal VLC (UVLC) table that uses an infinite-extend codeword set (Exponential Golomb). Instead of multiple VLC tables as used by other video codecs, only the mapping to the single UVLC table is customized according to statistics.

For transform coefficients, which consume most of the bandwidth, H.264 uses Context-Adaptive Variable Length Coding (CAVLC). Based upon previously processed data, the best VLC table is selected.

Additional efficiency (5–10%) may be achieved by using Context Adaptive Binary Arithmetic Coding (CABAC). CABAC continually updates the statistics of incoming data and real-time adaptively adjusts the algorithm using a process called context modeling.

11.10.4 Network Abstraction Layer (NAL)

The NAL facilitates mapping H.264 data to a variety of transport layers including:

- RTP/IP for wired and wireless Internet services
- File formats such as MP4
- H.32X for conferencing
- MPEG-2 systems

The data is organized into NAL units, packets that contain an integer number of bytes. The first byte of each NAL unit indicates the payload data type and the remaining bytes contain the payload data. The payload data may be interleaved with additional data to prevent a start code prefix from being accidentally generated.

When data partitioning is used, each slice is divided into three separate partitions, with each partition using a specific NAL unit type. This enables data partitioning to be used as an efficient layering method that separates the data into different levels of importance. By partitioning data into different NAL units, it is much easier to use different error protection for various parts of the data.

Bibliography

H.264 Advanced Video Coding: A Whirlwind Tour, by PixelTools, 2003.

H.264 Coding Efficiency Has a Price, by Eric Barrau, Philippe Durieuz, and Stephane Muta, Sophia Antipolis Micro-Electronics Forum, January 2003.

ISO/IEC 14996–1, Information Technology—Coding of Audio-Visual Objects—Part 1: Systems.

ISO/IEC 14996–2, Information Technology—Coding of Audio-Visual Objects—Part 2: Visual.

ISO/IEC 14996–3, Information Technology—Coding of Audio-Visual Objects—Part 3: Audio.

ITU-T H.264, Advanced Video Coding for Generic Audiovisual Services, May 2003.

Technical Overview of H.264/AVC, by R. Schafer, T. Wiegand, and H. Schwarz, EBU Technical Review, January 2003.

The H.264/AVC Video Coding Standard for the Next Generation Multimedia Communication, by M. Mahdi Ghandi and Mohammad Ghanbari, IAEEE invited paper.

CHAPTER 12

Liquid Crystal Displays

K.F. Ibrahim

Liquid crystal displays have come a long way since the first monochrome versions appeared in digital quartz wrist watches in the 1970s. Today TVs with LCD displays far outsell their CRT counterparts, and almost all portable consumer electronics devices—including well over a billion cell phones sold each year—use them. This is clearly a technology that works.

But how does it work? To start with, liquid crystals are neither liquid nor crystals, though they display the characteristics of both. They are large, elongated molecules whose shape can be altered by an electrical field, which in turn determines how much light they absorb. Liquid crystals occur in nature, some of which are used in organic light-emitting diodes (OLEDs), which are commonly found on the outer display on clamshell-type handsets. This phenomenon was discovered as far back as 1888 when Viennese chemist Friedrich Reinitzer noted the crystalline nature of cholesterol he extracted from carrots (which I previously thought were healthy). Most LCDs today are based on good old biphenyl, cyclohexane and ester.

The most common LCD displays place a liquid crystal emulsion between two plates of glass; one plate is scored with horizontal lines, the other with vertical ones. The crystals line up along the surface of the ridges, forcing the intervening molecules to twist into a helical shape. Light passing through the panel changes polarization by 90 degrees. Passing a current through the plate forces the crystals to change shape, the degree determined by the applied voltage. Such twisted nematic displays enable fine control of the amount of light passing through them. Super-twist nematic screens can twist the LC molecules by 180–270 degrees, improving the contrast ratio. Most displays today are this type.

Things quickly get more complicated when you go to active-matrix (AMLCD), full-color, high-resolution displays. There are several types of AMLCD screens, depending on the type of switching used. In this chapter K.F. Ibrahim clearly explains the nature and structure of AM TFT LCD screens, from both the silicon and electrical points of view.

The physics of TFT LCD cells imposes some limitations that designers have to work with. For example, the lack of the phosphor persistence found in CRTs dictates the need for a pixel-cell "memory"; this is implemented in a separate capacitor for each pixel that maintains its luminence while the other lines are being scanned. The need to charge and discharge these capacitors detracts from AMLCD's response times, which are nowhere near as fast as CRTs. Ibrahim explains a number of techniques for dealing with the response time problem, none of which involve painless trade-offs.

LCDs are popular in portable devices in large part because they consume so little power; the same can't be said for their backlights, which are the biggest energy hogs in handsets. Cold-cathode fluorescent tubes (CCFTs) are more efficient than the alternatives, but anything that operates at 2500–3500 V_{P-P} and causes flicker and RF interference is not without its problems. Ibrahim discusses methods to address these problems as well as how to regulate brightness by combined backlight control and transmissive scaling—which in turn induces other problems that need to be corrected.

This chapter provides a clear introduction to both the technology of TFT LCD screens and the problems you'll encounter in incorporating them into your next portable design.

—**John Donovan**

Liquid crystal display (LCD) units used for the purposes of moving picture reproduction are some of the more popular flat panel displays. Like all flat panel displays, LCDs employ a matrix structure in which the active element, in this case a liquid crystal (LC), forming the pixel cell is located at the intersection of two electrode buses.

So, what is a liquid crystal? An LC is neither crystal nor liquid. It exhibits liquid-like as well as crystal-like properties. This feature is a result of the LC's comparatively elongated molecules and their structure. Though an LC is a natural material, the liquid crystal which is used for LC displays is a multi-component mixture that is artificially created by blending of biphenyl, cyclohexane, ester and the like.

12.1 Polarization

Light is a transverse electromagnetic (EM) wave composed of an electric and a magnetic field. The two fields are at right angles to each other traveling at the speed of light. In an EM wave, the electric field defines the designation of the wave in terms of its polarization: if the electric field is vertical, the wave is said to be *vertically polarized* or "*p-polarized*" (Figure 12.1) and conversely if the electric field is horizontal, the wave is said to be *horizontally polarized* or *"s-polarized"*. Natural light from the sun or any other light source such as a lamp is unpolarized. It contains both vertical and horizontal polarization. Light may become polarized if the vertical or horizontal polarization is reduced or removed completely by, for instance a polarizing filter, the type used in Polaroid sunglasses. If the horizontal polarization is removed, the light would be vertically polarized and vice versa. This simple principle is used to control the brightness of an LC cell.

12.2 Principles of Operation of LC Cell

By themselves, the molecules in an LC are arranged in a loose order. However, when they come into contact with a finely grooved surface, the molecules line up in parallel along grooves of alignment layer as shown in Figure 12.2.

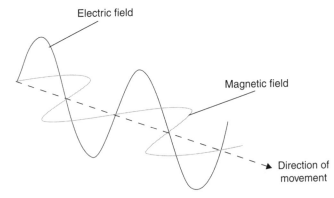

Figure 12.1: Vertically polarized electromagnetic (EM) wave

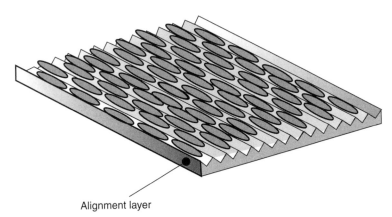

Figure 12.2: Alignment of LC molecules along tiny grooves of glass plate

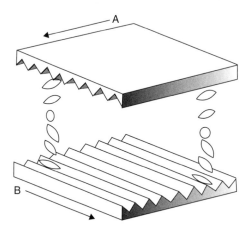

Figure 12.3: Twist-Nematic (TN) liquid crystal (LC) with a 90° twist

Furthermore, the application of an electric field across the LC causes a change in the molecular structure. This change affects the optical properties of the crystal in the way light is reflected off it or passes through it.

The most popular type of LC is the *Twist-Nematic* (*TN*). In this mode, the grooves in the two plates are at right angles as shown in Figure 12.3. The molecules along the upper plate point in direction "A" and those along the lower plate in direction "B" thus forcing the molecules of LC to arrange themselves in a helical form. The helix has the effect of twisting the EM wave (light) passing through it by 90°. Thus, if a vertically polarized light is forced through such a crystal, it will suffer a 90° twist and become horizontally polarized. However, what is special about LCs is that if an electric field is applied across it (Figure 12.4), the helical structure begins to break down and with it the polarisation of light resulting in a smaller twist than the natural 90°. The voltage level determines the extent to which breakdown occurs and the amount of twist. If a high enough voltage, in the region of 8 V, is applied, the twist is eliminated completely.

12.3 Reflective and Transmissive

An LC does not produce light, so the technology is *non-emissive* and therefore does not give off a glow like a cathode ray tube (CRT) or a plasma panel does. An external form of light is therefore necessary which may be provided in two ways for two types of LC displays: *reflective* and *transmissive*. In the reflective type, the change in the molecular structure controls the reflected light while in the transmissive type it controls the light passing through

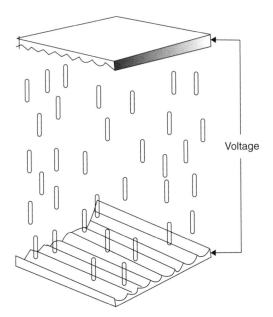

Voltage

Figure 12.4: TN crystal molecules rearranged removing the twist when a voltage is applied

it. The former is dependent on an external or ambient light for its brightness while the transmissive type has its own backlight and is not dependent on ambient light. For this reason, the transmissive type is the more popular of the two. In either case, the voltage across the LC controls its luminance.

12.4 The TN Transmissive LCD

Consider two differently polarized filter plates placed opposite each other with a backlight unit as shown in Figure 12.5. Plate A allows only vertically polarized light through while plate B permits only horizontally polarized light. The effect of the two glass plates is to block the unpolarized light emanating from the backlight completely. Now, consider the same arrangement with an LC placed between the two polarized glass plates as illustrated in Figure 12.6. The unpolarized light from the backlight passes through plate A, becomes vertically polarized and goes through the LC which, without any voltage across it, forces a 90° twist changing its polarisation from vertical to horizontal which pass through the second plate B without any hindrance. If now a voltage is applied across the LC, the 90° twist would be removed and light would be blocked. If a smaller voltage is applied, a twist angle less than 90° is introduced by the LC and low-intensity light would appear at the other end. Since the

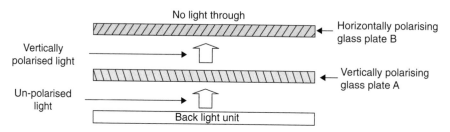

Figure 12.5: Light is blocked by two opposite polarized filters

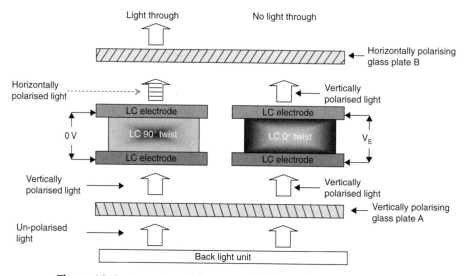

Figure 12.6: Amount of light controlled by voltage across LC cell

applied voltage determines the twist angle of the LC, it follows that the voltage now controls the intensity of light appearing at the other end and a greyscale may thus be obtained by varying the voltage applied across the LC.

12.5 Normally White and Normally Black

The LC display may be used in two different modes: *normally white* (or bright) and *normally black* (or dark). The former allows the backlight through while the latter blocks the backlight when the voltage across the LC cell is zero. The arrangement in Figure 12.6 is that for the more popular normally white LCD. A normally black LCD would have only one polarising plate.

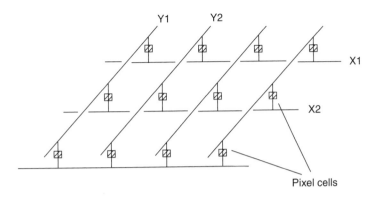

Figure 12.7: Passive-matrix LCD (PMLCD)

There are several types of TNLC cells depending on the angular twist the molecules are subjected to. In the simple TN type, the molecules are twisted by 90° resulting in a drop in contrast when used with large screens. The *Super Twist-Nematic* (*STN*) has its molecules twisted from 180 to 260° to improve the contrast ratio. Finally, the *Film Super Twist-Nematic* (*FSTN*) twists the molecules by 360°. This is used for very high quality black and white LCDs.

12.6 Passive- and Active-matrix LCDs

There are two matrix LCD technologies: passive-matrix LCD (PMLCD) and active-matrix LCD (AMLCD). In the PMLCD, pixels are addressed directly with no switching devices involved in the process as illustrated in Figure 12.7. The effective voltage applied to the LC must average the signal voltage pulses over several frame times, which results in a slow response time greater than 150 ms and a reduction of the maximum contrast ratio. The addressing of a PMLCD also produces a kind of cross-talk resulting in blurred images because nonselected pixels are driven through a secondary signal-voltage path. This places a limit to the number of pixels that may be used in a display and with it a limit on the maximum resolution.

In the AMLCDs, on the other hand, a switching device is used to apply the voltage across the LC (Figure 12.8) and hence a better response time becomes possible. In contrast to PMLCDs, the active type, AMLCDs has no inherent limitation in the number of pixels, and they present fewer crosstalk problems.

There are several kinds of AMLCD depending on the type of switching device used. Most use transistors made of deposited thin films, which are accordingly called *thin-film transistors* (*TFTs*). The most common TFT semiconductor material is made of *amorphous silicon* (*a-Si*).

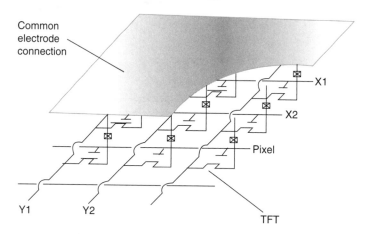

Figure 12.8: Active-matrix LCD (AMLCD)

a-Si TFTs are amenable to large-area fabrication using glass substrates in a low temperature (300–400°C).

An alternative TFT technology, polycrystalline silicon, normally known as *polysilicon* or *p-Si* is costly to produce and especially difficult to fabricate when manufacturing large-area displays. Nearly, all TFT LCDs are made from a-Si because of the technology's economy and maturity, but the electron mobility of a p-Si TFT is 100 times better than that of an a-Si TFT. This makes the p-Si TFT a good candidate for a TFT array containing integrated drivers, which is likely to be an attractive choice for small, high definition displays such as view finders and projection displays.

12.7 TFT Cell Drive

In the TFT LCD, switching transistors are provided for each pixel cell as shown in Figure 12.9. One side of each LC cells is connected to its own individual TFT while the other side is connected to a common electrode which is made of transparent *indium tin oxide* (*ITO*) material. This is necessary to ensure high aperture ratio. *Aperture* ratio is the ratio of the transparent area to the opaque area of the panel. A cross-section of a TFT is shown in Figure 12.10.

Unlike the CRT in which the phosphor persistence allows for continued luminance of the picture even after the electron beam has moved to scan other lines, in flat display applications, no such persistence exists and refreshing of pixels to produce natural moving pictures becomes difficult as the number of pixels increases. Hence, the need for a pixel cell

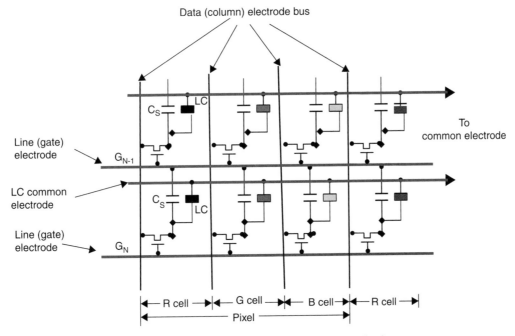

Figure 12.9: Equivalent circuit for TFT-LC display

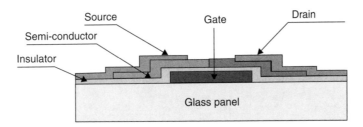

Figure 12.10: TFT cross-section

"memory". A charge on a storage capacitor C_S is used for this purpose as illustrated in Figure 12.9. Each cell consists of three sub-pixels (RGB) normally referred to as cells. Each cell contains an LC driven by a TFT acting as a switch. The LC is placed within two electrodes. One electrode is connected to the TFT's source electrode and the other goes to a common electrode.

The TFT-LCD panel is scanned line by line. Each line is selected in turn by a V_{SEL} pulse to the *line* (or *gate*) *electrode* bus. Once a line is selected, the pixel cells along that line can be addressed and their luminance levels set by a voltage applied via a source driver to their corresponding data (also known as *source* or *column*) *electrode*. The source driver supplies the desired voltage level known as the *greyscale voltage* representing the pixel value, i.e., the luminance of the pixel cell. The storage capacitor C_S is charged and this charge maintains the luminance level of the pixel cell while the other lines are being scanned. When all the lines have been scanned and all the pixel cells addressed, the process is repeated for the next frame and the picture is refreshed.

Figure 12.11 shows the operation of a TFT-LC cell where G_N is the currently selected gate line and G_{N-1} is the immediately preceding gate line. The TFT gate is connected to the line (or gate) electrode bus, also known as the gate bus and the drain is connected to the data (or column) bus, also known as the source bus. Storage capacitor C_S is connected between the current gate line G_N and the immediately preceding gate line (G_{N-1}). For this reason, C_S is known as C_S-*on-gate*. It forms the drain load for the TFT. The TFT turns fully on when its gate voltage is 20 V and turns off when its gate goes to at least -5 V. To select the pixel cell, a 20 V pulse, V_{SEL}, is applied to the gate. At the same time, data in the form of an analog positive voltage V_{DAT} is applied to the drain. For peak white, V_{DAT} is 0 V while for pitch black V_{DAT} is a maximum of about 8 V. With the TFT on, the source and drain are shorted and V_{DAT} is applied across the LC. The storage capacitor, C_S-on-gate charges up and this charge is sustained even when the TFT is turned off. This is then repeated for the next line and so on. The main function of C_S is to maintain the voltage across the LC until the next line select

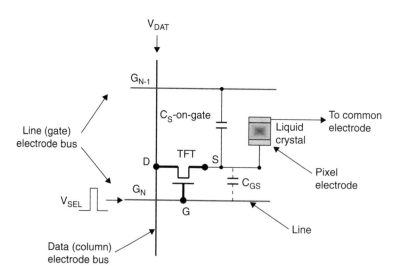

Figure 12.11: TFT cell equivalent circuit

voltage is applied when the picture is refreshed. A large C_S can improve the voltage holding ratio of the pixel cell and improve the contrast and flicker. However, a large C_S results in higher TFT load and lower aperture ratio. In determining the value of the storage capacitance, account must be taken of the stray capacitance between the TFT's gate and source, G_{GS} which is effectively in parallel with C_S.

12.8 Response Time

Response time is one of the few areas remaining where the performance of a traditional CRT still holds an advantage over LCD displays. CRTs have nearly instantaneous pixel response times, but LCDs tend to be much slower. The result is the user might see smearing, motion blur or other visual artefacts when there is movement on the screen.

A pixel's response time is the time it takes a pixel to change state. If it is a *rise-and-fall* response, then it is a measure of the time it takes a pixel to change state from black-to-white-to-black as illustrated in Figure 12.12. More specifically, it represents the pixel ability to change from 10% "on" to 90% "on" and then back from 10% "off" to 90% "off" again. Originally, this was the standard way of reporting response times of LCD TVs and computer monitors, and was normally listed as a *TrTf* (*time rising, time falling*) measurement. Some manufacturers started using a *grey-to-grey* (*GtG*) measurement for LCD response times which is different from TrTf. There are as yet no standards and manufacturers can state any figure that suits them.

One factor that affects the response time is the viscosity of the LC material. This means it takes a finite time to reorientate its molecules in response to a changed electric field. A second

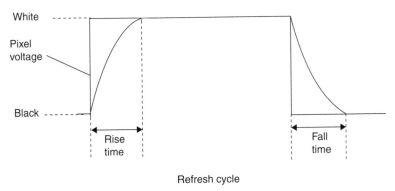

Figure 12.12: *Time rising, time falling* (TrTf) response of an LC

factor is that the capacitance of the LC material is affected by the molecule realignment which changes the TFT load and with it the brightness to which the cell ultimately settles.

A good response time starts at around 25 ms with some LCD TV manufacturers claiming a response time as fast as 16 ms or less. Short response times are required for fast moving images such as games. New techniques have been developed to reduce the response time. Such techniques include the use of lower viscosity LC material. Reducing the cell gap thickness is another technique which results in fewer LC material to reorientate giving a response time as little as 8 ms. Thinner cells make production more difficult with lower yields and hence more expensive. Another technique apply a drive signal for a brief duration in order to give the pixel cells a "jump start" and then reducing it to the required level as illustrated in Figure 12.13. This technique known as *amplified impulse* provides grey-to-grey transition to be completed up to five times faster than a typical LC display.

For TV images, two techniques have been developed, both attempt to hide the cell transition time. *Backlight strobing* involves flickering the backlight off momentarily and the *black frame insertion* introduces a black frame during the LC transition. Backlight strobing also helps improve motion blur caused by the *sample-and-hold* effect in which an image when held on the screen for the duration of a frame-time, blurs the retina as the eye tracks the motion from one frame to the next. By comparison, when an electron beam sweeps the surface of a cathode ray tube, it lights any given part of the screen only for a miniscule fraction of the frame time. It's a bit like comparing film or video footage shot with low- and high-shutter speeds. This type of motion-blur has come about as manufacturers moved from the traditional resistor type digital-to-analog converter (DAC) to the much more compact sample-and-hold type.

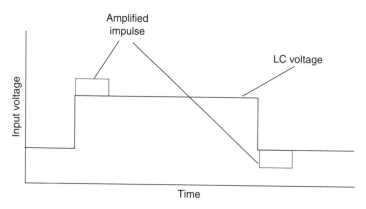

Figure 12.13: Amplified impulse LC cell drive technique

Motion blur originating from sample-and-hold in the display can become less of an issue as the frame rate is increased

12.9 Polarity Inversion

In LC cells, it is the magnitude of the applied voltage which determines the amount of light transmission. Such voltage may be DC or AC. Applying a voltage of the same (DC) polarity to an LC cell would cause electroplating of one electrode resulting in what is known as *DC stress* causing deterioration in image quality. To prevent polarization (and rapid permanent damage) of the LC material, the polarity of the cell voltage is reversed, a process known as *polarity inversion*. Polarity inversion may be implemented in three different ways: *frame inversion, line* (or horizontal) *inversion* and *dot inversion* (Figure 12.14). It will be noticed that line inversion incorporates frame inversion as well since a positive line in one frame becomes negative in the following frame and vice versa.

Unfortunately, it is very difficult to get exactly the same voltage on the cell in both polarities, so the pixel-cell brightness will tend to flicker. This flicker is most noticeable with frame inversion in which the polarity of the whole screen is inverted once every frame resulting in a 25-Hz and 30-Hz flicker for PAL and NTSC, respectively. Flicker may be reduced by having the polarity of adjacent lines using line inversion thus cancelling out the flicker. Better results may be obtained with dot inversion. In this way, the flicker can be made imperceptible for most "natural" images.

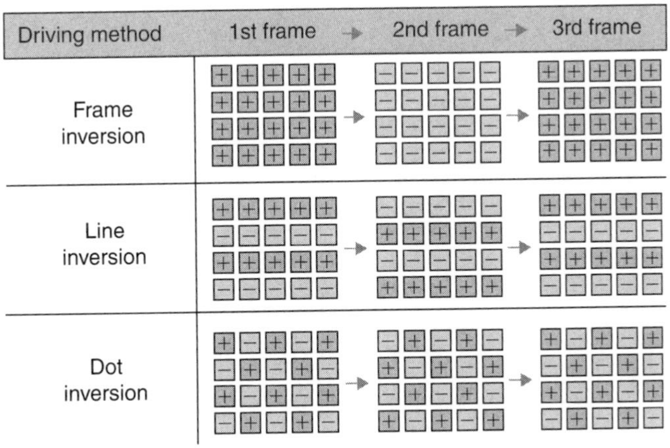

Figure 12.14: Polarity inversion

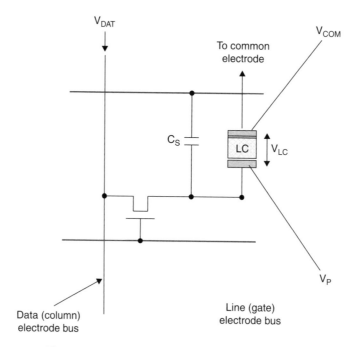

Figure 12.15: Voltage across LC, $V_{LC} = V_P - V_{COM}$

Polarity inversion is carried out by inverting both the pixel cell electrode V_P and voltage at the common electrode, V_{COM} frame-by-frame, line-by-line or dot-by-dot. Referring to Figure 12.15, the voltage across the LC cell, V_{LC} is the difference between V_P and V_{COM}:

$$V_{LC} = V_P - V_{COM}$$

When both V_P and V_{COM} are inverted,

$$V'_{LC} = -V_P - (-V_{COM})$$
$$V'_{LC} = -V_P + V_{COM}$$
$$V'_{LC} = -(V_P - V_{COM})$$

The line inversion sequence for three lines of a plain white screen is shown in Figure 12.16. For line n, the pixel voltage V_P for all the pixels on the line is high to remove the 90° twist of the LC and remains constant throughout "line n" since all pixels are at the same luminance level. They are then inverted for the following "line $n+1$" and so on. The pixel voltage V_{COM} is also constant over one whole line and inverted for the following line. For a 5-step greyscale display, the pixel values change along the line as shown in Figure 12.17 starting with V_{white}

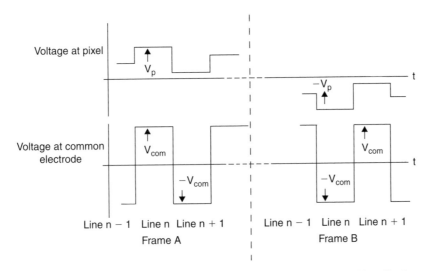

Figure 12.16: Line inversion pixel voltage sequence for plain white display

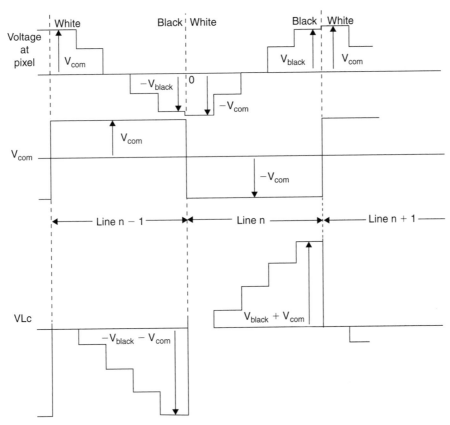

Figure 12.17: Line inversion waveform for a 5-step greyscale display

for peak white and $-V_{\text{black}}$ for black at the end of the line. This is then inverted and repeated for the next line and so on. V_{COM}, on the other hand, is constant over one line and inverted over the next. The LC voltage V_{LC}, being the difference between V_P and V_{COM} is that shown in Figure 12.17 inverted every line. For dot inversion, the pixel voltage and the common electrode voltage will invert on successive dots.

12.10 Greyscale and Color Generation

In general, there are two ways of generating greyscale: digital and analog. The digital method includes the sub-field coding known as *time ratio* technique employed in plasma panels. The other digital technique is the *area ratio* in which the pixel cell is divided into several smaller areas, e.g. 6, 8 or 10. Each area is energized independently. The sum total of the luminance of the separate areas is the luminance of the pixel cell. The *analog method* is by far the simplest technique by which luminance is determined by the instantaneous value of the voltage fed into the data bus and this is the method used in LCD panels.

With the video data being in digital format, the analog technique requires a digital-to-analog DAC. Traditionally, a resistor-DAC was used employing a resistor ladder to generate a set of binary-weighted reference voltages, which may be combined to produce an instantaneous voltage that represents the value of the signal. Given eight different reference voltages, $2^8 = 256$ different greyscale voltage values may be produced. In the case of LCD panel drives, two separate sets of reference voltages are needed because of the asymmetrical gamma characteristics encountered during polarity inversion. The resistor chains may of course be incorporated within the DAC chip itself which can also provide the necessary gamma correction. Current LCD panels use linear DACs and sample-and-hold architecture in which the DACs are shared by the column video data thus reducing their number and with it the chip size.

To produce color, each pixel is divided into three R, G and B sub-pixels or cells in the same way as a plasma pixel. While this arrangement can produce a greyscale, it cannot generate a color image. To do that, three filters, one for each of the three primary colors are provided on a *masked filter* substrate. The RGB elements of this color filter line up one-to-one with the pixel cells on the TFT-array substrate. Because the cells are too small to distinguish independently, the RGB elements appear to the human eye as a mixture of the three colors. Practically, any color can thus be produced by mixing these three primary colors. Pixel cells may be arranged in three formats: *vertical stripe*, *horizontal stripe* and the *triad* or *delta* as illustrated in Figure 12.18. The most popular with manufacturers is the vertical stripe which is sometimes described as $3\,\text{m} \times n$ format. Thus, for an SVGA-resolution LCD panel (resolution

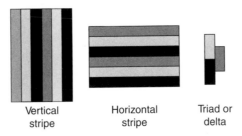

Vertical stripe   Horizontal stripe   Triad or delta

Figure 12.18: Color filter formats

800×600), the total number of pixel cells $= 800 \times 3 \times 600 = 1{,}440{,}000$ with the number of columns $= 800 \times 3 \times 2400$.

The number of colors provided by the LC display is determined in the same way as that for plasma panels namely by the number of combinations of R, G and B that may be produced given a particular greyscale levels. In general, the number of colors $= 2^n$ where n is the total number of bits allocated to the video which for a 24-bit video is 24 bits. Given this, the number of different colors $= 2^{24} = 16.78$ million. Alternatively, the same figure may be produced by multiplying the greyscales levels of each color, namely $256 \times 256 \times 256 = 16.78$ million.

12.11 Panel Drive

In a TFT-LCD panel, each pixel cell may be addressed by selecting two electrodes: the line (or gate) electrode and the data (or source) electrode with respective line and data drives as illustrated in Figure 12.19. 24-bit video is fed into the LCD controller which addresses each line in turn by a line select voltage which turns all the TFTs on that line on. At the same time, the controller feeds the corresponding RGB video data for the pixel cells of the selected line to the greyscale generator. The greyscale generator consists of a DAC and a sample-and-hold circuit. It converts the digital greyscale value of each cell into an equivalent analog voltage that when fed into the LC electrode produces the correct luminance. The sample-and-hold ensures the analog value remains steady long enough for the cell to respond to any change in value. Before going into the cell electrode, the voltage is inverted by a line sync for line inversion (shown) or by a dot frequency pulse for dot inversion. Similarly, the common electrode voltage V_{COM} is inverted by a line sync pulse of a dot-frequency pulse. The backlight which forms a part of the panel assembly is fed by a sine wave in the region of 2000 V from an external DC–AC converter.

The analog cell values are fed into the panel using a *tape carrier package (TCP)* in which an LSI chip is installed on a thermostabile film and sealed with plastic as illustrated in Figure 12.20.

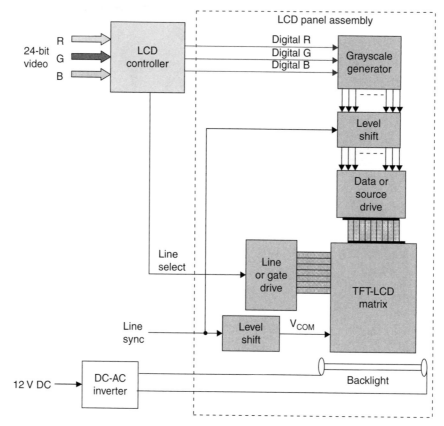

Figure 12.19: TFT-LCD panel drive

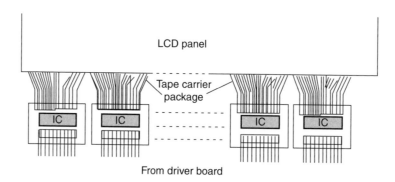

Figure 12.20: Tape carrier package connection to the LCD panel

Figure 12.21: Symptom of a faulty tape carrier package (TCP) chip or tape carrier itself

The chip may be a driver or a shift register or both delivering not an inconsiderable power to the cells, power that must be dissipated using a heat sink. The heat sink normally forms part of the panel frame. A faulty TCP IC would result in vertical lines along the section of the screen which is fed by the particular TCP as shown in Figure 12.21. This is an unrepairable fault and the whole panel must be changed. The same symptom would be observed if the tape carrier is damaged in any way.

12.12 The Backlight Assembly

There are two types of backlight formats: the *guided type* for screen of 20 in. or less and the *direct type* for larger screen sizes. Both use *cold cathode fluorescent tubes* (*CCFT*) for their low energy consumption and low cost. The guided type is slim and compact but suffers from complicated structure and low light efficiency. By contrast, the direct type is thick in structure but simpler in structure with high efficiency.

The assembly of the guided type consists of one CCFT on either side of the screen with a *light guide*, and a *reflector* behind the light together with one or more *microprisms* and one or more diffusers in front of it (Figure 12.22). The light guide is based on a methacrylate material and it is used to guide the light through the layers. The reflector is located in the back of panel and it is used to improve the reflection of light. It is made of a material called polyethylene (PET). The diffuser layer has two functions, to diffuse and to collimate the light to make it parallel and uniform. Using the same material PET as the reflector, the diffuser

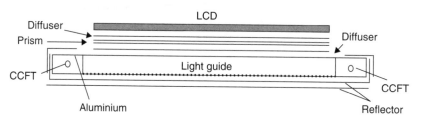

Figure 12.22: Backlight assembly (light guide type)

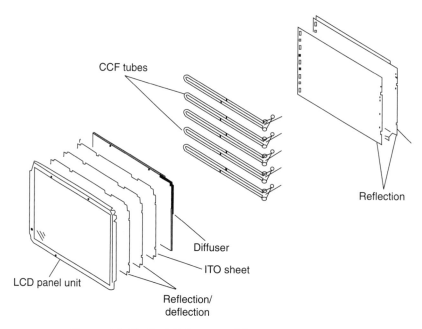

Figure 12.23: Backlight assembly (direct light type)

improves brightness by 20%. The light from the diffuser is collimated by the next microprism layer which improves brightness by 40%.

The direct light assembly has more than two lamps as shown in Figure 12.23. The same layers are used as for the guided light type performing the same functions. Direct light diffuser are used to diffuse the light and to avoid seeing the backlight. A transparent indium tin oxide (ITO) sheet connected to ground is used to filter out the noise produced by the lamps.

When servicing a backlight assembly, care must be taken to ensure that the layers are replaced in the correct order as well as orientation otherwise a permanent pattern may appear on the screen. Other precautions that must be observed when servicing the LCD panel include keeping the surfaces clean and avoiding scratching their surfaces, keeping the LCD dry as water could cause electrical shorts and corrosion, avoiding swift temperature changes as dew and ice could cause non-conformance and malfunction, avoid electrostatic discharges by ensuring proper body earth before handing the LCD and do not operate the LCD for a long time with the same pattern as this would cause image persistence which may result in permanent damage.

12.13 CCFT parameters

The following are the main parameters of a cold cathode fluorescent tube that have to be considered when designing the driving DC–AC inverter or in replacing the tube:

- *Starting voltage* (typical values 2000–3000 V_{peak}): Also known as the discharge voltage, the starting voltage is the minimum voltage required to ignite, i.e. start the tube. The starting voltage is usually 50% higher than its operating voltage. The starting voltage is the primary parameter which determines the "end of life" for the tube. The older is the tube, the higher is its starting voltage.

- *Operating voltage* (typical values 2500–3500 V_{P-P}): This is the voltage across the tube when it has been lit. It is a key parameter in the design of the DC–AC inverter.

- *Tube current*: The current through the tube determines, to a large extent, its brightness. It also indirectly determines the tube's useful life. In general, the tube's life is the square of its current. If the current increases by 20% above its normal value, its life span decreases by 40%. Higher than normal current also results in excessive heat.

- *Frequency* (typical value: 40–60 kHz): Frequency generally has no effect on the brightness of the tube, its efficiency or its useful life. However, it does have an impact on the compatibility between the tube, the display itself and the graphic information displayed by the tube.

- *Waveform*: An undistorted current and voltage sinusoidal waveforms are required to avoid radiated electric noise that may impact on the system and surrounding environment introduced by a distorted sine wave. Although the AC–DC inverters produce pure sine waves, the dynamic nature of the tube distorts both the voltage and the current waveforms.

- *Impedance*: A high impedance is presented by the CCFT assembly which is in the region of 50–70 kΩ.

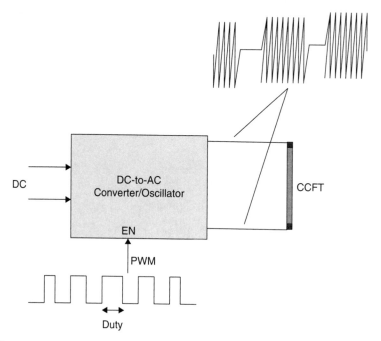

Figure 12.24: Cold cathode fluorescent tube (CCFT) brightness control

12.14 Tube Brightness Control

The CCFT requires a sine wave with an amplitude of few thousand volts and a frequency in the region of 50–70 kHz. This is provided by a DC-AC converter which is essence an oscillator. While maximum brightness may be obtained by turning the tube fully on, in most application, there is a need to reduce the lamp's brightness. There are two basic methods for dimming the CCFT. The first simply reduces the tube current either directly or indirectly by reducing the voltage applied to it. The second method maintains a constant current but turns the lamp on and of to control its brightness (Figure 12.24). If the inverter is turned on for longer periods than it is off, a brighter light is produced and vice versa. This technique employs a *pulse-width modulated* (*PWM*) waveform to turn the inverter oscillator on and off. A typical tube drive and control signals are shown in Figure 12.25.

The frequency of the PWM waveform has to be chosen carefully to avoid interaction with the frame rate. Typical values are 270 and 330 Hz and in the case of PC monitors, it is varied with the frame rate itself to avoid interference with the graphics.

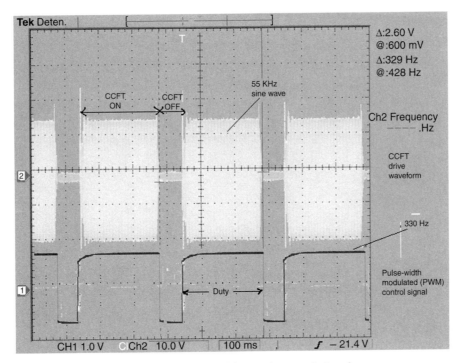

Figure 12.25: CCFT drive and control signals

12.15 The DC-AC Inverter

Essentially, the DC-AC inverter is a tuned collector oscillator (Figure 12.26). When power is switched on, the transistor conducts feeding energy into inductor L. When the inductor saturates, current ceases and the back e.m.f. forces the current to reverse. Energy in L is now transferred to C. When C is fully charged, charging current ceases causing an opposite back e.m.f. across the inductor and the capacitor discharges into L with its energy transferring back to L until the inductor saturates and so on. The output across the secondary of the transformer is a pure sine wave.

The basic elements of a DC-AC inverter are illustrated in Figure 12.27. Capacitor C_P is the primary tuned capacitor which resonates with the inductance of the primary winding of transformer T1. Capacitor C_S is connected in series with the tube to ensure constant current operation. At the frequency of operation, the impedance of the tube assembly together with the ballast capacitor is very high making the inverter act as a constant current source. The output is a sine wave with a slight distortion caused by the reactance of the tube. Because of

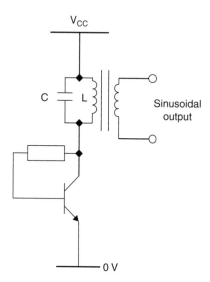

Figure 12.26: Tune-collector oscillator as a DC-AC inverter

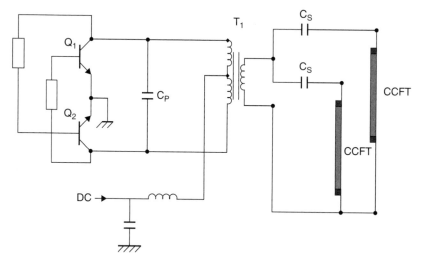

Figure 12.27: The essential elements of a DC-AC inverter

the very high impedance, measuring devices such as a DVM or a CRO would load the output so much as to render the readings almost meaningless. A current probe should be used to observe the shape of the waveform on an oscilloscope.

Figure 12.28 shows practical backlight inverter driving two CCFTs together with the control chip as used by a Panasonic 15-in. LCD receiver. The control chip provides the PWM signal

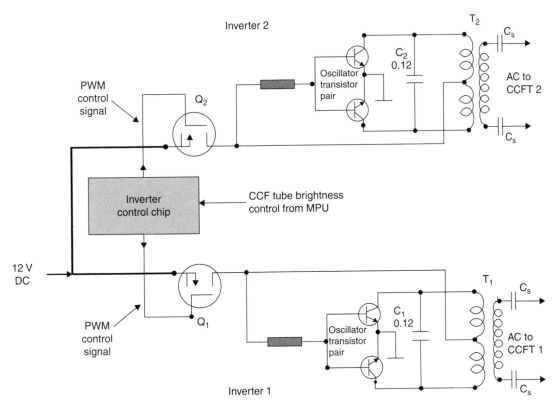

Figure 12.28: The essential elements of a practical DC-AC inverter

to drive two separate inverters, one for each CCFT. The control chip itself is controlled by a signal from the microprocessor. For inverter 1, a center-tapped step-up transformer is used to feed 700 V_{rms} to drive CCFT1. The tuned circuit is formed by C1 and the primary of the transformer and a pair of transistors is used to oscillate back-to-back. The DC power to the oscillator is obtained from switching transistor Q1. While Q1 is on, oscillation takes place and the tube lights up. However when Q1 is off, the oscillator turns off and with it the tube itself. Switching transistor Q1 is controlled by a PWM signal from the control chip. The width of the pulse controls the ON/OFF ratio of Q1 and with it the brightness of the tube. For inverter 2, Q2 is the switching transistor and C2 is the tuning capacitor. The tuning capacitors may be recognized by their non-nominal values, in this case 0.12 μF. Capacitors marked C_S are the series capacitors that ensures the lamp presents a very high impedance to the inverter. It is normal to include over voltage/over current protection as well as a panel enable from the microprocessor controller.

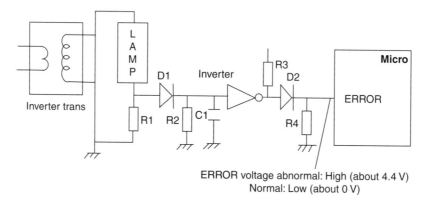

ERROR voltage abnormal: High (about 4.4 V)
Normal: Low (about 0 V)

Figure 12.29: Lamp error detection circuit

12.16 Lamp Error Detection

Invariably, LC displays go into standby if one or more lamps fail to light up. This may be caused by a actual malfunction lamp or a faulty inverter circuit. A typical lamp error detection circuit is shown in Figure 12.29. When the lamp is functioning normally, lamp current flows through R1 and turns D1 on feeding a positive voltage to the inverter. The output from the inverter is negative which turns D2 off producing an zero error voltage. If the lamp fails to light up, current through R1 ceases, D1 turns off, inverter output is positive and D2 turns on resulting in a positive (~4 V) error voltage which will interrupt the microcontroller and the set will be forced into standby. Smoothing is provided by R2C1.

12.17 Adaptive Transmissive Scaling

Light, which is emitted from the LCD panel, is a function of two parameters: the intensity of the backlight and the *transmissiveness* of the LC cells. The latter is the amount of polarising twist the LC imposes on the light passing through it. Therefore, by carefully adjusting these two parameters, one can achieve the same perception of brightness at different values of the backlight intensity and the LCD twist. Since the changes in energy consumption of the backlight lamp are higher than that of the LCD cells, energy may be saved by simply dimming the backlight and increasing the LC cells' transmissiveness. This is known as *brightness enhancement* which may be carried out dynamically, frame-by-frame. Apart from power savings, a dramatic improvement in brightness and contrast ratio as well as black reproduction is achieved.

Figure 12.30 shows a block diagram for an adaptive transmissive scalar. The video information is assessed by the brightness detector to obtain three-frame parameters: average

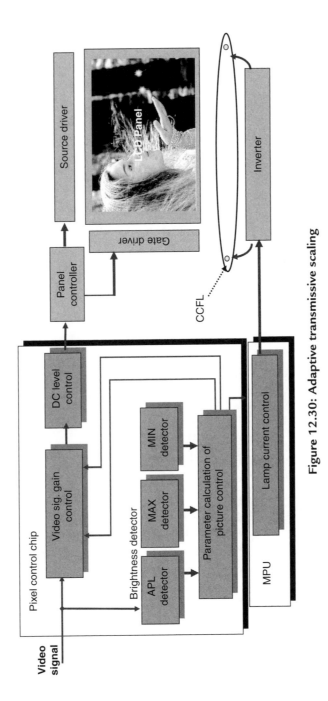

Figure 12.30: Adaptive transmissive scaling

picture level (APL) and maximum and minimum luminance. These parameters are then used to establish the optimum combinations of pixel voltage and the backlight brightness. Signals are sent to the video gain control and the lamp current control as shown.

Transmissive scaling invariably introduces distortion which may be minimized by the use of a histogram. A histogram is used to calculate or estimate the distortion produced by transmissive scaling and if it is above a specified threshold, both the backlight and the LC pixel values are readjusted to bring the distortion down to an acceptable level.

12.18 LCD Panel Faults

The first step in fault finding on LCD panels is to ascertain if the fault is a panel malfunction, in which case it has to be replaced, or if the cause of the fault is outside the panel in which case a repair is possible. Some of the symptoms point clearly to a panel fault, others may be ambiguous. One of the main symptoms of a faulty panel is pixel defect. This may be a single pixel, several pixels or cluster of pixel failure in the same way as pixel failures in a plasma

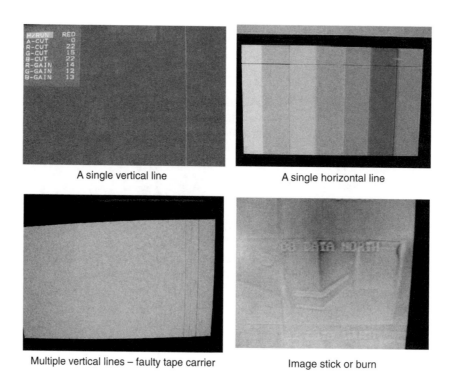

A single vertical line

A single horizontal line

Multiple vertical lines – faulty tape carrier

Image stick or burn

Figure 12.31: Symptoms of faulty panel

Figure 12.32: "Negative picture" effect due to panel malfunction

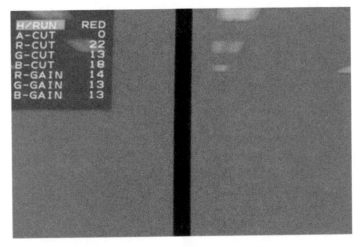

Figure 12.33: Source or column drive fault

panel. Again, in the same way as plasma displays, manufacturers allow for a certain number of pixel defects before the panel is rendered obsolete. The display panel is divided into a number of areas and the number of bad pixels permitted depends on the area. Other classic symptoms pointing to faulty LCD panel are shown in Figure 12.31. A horizontal line across the screen may also be caused by a bad connection between gate driver board and the panel. The "negative picture" effect in Figure 12.32 is also a result of panel failure.

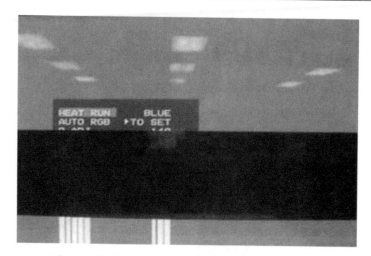

Figure 12.34: Symptom of a line gate driver fault

12.19 Drive faults

The classic symptom for a source (or column) drive is a vertical band across the screen as shown in Figure 12.33. A horizontal band (Figure 12.34) would indicate a faulty line scan drive.

Index